4th International Scientific Conference SEA-CONF 2018

IOP Conference Series: Earth and Environmental Science
Volume 172

Constanta, Romania
17 – 19 May 2018

ISBN: 978-1-5108-6746-8
ISSN: 1755-1307

Printed from e-media with permission by:

Curran Associates, Inc.
57 Morehouse Lane
Red Hook, NY 12571

CURRAN ASSOCIATES INC.
proceedings.com

Some format issues inherent in the e-media version may also appear in this print version.

Printed by Curran Associates, Inc. (2018)

For permission requests, please contact the Institute of Physics
at the address below.

Institute of Physics
Dirac House, Temple Back
Bristol BS1 6BE UK

Phone: 44 1 17 929 7481
Fax: 44 1 17 920 0979

techtracking@iop.org

Additional copies of this publication are available from:

Curran Associates, Inc.
57 Morehouse Lane
Red Hook, NY 12571 USA
Phone: 845-758-0400
Fax: 845-758-2633
Email: curran@proceedings.com
Web: www.proceedings.com

Table of contents

Volume 172

4th International Scientific Conference SEA-CONF 2018

17–19 May 2018, Constanta, Romania

Accepted papers received: 20 June 2018
Published online: 12 July 2018

Preface

Papers

The 4th International Scientific Conference SEA-CONF 2018

(The 4th SEA-CONF 2018)

MAY 17-19, 2018 — CONSTANTA, ROMANIA

Accepted, peer reviewed papers from the 4th SEA-CONF 2018

Edited by

Naval Academy Press
Constanta – Romania

2018

IOP Conf. Series: Earth and Environmental Science **172** (2018) 011001 doi:10.1088/1755-1315/172/1/011001

PREFACE

The objectives of the International Scientific Conference SEA-CONF are to bring together stakeholders from marine industry, research and education, and to provide a networking opportunity for academics, professionals, and public authorities active in this domain.

The International Scientific Conference SEA-CONF is a place of debate and sharing research ideas and results related to the topics of the conference.

Sea-Conf 2018 aims to be a hub for sharing research ideas and results, built on the scientific contributions of the participants, in accordance with the conference sections.

Sea-Conf 2018 brings together researchers, experts and representatives from maritime academia, companies, and public authorities, eager to review, discuss, and explore the latest opportunities and strengths in the maritime domain.

IOP Conf. Series: Earth and Environmental Science **172** (2018) 011001 doi:10.1088/1755-1315/172/1/011001

ORGANIZING COMMITTEE

CHAIRMAN:
Catalin-Paul CLINCI, Ph.D.

MEMBERS:
Dinu-Vasile ATODIRESEI
Paul BURLACU
Ionut-Cristian SCURTU
Filip NISTOR
Florentiu DELIU
Sergiu LUPU
Ovidiu CRISTEA
Alexandru COTORCEA
Petrica POPOV
Andrei BAUTU
Ligia SPORIS
Sergiu SERBAN
Corina SANDIUC
Florin POSTOLACHE
Mihai BALACEANU
Andrei POCORA

SCIENTIFIC COMMITTEE

CHAIRMAN:
Octavian TARABUTA, Rector of "Mircea cel Batran" Naval Academy

MEMBERS (in alphabetical order):

- Avni Zafer ACAR (Turkey)
- Taner ALBAYRAK (Turkey)
- Beazit ALI (Romania)
- Dinu -Vasile ATODIRESEI (Romania)
- Laura BACALI (Romania)
- Nicolae BADARA (Romania)
- Dana BADAU (Romania)
- Alina BALAGIU (Romania)
- Ghita BARSAN (Romania)
- Andrei BAUTU (Romania)
- Blagovest BELEV (Bulgaria)
- Hristo BELOEV (Bulgaria)
- Mircea BOSCOIANU (Romania)
- Romeo BOşNEAGU (Romania)
- Paul BURLACU (Romania)
- Gheorghe CALOPAREANU (Romania)
- Maria Filomena CAPUCHO (Portugal)
- Ionel CHIRICA (Romania)
- Carmen COJOCARU (Romania)
- Irina CRISTEA (Slovenia)
- Florentiu DELIU (Romania)
- Vasile DOBREF (Romania)
- Leonard DOMNISORU (Romania)
- Carmen ENE-VOICULESCU (Romania)
- Virgil ENE-VOICULESCU (Romania)
- Oral ERDOGAN (Turkey)
- Jean-Claude FRERY (France)
- Mihai GIRTU (Romania)

IOP Conf. Series: Earth and Environmental Science **172** (2018) 011001 doi:10.1088/1755-1315/172/1/011001

- Rosen IVANOV (Bulgaria)
- Ricardo IZZO (Italy)
- Nicolae JULA (Romania)
- Kalin KALINOV (Bulgaria)
- Dorin LUCACHE (Romania)
- Delia LUNGU (Romania)
- Sergiu LUPU (Romania)
- Liviu MARSAVINA (Romania)
- Boyan MEDNIKAROV (Bulgaria)
- Lucia MELNIC (Romania)
- Florin NICOLAE (Romania)
- Filip NISTOR (Romania)
- Suleyman OZKAYNAK (Turkey)
- Dimitrios Nikolaos PAGONIS (Greece)
- Cristian Adrian PAPARI (Romania)
- Radu-Dumitru PENTIUC (Romania)
- Adrian POPA (Romania)
- Catalin POPA (Romania)
- Mihaela POPA (Romania)
- Dorin-Mircea POPOVICI (Romania)
- Mihail PRICOP (Romania)
- Anastase PRUIU (Romania)
- Anca Alexandra PURCAREA (Romania)
- Serban RAICU (Romania)
- Sorin RUGINA (Romania)
- Gheorghe SAMOILESCU (Romania)
- Ionut Cristian SCURTU (Romania)
- Tomasz SZUBRYCHT (Poland)
- Alecu TOMA (Romania)
- Deniz UNSALAN (Turkey)

Peer review statement

All papers published in this volume of *IOP Conference Series: Earth and Environmental Science* have been peer reviewed through processes administered by the proceedings Editors. Reviews were conducted by expert referees to the professional and scientific standards expected of a proceedings journal published by IOP Publishing.

IOP Conf. Series: Earth and Environmental Science **172** (2018) 012001 doi:10.1088/1755-1315/172/1/012001

Three-layer informatics solution for demand response management in the smart grid context

Simona-Vasilica Oprea[1], Adela Bâra[1] and Osman Bulent Tor[2]

[1]The Bucharest University of Economic Studies, Romania
[2]EPRA Engineering Procurement Research Analysis, Ankara, Turkey
E-mail: simona.oprea@csie.ase.ro

Abstract. In consistency with the European Union (EU) vision of creating an electric power system that integrates renewables and enables flexible consumption and generation technologies by means of smart grid (SG) implementation, the main aim of the paper is to develop different level of aggregation mechanisms design that facilitates demand response (DR) and energy systems to utilize up to 100% renewable generation.
In this paper, we will present a three-layer informatics solution for demand response management in the smart grid context. The data model will be developed as a scalable and customized framework using open solutions based on cloud computing.

1. Introduction

Main challenge is to increase flexibility of the grid, complex interactions among large and local or regional energy stakeholders with massive volumes of heterogeneous data (big data challenge). Advancements in SG technologies make possible to apply various strategies for optimization of DR. Its aggregation would essentially accumulate potential consumption schedules and constraints given by small/medium-size consumers for the participation in wholesale electricity markets. However, there are still barriers for widespread adoption of DR aggregation mechanisms at the EU level. According to the *Final 10-year ETIP SNET R&I roadmap 2017-26*, these include: differences between market structures and legislation; limited awareness and involvement of electricity consumers/prosumers; less motivating incentives for participants; ICT infrastructure at an affordable and reasonable cost compared with benefits; requirements of big data analytics for DR; interactions between aggregators and Distribution System Operators (DSOs); lack of business models for market trading [1].

Implementations of DR aggregator solutions vary over EU countries due to different market structures and development. Advancements in SG technologies enable various options/strategies for the DR optimization in electricity markets. However, there are still barriers for the widespread adoption of DR aggregation mechanisms at the EU level, as emphasized in *Final 10-year ETIP SNET R&I Roadmap (2017-26)*. The community aggregator (CA) concept has been implemented in Nordic countries and can essentially be implemented in other countries (Turkey, Romania), where market development (legislation, rules & policy) is behind compared to Nordic countries. Another concern is about reliability of the ICT infrastructure coupling the consumers, market players and DSOs at an affordable/reasonable cost compared with benefits. In response, our approach demonstrates that DR can be categorized into different levels, from simple CA solution, which can be implemented in any country in EU, to more complex wholesale electricity market aggregation concept for developed markets. Our approach uses the hands-on experience in the CA concept, while recommending market based DR solutions to expand in other countries. The idea of multi-layer aggregation (MLA) solutions integrates a wider range of requirements to a scalable prototype that leads to better scalability and transferability among the EU countries. Thereby, it has a better chance to disseminate the innovative technologies based findings to a wider audience and markets across EU countries.

Energy Union Package 2015 emphasizes the requirement for moving away from an economy driven by fossil fuels, based on a centralized, supply-side approach, relying on old technologies and outdated business models [2]. This necessitates empowering consumers by providing information, choice and flexibility to manage demand as well as supply. The European Commission is preparing an ambitious legislative proposal to redesign the internal electricity market (IEM) to link wholesale and retail. The electricity market will include many new producers, including renewables (RES), as well as enable full participation of consumers via DR.

The authors of [3] approach the demand response management by proposing an optimal community schedule of hourly electricity consumption considering advanced electricity tariffs such as real-time tariff that flattens the daily consumption curve. The proposed algorithm minimizes the electricity payment by considering the characteristics of electrical appliances and consumption preferences of the residential consumers. It is based on several steps, taking into account adjustable/programmable appliances, local generation sources, storage and common consumption at the community level.

In [4], the authors propose a heuristic-based evolutionary approach based on shifting techniques of several types of appliances at the communities level that treats different consumption sectors such as: householders, commercial and industrial consumption.

Decision-making models for demand response aggregators that act on the wholesale electricity markets are developed in [5]. The aggregators provide different types of contracts that aim to reduce and optimize the electricity consumption. The hourly schedule of the appliances depends on the electricity tariff rate and other provisions of the contracts.

In [6], the authors propose a demand response model bidding framework based on mixed-integer linear programming at the Independent System Operator level that considers consumption preferences and characteristics of the appliances. Demand response management includes measures such as load shifting, curtailments, local generation and storage devices usage.

Comparing with these approaches related to demand response management, the informatics solution proposed in this paper will be developed as an informatics prototype that aims to demonstrate the implementation of DR aggregator solutions across EU countries which have different electricity market structures, regulations and SG implementation stage. The prototype will manage massive data collected from smart meters (SM) and appliances (IoT) for advanced analytics and will be developed on a cloud computing (CC) platform which will enable ubiquitous access to distributed configurable computing resources.

2. Community aggregator multi-layer solution

Due to various stages of development of SG implementation in different countries or regions, our proposal addresses DR aggregating informatics solution in multi-layers beginning from the simplest community aggregator (CA Layer 1) solution. Motivation of the CA will be optimal load scheduling to minimize the community-aggregated electricity payment and consumption peak shaving considering the convenience of individual residents and hourly community load characteristics. Control of demand (i.e., active DR) including electric vehicle (EV) charging stations, distributed renewable energy sources (RES) generation and storage devices (SD) will be addressed in Layer 2. Finally, in Layer 3, aggregation of DR contracts will be considered in the price-based self-scheduling optimization problem to determine optimal DR schedules for participants in day-ahead energy markets as in Figure 1.

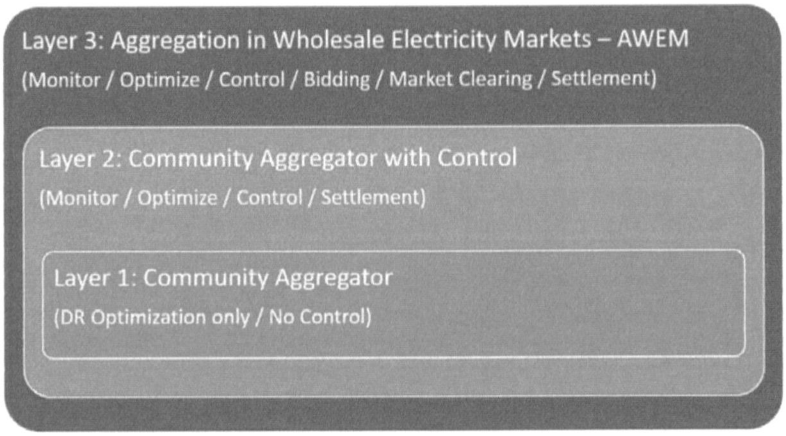

Figure 1. Multi-layer aggregation short description

Layer 1 can be the initial DR solution given its applicability to any community even without SG implementation. Layer 2 may include several sub-layers depending on SG implementation level. However, Layer 3 requires a DR bidding framework in day-ahead electricity markets, which integrates consumers' DR preferences and characteristics in the DSO's market clearing process. ENTSO/E and EDSO outline the main challenges connected with the DR from the grid perspective. *ENTSO/E R&D & Innovation Roadmap 2017–2026* highlights the need to activate the demand as a new source/tool for system operation and to integrate it in the planning and operation stages and market design; however it does not accentuates the tasks of activation of the source, leaving to DSOs, aggregators or other players the possibility to do it [7]. In any case, integration of DR has been identified as one of the outstanding issues to be dealt jointly with DSOs. Layer 3 addresses this issue.

The multi-layer aggregation approach covers different structures of markets and regulations as in Figures 2, 3 and 4. It is more applicable since in Europe, the development of SG implementation can significantly vary. For less developed regions, we propose a simple solution that monitors the consumption and provide recommendations (Layer 1), while for more developed markets where aggregators are involved in market clearing, we propose optimization and control solutions for DR based on interaction between aggregators and DSOs (Layers 2 and 3) utilizing the block-chain based smart contracts.

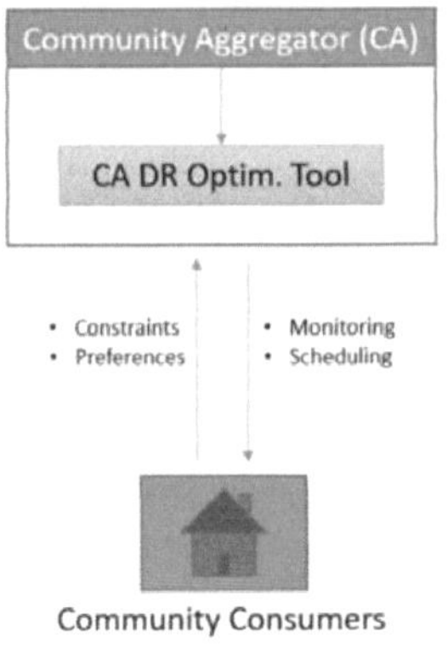

Figure 2. Layer 1 CA detailed concept

IOP Conf. Series: Earth and Environmental Science **172** (2018) 012001 doi:10.1088/1755-1315/172/1/012001

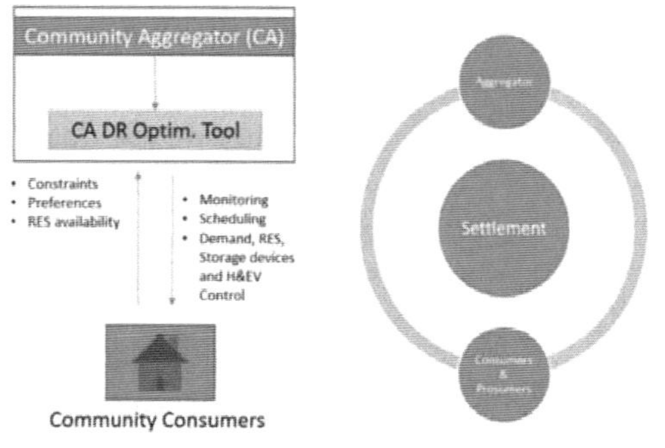

Figure 3. Layer 2 CA detailed concept

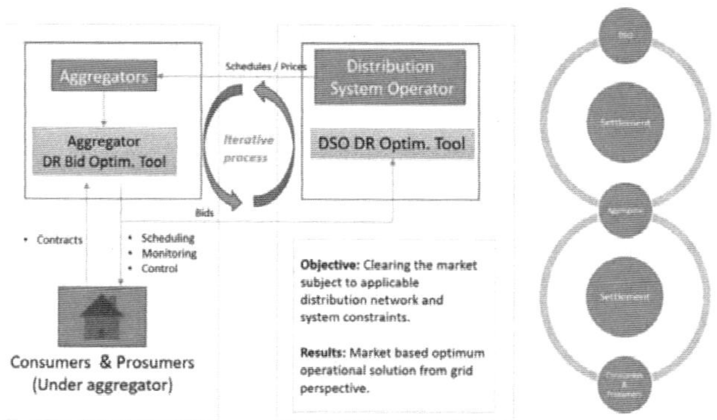

Figure 4. Layer 3 CA detailed concept

Our approach will develop a scalable cloud based platform to provide access to consumers/prosumers and aggregators/DSOs to a complex data framework and models allowing big data and real-time analytics including forecasting, optimization, profiling, and business models. We will also approach the big data and CC challenges related to SG in order to increase consumer awareness and improve the DR mechanisms. We will investigate and verify the high-reliability and low-latency ICT technologies to support real-time monitor and sensing. Block-chain technologies will be investigated to provide smart and flexible contracts. These aspects are among the most prevailing barriers for the widespread adoption of DR aggregation mechanisms at the EU level, as underlined by *Final 10-year ETIP SNET R&I Roadmap* and *EC Energy Union Package 2015*.

3. Data framework

The data framework contains several tiers that can be customized based on each layer. These tiers will allow complete customization, using different technologies for interconnecting sensors through smart meters (SM-Tier); fast processing and real-time analyses through big data (BG-Tier); high-reliable, low-latency, secure and performant data transmission and management with relational databases (RB-Tier); historical and advanced analytics through data warehouse (DW-Tier) as in Figure 5.

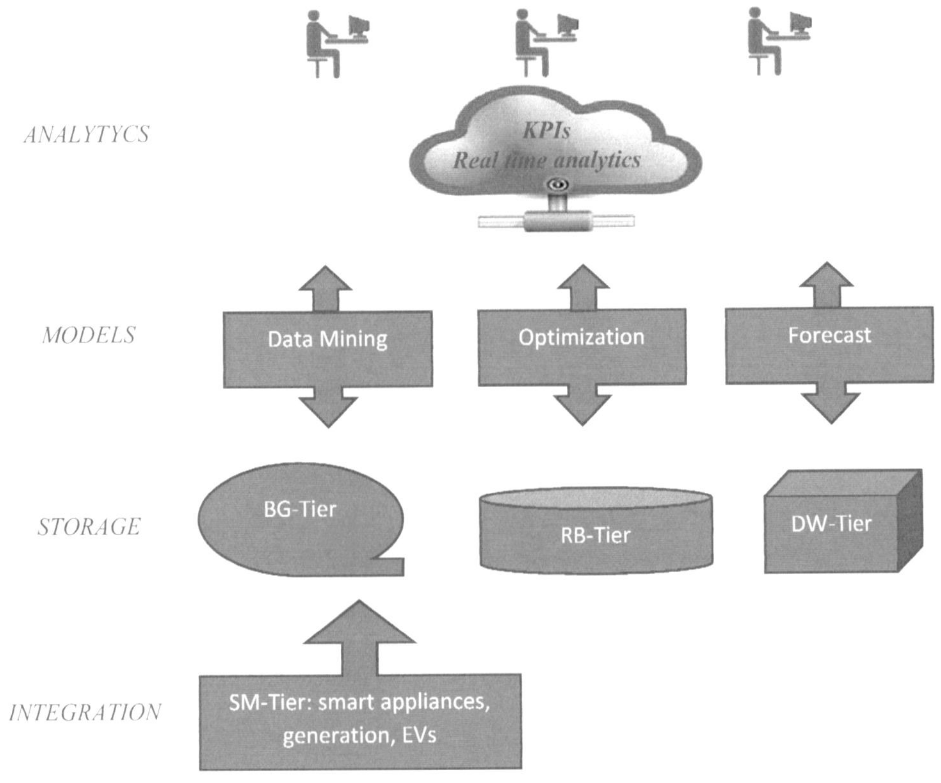

Figure 5. Tiers for each layer

The data framework will be developed on multi-tier architecture for high scalability, using open standards for wider integration and adoption. For SM-Tier, we'll consider OpenIoT, FIWARE & IoT-A ARM platforms currently adopted and standardized in EU. To achieve real-time information sensing and collection, we'll develop and demonstrate ultra-reliability low-latency communications based on short-packet. For BG Tier, we'll consider optimization methods, such as: hash partitioning, parallel execution, dimensionality reduction, Map Reduce. For load profiles (LP), we will use artificial neural networks (ANN) algorithms (Self Organizing Maps, k-means, Fuzzy C-Means) for clustering data across multiple nodes, with different type of SM, preserving consumers' location, weather, day type and dynamically adding particular characteristics. For forecasting, ANN algorithms will be developed/run on both historical and online data for short term load forecast (LF).

Optimization algorithms will be developed for each layer. Layer 1 provides optimal load scheduling of the community to minimize the electricity payment considering the convenience of individual residents and hourly community load characteristics. Inputs include: LP, LF, cost of electricity consumption, and preferences of the consumers in the community as constraints included on the hourly utility load (defined as CA load minus the local generation). Lagrangian relaxation (LR) will be applied to decouple the utility constraint and provide tractable sub-problems formulated as mixed-integer programming (MIP).

Optimization in Layer 2 includes various contract offers for the customers for load curtailment and shifting, local generation, managing electrical vehicles (EV) charging stations and storage systems as possible strategies for hourly load reductions. The aggregation of DR contracts is considered in the proposed price-based self-scheduling optimization model to determine optimal DR schedules for participants in day-ahead energy markets. We'll use block-chain to develop smart and flexible contracts.

In Layer 3, a hierarchical DR bidding framework in the day-ahead energy markets, which integrates consumers' preferences and characteristics in the DSO's market clearing process, will be developed. Aggregators can submit offers to the DSO which will optimize final decisions on aggregators' DR contributions to wholesale market. The DSO applies MIP to the solution of the proposed DR model in the day-ahead market clearing problem.

As an innovative concept, the prototype will be developed as a scalable-customized cloud computing web-services using open standards, models and connectivity as in Figure 6.

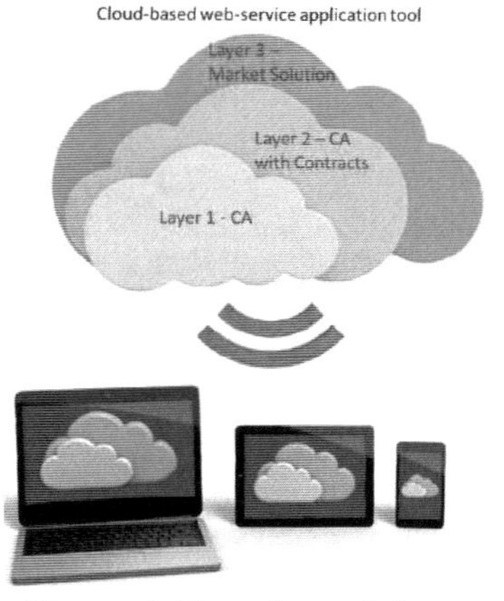

Figure 6. Tiers for each layer

The prototype will be developed modular and will be configured based on the requirements of stakeholders. Each proposed layer of the solution will have its innovative and original elements. The prototype's models will be developed on a data framework with multiple tiers (SM, BG, RB, DW Tiers) providing reliable and consistent data for the models and algorithms including LP, LF and electricity price from the utility, feed-in tariffs for RES, etc.

For integrating SM data, we'll develop agents to connect sensors with different communication protocols (ZigBee, LORA, WiFi, Wi-SUN) in order to load custom data in the BG-Tier. One of our key innovations in is to support ultra-reliable low-latency communications based on short-packet for supporting real-time SM data transmission and corresponding actuation, which is very important to multi-layer systems. The platform can be extended to connect sensors from other utilities (gas, heating) and enabling Advanced Metering Infrastructure (AMI).

The SM and BG Tiers will be used for real time analyses and for developing the algorithms for controlling, monitoring & online forecasting and optimization. The SM-Tier will be implemented on an edge computing architecture to perform data processing at the edge of the network, thus reducing the communications bandwidth between SM and BG-Tier. Relational databases and big data are complementary technologies that process data at an impressive rate of more than 10TB (terabytes) per hour. Through our integration framework, we'll first process big data in the BG-Tier on a distributed, scalable file system, such as HDFS (Hadoop File System) and develop query patterns with processing engines (Hive, Drill, Impala or Presto). For data processing, we'll develop algorithms that are executed in parallel and extract relevant data based on a pattern matching criteria. These patterns will be defined based on the requirements provided by the aggregator/DSO concerning their analytical perspective. At the BG-Tier, we'll develop forecasting algorithms based on ANN (feed-forward, deep-learning). They will be developed online, receiving immediate feedback and improving subsequent predictions in contrast to statistical methods. Data dimensionality will be adjusted for different time horizons, for example for very and ultra-short term forecasting the algorithms will perform on detailed data (minutes), since for STLF data will be aggregated hourly over locations or consumers' profiles. For LP we'll use distributed data mining with multi-node processing based on

Self Organizing Maps, k-means and fuzzy c-Means. Thus, data regarding type of consumption, tariffs schemas' influence, injection of RES, EV usage is transformed into a valuable information for aggregators/DSO. Optimization algorithms will be developed based on LR and MIP techniques to formulate the complex dynamic optimization problems that include several constraints at each layer. In Layer 1 LR will be applied to decouple the utility constraint and provide tractable sub-problems formulated as MIP model. A hierarchical DR bidding framework in the day-ahead markets, which integrates customer DR preferences & characteristics in the DSO's market clearing process, will be developed in Layer 3. The DSO applies MIP to solution of the proposed DR model in the day-ahead market clearing problem.

From BG-Tier, some aggregated data is loaded into RB-Tier to allow integration with on premises Energy Management Systems (EMS) already installed at aggregator/DSO. For example, the LP data will be partitioned by locations and consumers' profiles and stored in RB-Tier for integration with EMS. The DW-Tier will be developed on top of BG and RB tiers and will be subject oriented towards the business analyses regarding consumers' behaviour and wholesale electricity markets. Analytic solutions require data governance, data quality and stewardship that are absolutely critical and are achieved only through the DW-Tier. At this layer we'll developed a set of algorithms for analytical purpose and key performance indicators (KPIs) reporting. The framework will offer scalable web-services for targeted users. The aggregators/DSO will analyse data through an online portal available as web-services in CC platform. For consumers, the web content will be customized based on their preferences & requirements (monitoring/controlling/scheduling the appliances, real time billing information, micro-generation and SD). A section of the portal will be developed as a simulator similar to a computer game to motivate consumers to optimize their electricity consumption and to increase their awareness towards energy efficiency.

4. Conclusion

The full liberalization of the markets will lead to competitive prices and services offered by electricity suppliers including aggregators. Through business models, the prototype will allow suppliers to configure their tariffs and to attract new consumers groups as well. This will bring a significant benefit to prosumers, who will obtain several incentive options. They will play an active role in DR, having the possibility to gain profit, apart from saving money.

There are several impacts the consortium find relevant and beneficial for society. The multi-layer aggregator services encourage active participation of prosumers through differentiated tariffs between locally and centrally produced power, and through a market, handle storage and DR. This enables local markets to solve many of the issues regarding implementation of SG, especially around incentives, business models and adoption. This active participation will allow consumers to reduce their electricity bill maintaining their comfort. The solution opens up for new innovative business models where companies can develop new products combining wholesale and local markets, and energy services such as flexibility management. Again, this increases the economic benefits for participants willing to take active roles in local energy markets.

In addition, by aggregating and increasing the share of local RES, the costs for upgrading regional and national grid is lowered. By balancing the consumption and production in a local area, especially during peak hours, less energy is imported from the centralized grid and thereby reducing the amount of energy transported through the transmission grid. As a consequence, grid costs for end-users will diminish.

The proposed solution will also contribute in reducing the prices by making a market for the local energy produced, and thus increasing the incentives for buying local energy. In a functioning market, the price difference between locally produced energy and wholesale energy due to transportation costs, also creates incentive for consumers to become prosumers and sell energy to the local market.

Multi-layer aggregator concept needs new market polices, legislation and/or incentives schemes for SG infrastructure. By demonstrating that the solution can, under the right circumstances, give the desired effects, the pilot trials will demonstrate how such regulative bodies can be structured for the desired impact for adoption of SG technologies.

5. References

[1] Final 10-year ETIP SNET R&I roadmap covering 2017-2026. Available online: https://etip-snet.eu/pdf/Final_10_Year_ETIP-SNET_R&I_Roadmap.pdf.

[2] EC Energy Union Package 2015. Available online: http://eur-lex.europa.eu/resource.html?uri=cellar:1bd46c90-bdd4-11e4-bbe1-01aa75ed71a1.0001.03/DOC_1&format=PDF.

[3] Khodaei A., Shahidehpour M., Choi J. Optimal Hourly Scheduling of Community-Aggregated Electricity Consumption, in *JEET,* 2013, Vol. 8, Issue 6, pp. 1251-1260; DOI: 10.5370/JEET.2013.8.6.1251.

[4] Logenthiran T., Srinivasan D., Shun Z. Demand Side Management in Smart Grid Using Heuristic Optimization, in *IEEE Transactions on Smart Grid*, 2012, Vol. 3, Issue 3, pp. 1244-1252; DOI: 10.1109/TSG.2012.2195686.

[5] M. Parvania, M. Fotuhi-Firuzabad, M. Shahidehpour, Optimal demand response aggregation in wholesale electricity markets, IEEE TRANSACTIONS ON SMART GRID, 2013, Vol. 4, No. 4., pp. 1957-1965.

[6] M. Parvania, M. Fotuhi-Firuzabad, M. Shahidehpour, ISO's Optimal Strategies for Scheduling the Hourly Demand Response in Day-Ahead Markets IEEE TRANSACTIONS ON POWER SYSTEMS, 2014, Vol. 29, No. 6., pp. 2636-2645.

[7] ENTSO-E R&D & Innovation Roadmap 2017–2026. Available online: https://www.entsoe.eu/publications/research-and-development-reports/rd-roadmap/Pages/default.aspx.

Acknowledgments. This paper presents some results of the research project: Informatics solutions for optimizing the operation of photovoltaic power plants (OPTIM-PV), project code: PN-III- P2-2.1-PTE-2016- 0032, 4PTE/06/10/2016, PNIII - PTE 2016.

A wave energy system based on converting wave-induced motions into peristaltic flow in a rectangular channel

D Ünsalan[1], K Izet- Ünsalan[2] and K T Gürsel[1]
[1] Dokuz Eylül University, Institute of Marine Sciences and Technology, İzmir, Turkey
[2] Ovidius University, Faculty of Industrial, Mechanical and Maritime Engineering, Constanta, Romania
denizunsalan@yahoo.com

Abstract. Energy from the waves at sea is one of the areas that has been an area of interest of the engineering communities worldwide. Recent decades have witnessed the development of various wave energy systems, using various schemes and principles. The main challenge for most systems in efficient energy converting is posed by converting energy into a usable form from the reciprocating wave induced motions due to the harmonic motion of water particles. The authors have conceived a scheme to convert the harmonic motions of water particles which manifest themselves as the near-sinusoidal motion of the free water surface into a unidirectional water flow that can be exploited for harvesting energy: An array of buoys that are arranged in a line perpendicular to the wave direction, that are allowed to oscillate in heaving motion. The heaving motion enforces a series of actuators to induce a sinusoidal motion to a diaphragm sheet placed inside a rectangular channel, which is placed beneath the free surface at a sufficient depth. The motion of the diaphragm sheet shall result in the peristaltic motion of the otherwise stagnant water. The mass of flowing water peristaltic-pumped then can be channelled to a turbine from which energy can be obtained.

1. Introduction

The waves at sea, especially in heavy weather, transmits an enormous amount of energy as witnessed by their devastating effects on the coasts and coastal structures, on floating platforms and ships. Harnessing this energy for the benefit of mankind is one of the challenges that has intrigued engineering communities worldwide, especially in the past few decades, as can be observed from the various schemes depending on various physical phenomena. Each of the proposed/designed systems have their merits and inherent problems. Due to the harshness of the marine environment, such a system should be robust, cost effective and not be far away from the electrical grid. Summary of the existing wave energy systems and their classification can be found in [1] and [2].

A floating object on the surface of sea experiences a six degree of freedom motion under the effect of external disturbances, mainly that of sea waves. For a moored floating body, one of the most obvious is the heaving motion. If the natural frequency of the floating body is significantly higher than that of the sea wave in a monochromatic sea state, the body shall follow the wave contour. In a mixed sea state, the physical dimensions of the body shall act as a filter to eliminate the higher frequency components of the waveform.

The authors have conceived a system to convert the near-sinusoidal heaving motion of an array of buoys arranged in a direction perpendicular to the wavefront into peristaltic pumping of the seawater at a depth where the environment is otherwise stagnant. The buoys have a hydraulic linkage to the wave conversion device, with a reciprocating pump and a reciprocating hydraulic ram using seawater as the hydraulic fluid. The peristaltic pump element is a flexible diaphragm sheet, which is actuated by the hydraulic ram and is housed in a rectangular channel sufficiently below the free surface (figure 1). As the buoys oscillate, they shall induce the diaphragm inside the rectangular channel below to a sinusoidal motion that mimics the waveform on the free surface, which in turn shall result in peristaltic pumping action of seawater. The energy from the pumped seawater can be converted into electrical energy via a turbine (figure 2).

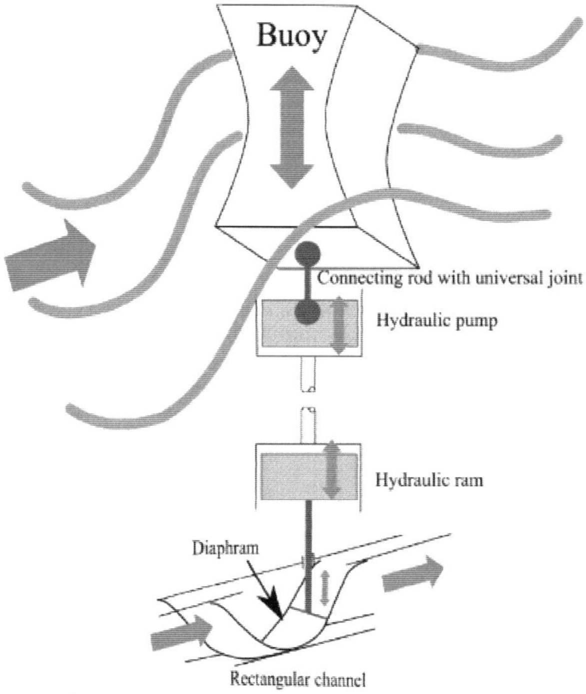

Figure 1. Buoy and peristaltic pump element

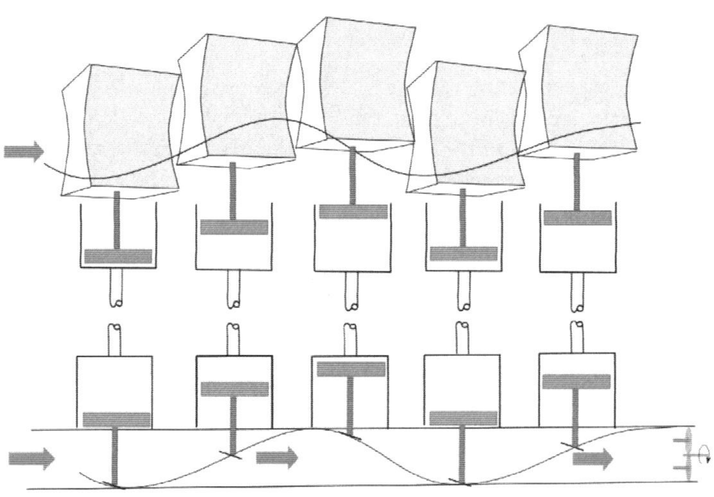

Figure 2. The proposed system- array of buoys and peristaltic pump in the rectangular channel

IOP Conf. Series: Earth and Environmental Science **172** (2018) 012002 doi:10.1088/1755-1315/172/1/012002

2. Peristaltic Pumps

The words "Peristaltic motion" or "peristaltic pump" are derived from the word "peristaltis" as defined by the Webster dictionary as "*successive waves of involuntary contraction passing along the walls of a hollow muscular structure (such as the esophagus or intestine) and forcing the contents onward*". Peristaltic pumps are bio-mimetic devices that are a class of positive displacement pumps by which the fluid is induced to move by the sinusoidal motion of the walls of the pump channel. Rotary peristaltic pumps are used in a variety of applications, especially in cases where the fluid that is handled should be isolated from the environment. A linear peristaltic pump is a device where the fluid is pumped by the sinusoidal motion of either of the surrounding walls. The geometry of a two-dimensional ideal peristaltic pump is shown below in figure 3. An analysis of linear peristaltic pumps has been given in [3]. The ideal flow rate, Q_{ideal} through the pump is given by:

$$Q_{ideal} = f \cdot w_p \int_{-\frac{\lambda}{2}}^{\frac{\lambda}{2}} y \cdot dx = f \alpha w_p \int_{-\frac{\lambda}{2}}^{\frac{\lambda}{2}} \left(1 + \cos\frac{2\pi x}{\lambda}\right) \cdot dx = f \alpha w_p \lambda \qquad (1)$$

where α is the "design amplitude of peristaltis" determined by the "design wave amplitude" and λ is the wavelength and f is the frequency of the peristaltic motion of the upper wall, w_p being the width of the rectangular pump channel.

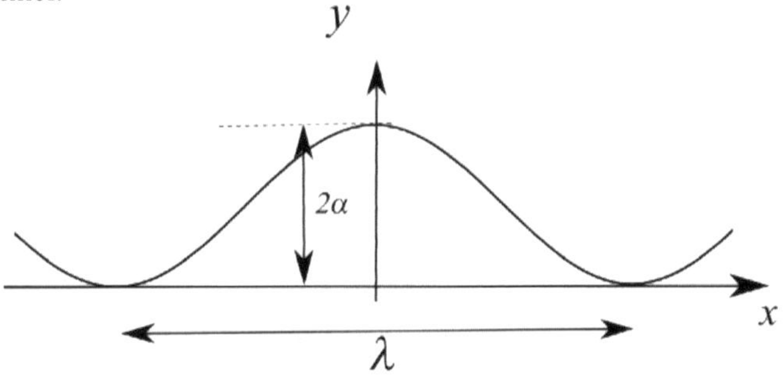

Figure 3. Waveform in peristaltic flow

For a sea wave which excites the system described, wavelength and wave period are related, thus, using linear, deep water approximations [4],

$$\lambda = \frac{gT^2}{2\pi} \qquad (2)$$

the ideal flow rate shall be:

$$Q_{ideal} = \frac{g}{2\pi} a w_p T = 1.56 a w_p T \qquad (3)$$

α is the amplitude of the waveform, w_p is the breadth of the peristaltic diaphragm and T is the period of wave. The pressure expected to be obtained from the peristaltic pumping configuration above is limited due to leakages in the diaphragm and shall be pulsed. Therefore, it is conceived that more than one sets of units with different phases, feeding an axial flow turbine shall be appropriate.

In waves with wave heights greater than designed height of the peristaltic pump's stroke amplitude, together with wavelengths greater than the length of the pump channel, i.e., in heavy seas, the pump shall function as a "travelling wave pump" (figure 4), rather than an ideal sinusoidal peristaltic pump.

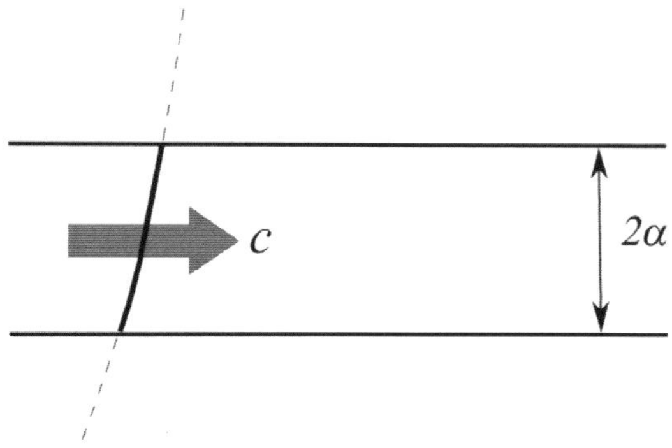

Figure 4. Peristaltic pumping in large amplitude/long waves

In this case, the flow rate shall be:

$$Q_{longwave} = 2\alpha w_p \cdot c = \frac{g}{\pi}\alpha w_p T = 3.12\alpha w_p T \quad (4)$$

3. Buoy system

The buoy shall be able to oscillate with an almost fixed amplitude of oscillation. In order to limit the amplitude of oscillations to the height of the peristaltic pump amplitude, a buoy with geometry indicated in figure 4 is adopted. The buoy shall have a waterline area defined by:

$$A_{wp} = w\left(b_0 + b_1 z^n\right) \quad (5)$$

where w is the length of the buoy, b_0 is the minimum breadth, and the breadth $b(x)$ expands away from the waterline as can be seen in Figure 5. By opting for such a buoy, a "hardening spring" type oscillation mechanism for heaving is aimed, which shall oscillate with a near-constant amplitude in the low excitation frequency regime [5].

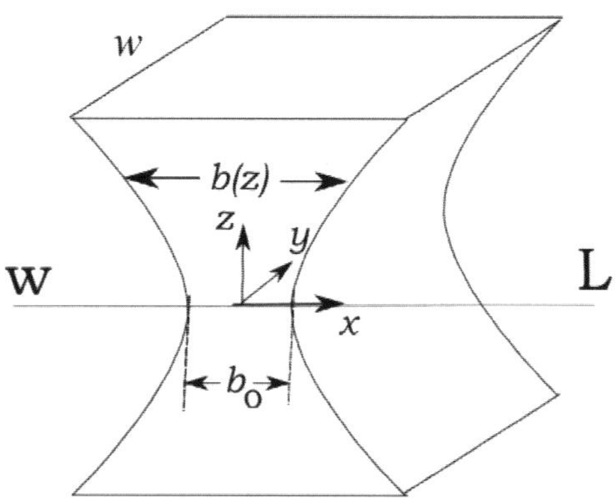

Figure 5. Geometry of a buoy

The governing differential equation of heave shall be [6]:

$$\ddot{z} + \frac{c}{m + m_a}\dot{z} + \left(\omega_{n0} + K\left|z^n\right|\right)z = F\cos\left(\Omega t + \sigma\right) \quad (6)$$

where:

- z Amplitude of heave
- c Damping constant (including both equivalent viscous damping and damping due to power extraction)
- m Mass of the buoy system

IOP Conf. Series: Earth and Environmental Science **172** (2018) 012002 doi:10.1088/1755-1315/172/1/012002

m_a	Added mass of the buoy system
ω_{n0}	Natural angular frequency in small heave, $\omega_{n0} = \sqrt{\frac{\rho g b_0 w}{m+m_a}}$
K	Coefficient of buoy's flare, $K = \frac{\rho g b_1 w}{m+m_a}$
b_0	Minimum breadth of buoy (figure 5 and equation (5))
b_1	Described by equation (5)
w	Width of buoy (in the direction of wave propagation)
n	Exponent of buoy flare, n >1 (equation 5))
F	Wave excitation force, normalized by total mass, $(m+m_a)$
Ω	Wave angular frequency
σ	Phase angle

As a special value of flare exponent, $n = 2$ yields the well-known damped- forced form of Duffing's equation, used to model vibrations of "hardening springs":

$$\ddot{z} + c\dot{z} + \omega_n^2 z + \mu z^3 = f(t) \tag{7}$$

Although various methods of solving Duffing's equation exist in literature, as summarised by [7], the authors have opted for a numerical solution to equation (6) above to analyse the behaviour of the system. The pumping action of the buoy, and hence power extraction, is modelled as the equivalent linear damping.

4. Damping and power extraction

Total power (sum of power dissipated by natural damping and extracted hydraulic pump power from the system) from the system with N buoys can be evaluated from the linear damping term, assuming no interaction related to damping between individual buoys:

$$P = N \cdot \frac{1}{T}\int_0^T F_D \dot{z} \cdot dt = \frac{N}{T}\int_0^T (m + m_a) \cdot (c\dot{z}^2) \cdot dt \tag{8}$$

where F_D is the damping force and T is the period of motion of heave being equal to the period of main component of the waveform. Assuming the heave motion is periodic with the wave period, T,

$$z = Z_0 \cos \Omega t \tag{9}$$

the total power of the system shall be:

$$P = (m + m_a) N \xi Z_0^2 \Omega^2 \omega_{n0} \tag{10}$$

where the classical assumption $c = 2\omega_{n0}\xi(m + m_a)$ was made, ξ being the nondimensional damping coefficient.

5. Heaving motions of the buoy- a numerical example

As an illustrative example, a system with six buoys, each 1 m apart, and connected to the peristaltic pumping system described above at 5 m. depth is conceived. The characteristics of an individual buoy is below:

Draught, T	:	1.0 m
Breadth	:	$b(z) = 0.50 + 0.25z^2$
Width, w	:	2.0 m
Space between two buoys	:	1.0 m
Mass	:	1367 kg

Nondimensional damping coefficient (ξ) was arbitrarily taken to be 0.40. Added mass (m_a) was taken to be 0.20×m, also arbitrarily. Natural frequency of heave for limitingly small amplitudes (ω_{n0})

of an individual buoy is $\omega_{n0} = \sqrt{\frac{\rho g b_0 w}{m+m_a}}$ =2.476 rad/s. The resonant frequency $\omega_r = \omega_{n0}\sqrt{1-\xi^2} =$ 2.269 rad/s. (Thomson and Dahleh, 2005). From linear wave approximation, Equation (5), this amounts to a wavelength of 12 m. With the well-accepted deep water approximation of wave steepness = a/λ = 0.025, a = 0.30 m, the differential equation, Equation (8) was solved numerically. Plot of wave motion is in Figure (5) and the initial transient behaviour in terms of phase plane is in Figure (6).

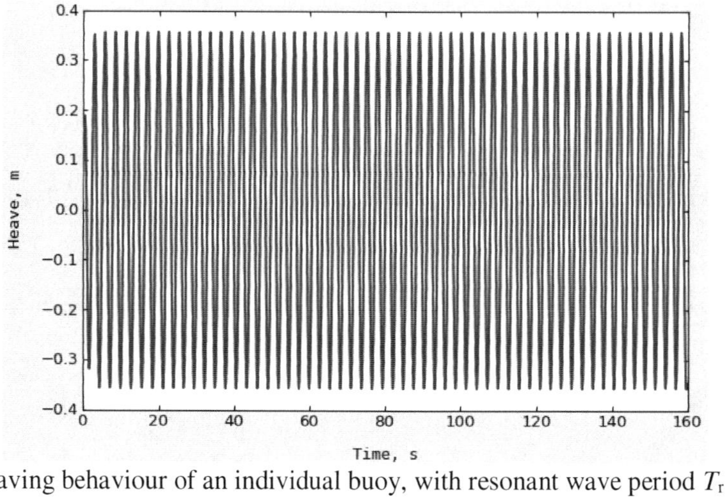

Figure 6. Heaving behaviour of an individual buoy, with resonant wave period T_r =2.77 s and corresponding waves with λ = 12 m, a = 0.30 m, initially at rest

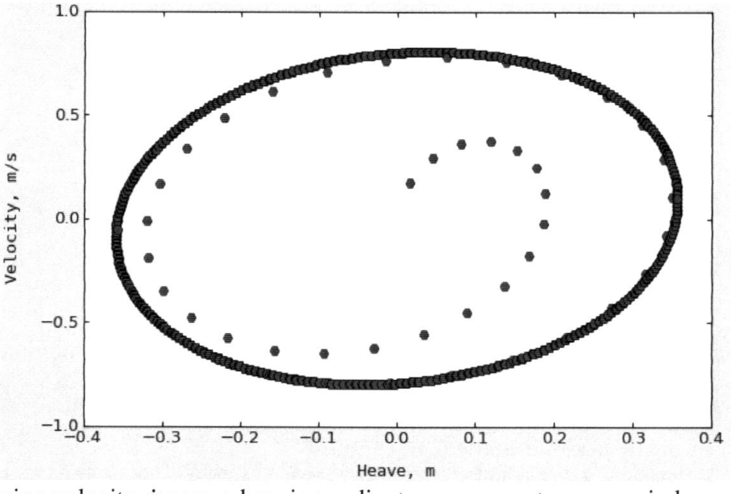

Figure 7. Heaving velocity $\dot{z}$ versus heaving ordinate, z, resonant wave period case of the case in Figure 6

As can be seen, the amplification factor, $\frac{z_{max}}{a} = 1.19$ for this resonant state. However, for a more efficient operation, heavier sea states should be assumed.

Next, a heavier sea state with wave period (T_w) = 5.06 s corresponding to wavelength (λ) = 40 m and wave amplitude a = 1.00 m was studied. The results for heaving versus time and heave velocity versus heave ordinate are exhibited in Figures 8 and 9 respectively below. It is noteworthy to observe that $\frac{z_{max}}{a} = 0.865$ for this case. The heaving behaviour of the buoy is shown in figure 9.

IOP Conf. Series: Earth and Environmental Science **172** (2018) 012002 doi:10.1088/1755-1315/172/1/012002

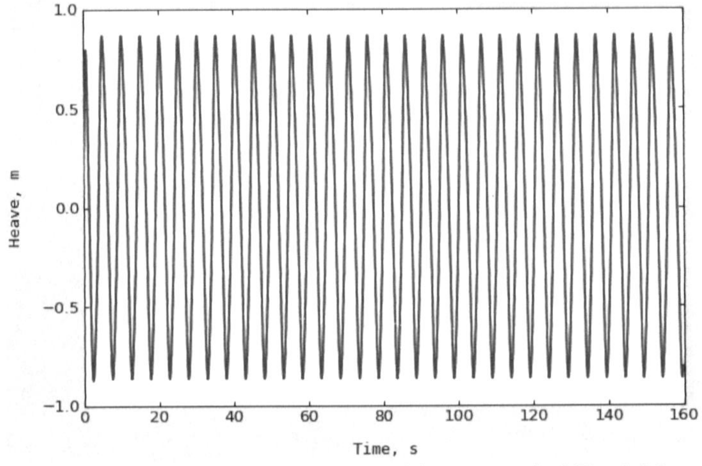

Figure 8. Heaving behaviour of an individual buoy, with wave period T_r =5.06 s and corresponding waves with λ = 40 m, a = 1.0 m, initially at rest

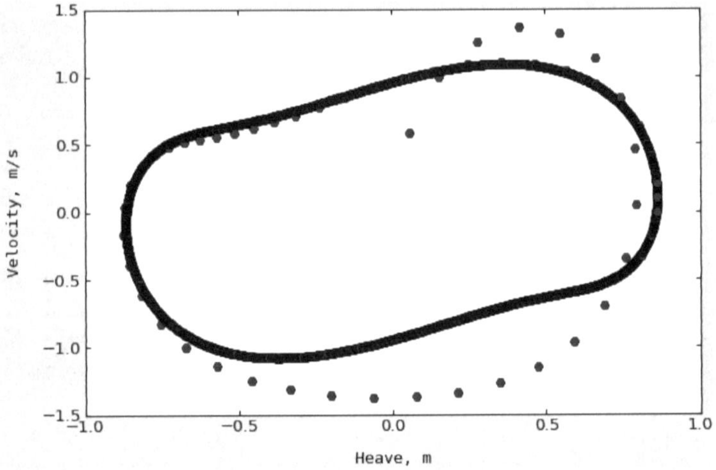

Figure 9. Heaving velocity $\dot{z}$ versus heaving ordinate, z, with wave period T_r =5.06 s and corresponding waves with λ = 40 m, a = 1.0 m of the case in Figure 7

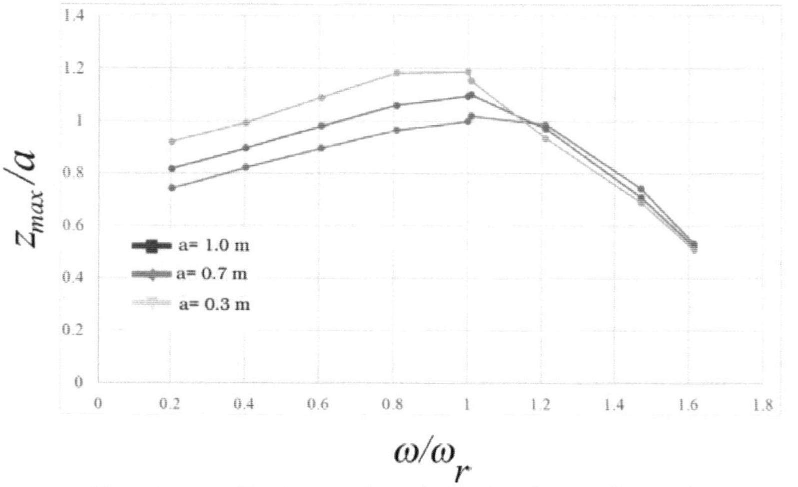

Figure 10. Response of heaving amplitude, z to the amplitude of monochromatic wave, a for a range of frequencies, normalized for the zero-amplitude resonant frequency, ω_r

Therefore, for this particular buoy configuration, based on 40 m design wavelength, a design with rectangular channel height = 2α =1.5 m shall be selected. Ideal pump capacity for this design case shall be, from Equation (3), 7.11 m^3/s. The power obtained, from Equation (10) was deducted to be approximately 7.5 kW, assuming power extraction was responsible for half of the dissipated energy. The pumping rate was found to be 23.68 m^3/s from Equation (4), for $\alpha = 0.75$ m and $w_p = 2$ m..

6. Some design considerations and conclusions

The proposed wave energy system has to be robust enough to endure heavy sea states and harsh marine environment, like all other wave energy systems. There should be another part above the water surface in addition to the buoys, where the electrical generation outfitting shall be located at. The peristaltic pump should be placed at a depth where the water is relatively stagnant, approximately at half the wavelength of the heavy seas expected.

The height of the peristaltic channel, 2α, is at the same order but also can be different from the waveheight, thanks to the hydraulic systems ratio of the pump piston area to ram piston area. In heavy seas, where both the hydraulic pump's and the hydraulic ram's amplitudes are restricted by maximum stroke lengths, the shape of the diaphragm's deflection shall approach to that of a square wave, rather than a sinusoidal one.

This study takes only the heaving motion of the buoys into account. However, the buoys also experience yaw and sway motions, as well as roll, pitch and yaw motions. Sway and surge motions shall also have an effect on the peristaltic pumping action. To ensure proper operation, the rod connecting the buoy to the hydraulic pump should be equipped with universal joints. Also, the buoys should be connected by a loose tether.

The proposed system with pump can either be bottom mounted or can be installed on a floating platform incorporating the peristaltic pump channel and apparatus, turbine and electrical generator and the buoys.

Like all other wave energy systems, it is proposed to manufacture a prototype system and test it at a location in sea where the waves are predictable and energic enough for an efficient operation. If further studies and model tests shall be made, a more complex non-linear and short-term motions analysis might be considered. The damping coefficient ξ and the added mass m_a were taken arbitrarily in this study, although more refined values, supported by tank tests can be taken. As a final word, other means of converting the surface wave shapes into peristaltic motion can be developed, although it is deemed that this is the mechanically simplest way.

IOP Conf. Series: Earth and Environmental Science **172** (2018) 012002 doi:10.1088/1755-1315/172/1/012002

References

[1] Falcao A 2010 *Wave energy utilization: A review of the technologies,* Renewable and Sustainable Energy Reviews 14 899–918

[2] Ganiev R T Ganiev S R Kasilov V P Pustovgar AP 2015 *Wave Technology in Mechanical Engineering,* Scrivener Publishing, 2015

[3] Latham T W 1966 *Fluid Motions in a Peristaltic Pump* M.S. Thesis, Massachusetts Institute of Technology, Boston, U.S.A.

[4] Le Mehaute B 1976 *An Introduction to Hydrodynamics and Water Waves*, Springer-Verlag, New York, U.S.A.

[5] Thomson,W T, Dahleh M D 2005 *Theory of Vibration with Applications*, Pearson Education Asia, Tsinghua, China

[6] Sabuncu T *Gemi Hareketleri, (2nd ed)*, 1993 İstanbul Technical University Press, İstanbul, Turkey

[7] Marinca, V, Herisanu N 2011 *Nonlinear Dynamical Systems in Engineering*, Springer Verlag, Berlin-Heidelberg

ESR Study of Irradiated Polysaccharides
ESR investigation of gamma irradiated pectin

Raluca Mihaela Slave [1], Maria Nicoleta Grecu[2] and Vlad Voicu Grecu[3]

[1]"Mircea cel Batran" Naval Academy, Constanta, Romania
[2] National Institute for Materials Physics, Bucharest, Romania
[3] Physics Faculty, University of Bucharest, Bucharest, Romania

Polysaccharides irradiation is usually used for their decontamination of fungi, pathogenic bacteria, toxigenic molds, parasitic organisms. ESR studies are performed just to evidence their irradiation. Therefore the accent is put on radical stability in time and ESR signal intensity dependence on irradiation dose. Less attention is given to radical nature and their dependence on plant nature. In this report a study of gamma irradiated pectin is given. Dose dependence of double integrated ESR signal is analyzed in terms of a recombination reaction during the radiation period. Radical stability dependence on water content is put in evidence.
Keywords: free radicals, gamma irradiation, pectin, ESR

1. INTRODUCTION

Polysaccharides, like pectin, corn and potato starch and cellulose are widely used in food industry. The pectins are heteropolysaccharides, having a heterogeneous chemical structure depending function on the plant origin and extraction method [1]. The main structure is a backbone chain made of uronic acid (galacturonic acid) residues linked through α-1-4 glicosidic bonds. Several side chains containing sugars like arabinose, glucose, fucose, mannose, xyulose are linked to the backbone chain. Pectins are soluble vegetable fibers, having the role of emulsifier, stabilizer and thickening agent in the food industry. It is extracted from the peel of some fruit. The richest fruits in pectin are citrus fruits (up to 30%) and apples (1-3%). Galacturonic acid has the structure:

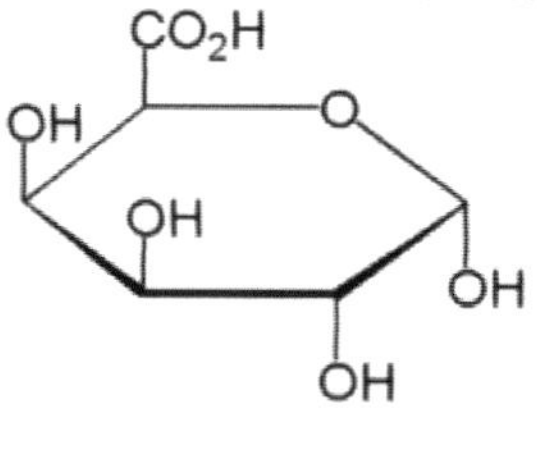

Carboxylic groups have poor acidic properties, so that:

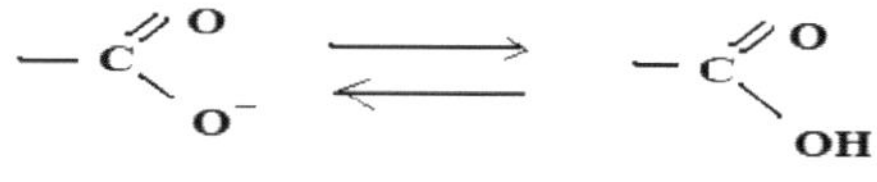

The pectins contain about 200 molecules of galacturonic acid, the acid units being linked to (1 → 4) α-D-galacturonic acid linkages, linked to α-L-rhamnose residues (simple saccharide structures).In the natural state most of the carboxyl groups are methylated (about 80%) with the structure:

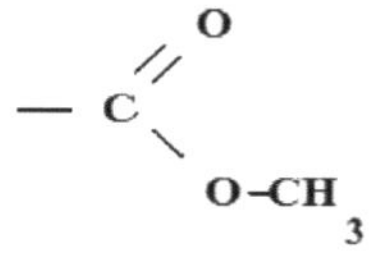

The pectins are extensively studied [2, 3].The apple pectins contain more sugars and have a molecule larger than the citrus pectins. Many studies have demonstrated that food irradiation with γ, X rays or high electrons (with energies below 10 MeV) and doses up to 10 kGy does not effect food quality, evidently does not induce radioactivity in it, being absolutely safe for consumers [4, 5]. Together with other methods like thermo and stimulated photoluminescence, viscosity measurements, chromatographic analysis of hydrocarbons, DNA comet assay screening, microbiological screening DEFT/APL, electron spin resonance ESR is accepted as a standard detection method for a large variety of irradiatedsolid food (meat with bones, sugars, spices and herbs, fruits, vegetables, fish and shellfish) [6]. This is confirmed by the European Standards emitted by European Committee for Standardization (CEN)[7]. Irradiated polisaccharides have been widely studied by ESR [8, 9, 13, 15]. Irradiation induces free radicals, stable for various time periods, which give measurable ESR spectra. The type and time stability of these signals are strongly dependent on the material, on temperature, humidity, the presence of oxygen and other factors [5, 7, 8, 11]. In irradiated fruits pectin radicals are significant, but are associated with radicals in other major constituents like cellulose, pulp, lignin [16]. Therefore it is important to identify the contribution of pectin free radicals to the whole ESR spectrum. This is one of the purposes of this paper. In the same time dose dependence of free radical concentration and their time stability will be investigated.

2. MATERIALS AND METHODS

The samples used in this paper are apple and citrus pectin, imported by Hofigal S.A., the main Romanian manufacturer of the natural products. All investigated specimens were presumed not be irradiated. Irradiation was performed in air, at room temperature, at a dose rate of 6.1 kGy/h, using a Co^{60} Gamma Chamber 5000 (BRIT, India) irradiation facility of Horia Hulubei National Institute of Physics and Nuclear Engineering Bucharest. Absorbed doses were evaluated by means of an ethanol-chlorobenzene dosimetry system with oscillometric readout method, andexpressed as absorbed dose in water. The target doses applied to samples were of 3, 7, 9 kGy, respectively, with an average uncertainty of 0.1-0.2 kGy. EPR measurements were performed at room temperature by using an X-band (9.5 GHz) Adani CMS 8400 spectrometer with the following characteristics: frequency range: 9.1-9.6 GHz; temperature: 83 K $< T <$ 480 K; magnetic field: 0.01 T$<$ B$<$ 0.7 T; sensitivity:2.5x 10^{10} spins/ Gauss. Irradiated and non-irradiated samples were introduced into quartz tubes (1 cm sample length, 3-4 mm inner tube diameter) and placed in the resonant cavity always in the same position. All measurements were performed at the same microwave power. Finally, the spectra were scaled at the same amplification and a standard sample weight. The numerical analysis of ESR spectra was performed with the Win-ESR Simpfonia, ESR spectra simulation program from Bruker

3. RESULTS AND DISCUSSION

Examples of irradiated apple pectin are given in fig.1

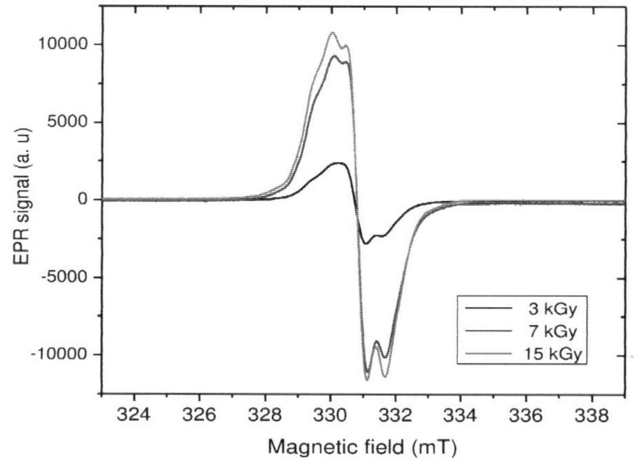

Fig. 1. EPR spectra of apple pectin samples irradiated

Irradiation increases their intensity, but does not induce noticeable changes in their shapes. The double integrated spectra, corresponding to the area below the absorption curve, show a monotonous increase with irradiation dose. These values are proportional to radical concentration in sample, therefore its dose dependence corresponds to radical concentration dose dependence. Their dependence is given in Figure 2.

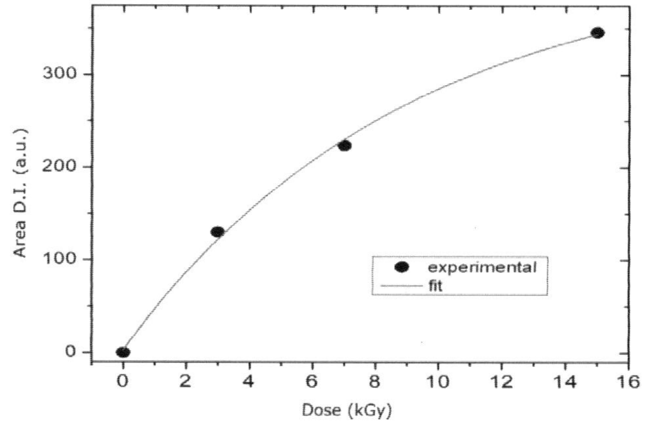

Fig. 2. Dose dependence of double integrated EPR spectra of apple pectin samples.

It can be fitted with the equation

$$A(D) = A_0 + A_{sat}\left(1 - e^{-\ln 2 \frac{D}{D_{1/2}}}\right) \qquad (1)$$

Where A_0 is the initial value, A_{sat} and $D_{1/2}$ parameters related to radical generation and recombination rate. As you can see from Fig. 2, saturation effect appears at doses:

$D_{1/2}$ = 6,193 kGy ± 1,169 kGy ; A_{sat} = 437±23, A_0 = 0, Adj.R-Square=0,99428.

Such a dependence is quite common for irradiated food and was reported for many materials [17]. It corresponds to a differential equation for time dependent radical concentration N(t),

$$\frac{dN}{dt} = Q - kN \qquad (2)$$

Where Q – rate of free radicals induced by irradiation in the sample (constant), k – reaction rate for a monomoleculaire reaction. The time dependence is changed in dose dependence

$$t = \frac{D}{D_o}$$

where D_o is the dose rate. The solution for N(D) (with initial condition N(0) = 0) is:

$$N(D) = N_{sat}\,[1\text{- }\exp(-\frac{ln2}{D_{1/2}}D)] \qquad (3)$$

Where $N_{sat} = \frac{Q}{k}$, $D_{1/2} = \frac{ln2}{kD_o}$ is the dose at which N = (½) N_{sat}. Relation (3) is equivalent to (1). The dose dependence for citrus pectin is similar, with $D_{1/2}$ = (5.47 ± 2.37) kGy

The complex spectrum of irradiated pectins is composed of at least three different signals, a single line in the center, a doublet component on each side of the single and a broad line whose structure can not be resolved in components. The parameters of these components are summarized in table1

Table 1

Sample	Irradiation Dose(kGy)	g-value	Hyperfine A-value(mT)	Line width (mT)	Component identification
Apple pectin	12	2.003	---	8	Single line
		2.001	13.5	9.5	Doublet component
		2.005	?	30-35	Broad line, unresolved structure

A single line of such width was identified in other polycarbonates containing uronic acids [16, 17]. It can be associated to a **C-O-O·** type radical, very likely to be produced in any uronic acid and monosaccharide sugars. The doblet component can be associated to a radical of the type

$$-C{\scriptstyle\genfrac{}{}{0pt}{}{/O}{\backslash OH}}$$

The hyperfine interaction is due to the hydrogen nucleus. The third line, very probably o superposition of different lines of a triple-quadruplet structure, is quite common in various irradiated monosccharides [17], but in such complex material it is impossible to resolve its structure using only ESR spectroscopy (ENDOR investigations could solve this problem).

Concerning the radical time stability it is evident that for samples with a higher water content, like citrus pectin, the stability is much lower. Such water efect on free radical stability was frequently reported in literature [18].

The following picture (Fig.3) shows the amplitude of the EPR signal of citrus pectin irradiated 7kGy after 2h, 4h respectively after irradiation. After 30 hours of irradiation, the signal disappears. In apple pectin, less wet, the time dependence is much slower. For citrus pectin free radical concentration decreases exponentially, with a half time $T_{1/2} = 3.20 \pm 0.75$

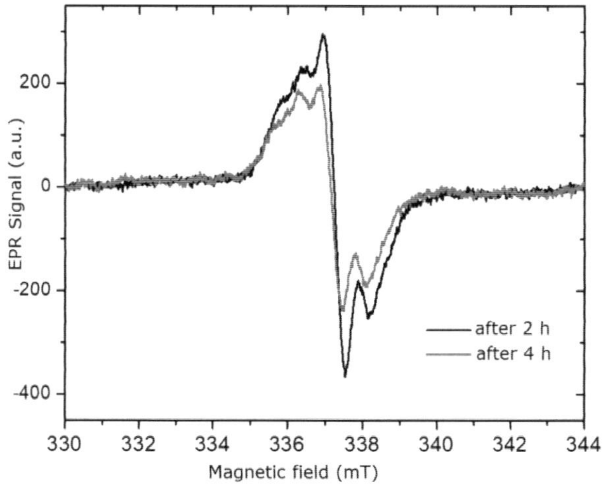

Fig.3- EPR signal of citrus pectin samples irradiated

4. CONCLUSIONS

- The gamma-irradiation induces stable radicals in apple and citrus pectin. They can be observed up to 30 hours after irradiation for wet samples and for much longer time periods for dry samples.
- The dose dependence of the double integrated area, i.e. proportional to free radical concentration, follows a generation with saturation mechanism.
- The gamma irradiated citrus pectin has a similar shape as apple pectin, but much less stable at room temperature, due to its higher water content.
- EPR spectrum is composed at least of three components. A qualitative analysis made possible assignments to most probably free radicals types.

5. REFERENCES

[1] Muller-Maatsh J, Bencivenni M, Caligiani A, Tedeschi T, Bruggeman G, Bosch M, Petrusan J, Van Droogenbroeck B, Elst K and Sforza S. *Pectin content and composition from different waste streams,* Food. Chem. 201, 37-45, 2016

[2] Lopes da Silva JA and Rao MA. *Pectins: Structure, functionality and uses.* Food Polysacharides and their Applications,. Stephen AM, Phillips GO and Williams PA, (eds) CRC Press, 2006, Bocca Raton

[3] Kravtchenko TP, Voragin AG,I and Pilnik W. *Analytical comparison of three industrial pectin preparations.* Carbohydr. Polym. 18, 17, 1992

[4] Diehl JF. *Assessment of wholesomeness of irradiated food*, Acta Alimentaria 23, 195-214, 1994

[5] Delincee H . *Analytical methods to identify irradiated food- a review*, Radiation Rhys. And Chem. 63,455-458, 2002
[6] Ahmit SK, Uddin MdM, Rahman R, Islam SMR and Khan MS. *A rewiew on mechanisms and commercial aspects of food preservation and processing,* Agric. And Food Secur. 6, 51-73, 2017
[7] EN 1787.*European Standard. Determination of irradiated food containing cellulose: Method by ESR Spectroscopy*, European Committee for Standardization, Brussels, 2000
FAO/IAEA/WHO 1999; *High-dose irradiation: Wholesomeness of food irradiation with doses above 10kGy*
[8] Raffi JJ, Yordanov ND, Chabane S, Donifi L, Gancheva V and Ivanova S. *Identification of irradiation treatment of aromatic herbs, spices and fruits by electron paramagnetic resonance and thermoluminescence*, Spectrochimica Acta A 56,409-416, 2000
[9] Yordanov ND, Gancheva V, Radicheva M, Hristova B, Guelev M and Penchev P. *Comparative identification of irradiated herbs by the methods of EPR and TL*, Spectroscopica Acta A 54, 2413-2419, 1998
[10] Yordanov ND and Aleksieva K. *Xand Q band studies in fine powders of irradiated plants. New approach for detection of their radiation history by using Q band EPR spectrometry*, Rad.. Phys. and Chem.,69,59-64, 2004
[11] Bercu V, Negut CD and Duliu OG. *ESR studies of free radical kinetics in gamma-rays irradiated pleurotus ostreatus oyster mushrooms,* Food Res.Int., 44, 1008-1011, 2011
[12] Slave RM, Grecu MN, Negut CD and Grecu VV. *ESR on some gamma- irradiated aromatic herbs.* Romanian Journal of Physics, 2014
[13] Thayer DW and Rajkowski K T. *Developments in irradiation of fruits and vegetables*, Food Technology 53 (11), 62-65, 1999
[14] Yordanov ND and Gancheva V. *A new approach for extension of the identification period of irradiated cellulose containing foodstuffs by EPR spectroscopy.* Appl.Rad. and Isotropes 52,195-198,2000
[15] Yu L, Reitmaier CA and Love M H. *Strawberry Texture and Pectin Content as Affected by Electron Beam Irradiation,* Journ. of Food Science, 61(6), 844-846, 1996
[16] Raffi JJ and Agnel JPL. *ESR identification of irradiated fruits,* Int. Journ. of Rad. Applic. And Instr., Part C: Rad. Phys. And Chem., 34(6), 891-894, 1989
[17] Nelson W.H., *Dose-Response Relationship for Radicals Trapped in Irradiated Solids*, Radiation Research163, 2005, 673-680; Roberts J., Simmons J. A., *Dose-response relationship in the yield of radiation induced free radicals in amino-acids*, Phys. Med. Biol., 7, 1963, 429-497, and references in these papers
[18] Ahmed ben Ahmed Salem Mamas. *Gum Arabic and Costituents Sugars Studied by ESR,* Ph.D Thesie, Brunel Univ., UK, 1991
[19] Leonor SJ, Gomez JA, Kinoshita A, Calandreli I, Tfouni E and Baffa O. *ESR spectroscopic properties of irradiated gum Arabic*, Food Chem., 141(3), 1860-1864, 2013

Port performance evaluation. Case study: Ports in the Black Sea basin

Florin Nicolae[1], Marius Bucur[2] and Alexandru Cotorcea[1]

[1]Navigation and Naval Management Interdisciplinary Research Centre, Mircea cel Batran Naval Academy of Constanta, Romania
[2]Politehnica University of Bucharest, Romania

E-mail: nicolae_florin_m@yahoo.com

Abstract. Water transport is one of the essential elements of international economic exchanges. Over 85% of international trade is achieved through this mode of transport, which remains the cheapest and at the same time an indispensable factor for meeting demand for goods and services. The development of a country's economy, which is open to the sea, requires, among other things, the stimulation of maritime and river transport and, implicitly, of port activity. Improving port system performance reduces transport costs, and indirectly can ensure increased competitiveness of the economic sectors at the level achieved by other EU countries. This is only possible through innovative management to ensure the identification and implementation of quality improvement solutions and optimization of port logistics processes. The paper summarizes the results of the research on: the main features of port services, the quality of port services, port system components and port competitiveness. The comparative analysis of port development priorities was based on a set of criteria accepted internationally. The research carried out included both the highly developed ports and those in the average performance area. The conclusions of the paper highlight several relevant aspects that could be detailed in order to find solutions for increasing port competitiveness.

1. Introduction

Ensuring a growing volume of goods transported by water is made by larger and more specialized ships that are subject to strong pressure to adapt to the new international trade performance requirements. Literature reveal that one of the medium and long-term priority objectives for port performance is associated with increasing the efficiency of port subsystems: cargo handling and transshipment, indoor transport, storage, delivery and reception, Figure 1.

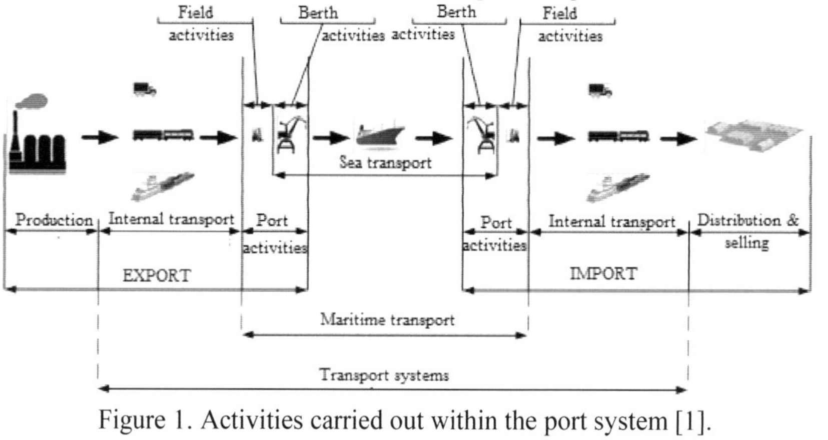

Figure 1. Activities carried out within the port system [1].

In this context, an analysis of the overall operating framework of the port system becomes a vital requirement for achieving a port-performance-enhancing strategy that will provide the port with a competitive position within the international logistics chain. Thus, the port has become, from the physical place where the cargo is transferred from the ship to the quay and vice versa, the logistics center that provides not only cargo handling operations but also a series of complex services provided to ships and cargo, such as: packaging and marking of goods, container repairs and industrial processing activities (such as petroleum refining), ship repair, bunkers etc. Step by step modern ports have become dynamic knots in international production and distribution networks over the last 20 years [2].

Port as a logistics center is a concept closely linked to the development of a country's foreign trade and is therefore associated with the expansion of the processing industry with added value for export [3]. In most cases, by increasing port productivity, maintenance of operating costs at lower levels is ensured and continuous management improvement for all subsystems of the transfer system transforms the port into an economic system that boosts economic growth at regional and national level.

The globalization process makes ports the main points of access to the international commerce network. That is why the world economy needs well-developed and efficient ports according to international standards. In the case of the European Union, while some of the ports are among the best performing in the world (Rotterdam, Antwerp and Hamburg), others have consistently recorded lower performance for some time. Lack of performance has become permanent, turning into a structural decline that slows regional economies. Consequently, the structural performance gap reduces the options available to maritime and logistics operators at international level. It is important to note that in 2011 19.8% of the goods arriving in Europe by sea crossed the three ports mentioned above. In 2016 this percentage increased to 20.3% [4].

The domino effect is anticipated in numerous scientific papers. In the case of high-performance ports in the western part of Europe, there is congestion and "suffocation" of the hinterland area with major repercussions on road and rail traffic and implicitly on increasing discomfort for citizens in these regions. Given that freight volumes are projected to reach more than 5.5 billion tonnes of freight by 2030 compared to 3.9 billion tonnes in 2012, it will be necessary that all ports in the trans-European network to contribute in order to cope this growth [5].

Figure 2. Main ports in the Black Sea Basin

Given that the ports in Western Europe are overburdened, statistics show that only 2.5% of international maritime trade has taken place in the Black Sea ports over the last 10 years [6]. This information shows that the advantages offered by the geographical opening of this area and its connectivity with transport routes in Europe, Asia and Africa are not yet capitalized, Figure 2. It is to be expected that in the medium and long term the freight volumes that will transit the ports in the Black Sea basin, such as Varna and Burgas (Bulgaria), Trabzon (Turkey), Poti and Batumi (Georgia) and Novorossiysk (Russia) to grow.

The analysis shows that these ports have experienced important economic and institutional changes, the globalization process, structural changes in logistics and distribution networks, as well as the fierce competition between them that influences their development. At the same time, countries in the Black Sea region have major plans to extend port and terminal specialization for certain types of freight, especially bulk and container cargo [7].

The development of ports in this area could lead to new models of regional development of port networks and other dynamics of trade relations. Given that the port has become a key component, dynamically integrated into Supply Chain Management, the authors highlight the need for a new approach to the issue of the general framework of the port system, whose international dimension is also recognized in European documents. Starting from the existing realities in the Black Sea basin, in the case of the provision of goods transhipment services there are unfair competition practices from the Black Sea ports [8]. Therefore, medium and long term sustainable development of this geographical area will not be possible without closer cooperation with neighboring countries, with a goal to achieve a sustainable transport system which will be beneficial for all EU partners.

Under these circumstances, to identify the port development priorities in the Black Sea basin, an analysis of the overall port operation framework and port facility and performance reporting is required based on a set of internationally accepted criteria.

The research aims determining a set of Priority Indicators for Port Development (PIPD) and assessing them for ports in the Black Sea basin. Under these conditions, 15 European ports were identified: developed European ports as reference models (Rotterdam, Antwerp, Hamburg), average performance ports (Marseilles, Algeciras and Valencia, stating that in EU records also the port of Constanta is in this area) and the main ports in the Black Sea basin.

2. Research methodology and results

An important concept in scientific research is that of research methodology, which describes the methods used and the steps taken on the road to knowledge. In the scientific approach, the authors combined a set of qualitative and quantitative methods to obtain a set of priority indicators for port development.

The conceptual framework for this research was structured in two parts. The first part summarizes the information resulting from the analysis of the literature regarding the priorities of the development of the port activity. The results were centralized in the form of a set of 10 indicators, which can ensure the assessment of the level of development of a port. The proposed indicators are: the advantages of geographic positioning, the quality of hinterland connections, the availability and efficiency of port services provided to goods and ships, the price of port products, socio-economic sustainability, port infrastructure, information and communication systems in port activity, flexibility and adaptability to changes in international trade. The literature with the highest relevance for the problems related to the development of the port system were selected, Table 1.

Table 1. Priority indicators for port development – Literature review

Identified indicator	Code	Reference
Geographic positioning	I1	[9, 10, 11]
Quality of transport connections with hinterland	I2	[12, 13, 14]
The availability and efficiency of port services provided to goods	I3	[2, 15, 16, 17, 18]
The availability and efficiency of port services provided to ships	I4	[2, 19, 20]
Prices of port products	I5	[2, 20]
Socio-economic sustainability	I6	[21, 22]
Port infrastructure	I7	[1, 23, 24]
Informational and communication systems in port activity	I8	[2, 25, 26]
The human resource involved in port activity	I9	[22, 27, 28]
Flexibility and adaptability to changes in international trade	I10	[29, 30, 31]

From the selected references, the main characteristics that resulted for each indicator were summarized and they were centralized in Table 2. The selection of characteristics was reported to the port activity framework and to the best practices implemented by the developed European ports, considered in the paper as reference models (each indicator having the maximum performance level for them).

For the second part of the research method, the identified indicators, along with their characteristics, formed the basis of the implementation of a port performance assessment questionnaire (PPAQ). The questionnaire was discussed and analyzed in exploratory interviews (expert group method) with the most relevant stakeholders in the maritime industry in Romania, representatives of: shipping companies, terminal operators, shippers, logistics groups and port administration. In the process of selecting the stakeholders, it was considered that the interviewees had connections through their activities with the ports and freight forwarding companies in central and western Europe as well as with ports and companies from the countries in the Black Sea basin. Individuals selected to participate in exploratory interviews have a high level of expertise and are familiar with port and port system issues.

Table 2. Details of priority indicators for port development

Indicator	Characteristics of priority indicators for port development
I1	A port located at the confluence of the major maritime routes and benefiting from favorable physical conditions (natural embankment, a area protected from the marine environment etc.) has a competitive advantage over other seaports. Examples of such ports: Singapore, Antwerp, Rotterdam, Hamburg, Constanta.
I2	Port activity is dependent on the quality of hinterland transport connections. These links, which ensure the relationship between the exporter and the consignee, are decisive for the international freight forwarding. A high quality level of hinterland connections contributes to the integration of services in the freight logistics chain.
I3	The main stakeholders involved in the work of a port are constantly represented by shipowners and shipboard operators (chargers) who subcontract services to maritime and transit agents. The services provided to the goods are diverse, of high importance for the logistics chain.
I4	The services provided to ships must provide: maneuvers for arrival / departure from the operational berth, binding / unbinding, power / water / fuel supply. Given the importance of total port staying time on its economic performance, these services must be delivered in time with maximum efficiency.
I5	The port, as an integrated system in the international logistics chain, must develop a commercial strategy tailored to regional and international competition. Each time the final price to be paid for the port product by importers / exporters is determined as the sum of the costs associated with the services provided by the shipping, transit and other port system components. For this reason, port administration, shipping agents, charterers and other stakeholders must ensure competitive prices in the maritime market.
I6	Socio-economic sustainability refers to solutions that prove to be difficult to control for a particular geographic area. Practice shows that unstable situations are mainly generated by political and military instability, economic boycott, incoherent and changing administrative policies, macroeconomic imbalances.
I7	Ports are socio-technical systems with a high degree of complexity, with a very important role in the international commodity exchange as a link for all forms of transport. Under these circumstances, port infrastructure must provide, through port systems and equipment, quality services for ships and freight. In the same context, ports are used as warehouses or distribution centers where value-added activities are carried out, such as labeling, packaging, packing, palletizing. The permanent connection to the international

	logistics chain often requires retechnology, the implementation of new and innovative solutions for all the components of the port system.
I8	Informational and communication systems have an increasingly important role in ensuring and maintaining an efficient management of commodity flows and relationships with partner organizations. The e-maritime concept is increasingly in port activity and in the activities of shipping companies.
I9	The port industry in recent years has focused on implementing advanced, high-productivity technological solutions that are less dependent on human effort and the immediate consequence is the rapid adaptation of professional training to these transformations. In recent studies, port personnel is grouped into two main categories: port workers and port executives. The port workers category includes stevedores, checkers, tallymen and clerical staff; executive staff includes supervisors of operations performed in the terminal. Special training programs are required for each category.
I10	Structural changes in international trade are permanent, generating significant quantitative and qualitative mutations in port activity. Flexible management in port administration, quick adaptation of port community members to change will be the most used tools for achieving port performance.

In the initial exploratory interview, the set of indicators identified in Table 1 that was converted into a set of Port Performance Measurement Indicators (PI) was validated and the assessment questionnaire corrected on the basis of the comments received. In the exploratory interviews that followed, the questionnaire was completed by all the interested stakeholders. The interviewees used information from various sources in which statistical data, technical information, annual activity reports on selected ports for analysis were presented.

The performance level for each indicator (PI) was measured using a five-point Likert scale ranging from 5 - excellent; 4- very good; 3- good; 2 - satisfactory; 1 - mediocre. Each indicator was evaluated for each port and the results are shown in Table 3.

As mentioned above, the selection of features was reported to the port activity framework and the good practices implemented by the developed European ports, considered in the paper as reference models (level 5 on the Likert scale), P* port. The average performance indicator PI_{med} shows the level of development of each port, level relative to the European reference model.

Table 3. Setting the indicators performance level of port development

Performance Indicator	**Evaluated port**									
	Odessa	**Illichivsk**	**Constanța**	**Varna**	**Burgas**	**Trabzon**	**Poti**	**Batumi**	**Novorosisk**	**P***
PI1	5.0	5.0	5.0	5.0	5.0	5.0	5.0	4.8	5.0	5.0
PI2	3.3	2.9	3.4	3.5	3.5	3.4	2.9	3.0	3.5	5.0
PI3	3.0	2.8	3.8	3.3	3.3	3.6	2.8	2.9	3.7	5.0
PI4	3.4	2.1	3.9	3.4	3.4	3.7	2.8	2.7	3.8	5.0
PI5	3.3	2.2	3.7	3.2	3.2	3.8	2.5	2.7	3.6	4.6
PI6	2.7	2.7	3.8	3.4	3.4	3.2	2.6	2.5	3.7	5.0
PI7	3.3	3.0	3.6	3.3	3.3	3.6	3.0	3.1	3.5	5.0
PI8	3.9	3.3	4.2	4.0	4.0	3.9	3.3	3.2	4.0	5.0
PI9	3.3	3.1	3.4	3.3	3.3	3.4	3.1	3.4	3.4	5.0
PI10	2.9	2.9	3.5	3.3	3.3	3.3	2.9	2.7	3.3	5.0
PI_{med}	**3.41**	**3.00**	**4.02**	**3.57**	**3.57**	**3.69**	**3.09**	**3.1**	**3.75**	**4.96**

The results presented in Table 3 show a major difference between the ports in the Black Sea basin, PI_{med}=3.118 and the reference ports P*, PI_{med}=4.96. Of the ports in this area, Constanta port has PI_{med}=4.02, followed closely by Novorosisk port PI_{med}=3.75. Port of Constanța has the advantage of its geographic position, at the intersection of the commercial routes linking the Central and Eastern

European markets with Transcaucasus, Central Asia and the Far East. This opportunity could be redeemed by providing a better connection with the hinterland. The main competitor of the port of Constanta is the port of Novorosisk. It is part of a group of ports alongside the Primorsk harbor, the Baltic Sea and the Baltisk harbor in the Kaliningrad region. This association can be an alternative route to North and Western Europe to the Western Black Sea ports. The analysis can be developed by a detailed research of each performance indicator.

3. Conclusions

Research shows that ports can play a decisive role in economic recovery and increased long-term competitiveness of a country or geographic economy. The paper proposes a set of priority indicators for port development and an assessment methodology through which port performance can be determined. The assessment methodology was applied to the representative ports of the Black Sea Basin in relation to a reference model corresponding to the highly developed European ports.

The results obtained have highlighted the performance of these ports in terms of the advantages of geographic positioning, the quality of transport connections with hinterland, the availability and efficiency of port services provided to goods and ships, the price of port products, socio-economic sustainability, port infrastructure, the quality of the human resource involved in port activity, flexibility and adaptability to changes in international trade. The obtained results highlight the structural disparities between these ports and the port reference model. The research can be continued in the following directions: the detail of each identified indicator on the types of terminals and sorted sorts of cargoes, the comparative analysis of port performance for various port entities, the identification of investment opportunities in certain sectors of port activity, the identification of training needs in certain domain/subdomain of port activity.

In conclusion, a correct assessment of a port activity is made when all the factors that influence its activity are considered. Port performance indicators should take account of these factors and reflect: how port facilities and resources are used to highlight the effect of corrective actions to prevent losses or improve operations; the way in which intensive activity factors are assessed so that planners can decide when and where additional resources are needed; the quality of services provided to shippers, carriers and other users of the port.

References

[1] Nicolae F., Ristea, M., Cotorcea A., Nistor, F., 2015. The Relationship Between Port Logistics and Global Logistics Performance. *Scientific Bulletin" Mircea cel Batran" Naval Academy*, **18(1)**, ISSN 1843-6749, pp. 83-88.
[2] Talley, W. K. (2017). Port economics. *Routledge*.
[3] Mangan, J., Lalwani, C., & Fynes, B. (2008). Port-centric logistics. *The International Journal of Logistics Management*, **19(1)**, 29-41).
[4] Review of Maritime Transport (Series), 2012-2017.
[5] European Comssion, (2017). Study on differentiated port infrastructure charges to promote environmentally friendly maritime transport activities and sustainable transportation.
[6] Ducruet, C. (Ed.). (2017). Advances in Shipping Data Analysis and Modeling: Tracking and Mapping Maritime Flows in the Age of Big Data.
[7] Grushevska, K., & Notteboom, T. (2014). An economic and institutional analysis of multi-port gateway regions in the Black Sea Basin. *Journal of International Logistics and Trade*, **12(2)**, 3.
[8] European Comission. (2014). Ports: an engine for EU growth.
[9] Chou, C. C. (2010). Application of FMCDM model to selecting the hub location in the marine transportation: A case study in southeastern Asia. *Mathematical and Computer Modelling*, **51(5-6)**, 791-801.

[10] Jacobs, W., Ducruet, C., & De Langen, P. (2010). Integrating world cities into production networks: the case of port cities. *Global networks*, **10(1)**, 92-113.
[11] Tongzon, J. L. (2009). Port choice and freight forwarders. *Transportation Research Part E: Logistics and Transportation Review*, **45(1)**, 186-195.
[12] Konings, R., Kreutzberger, E., & Maraš, V. (2013). Major considerations in developing a hub-and-spoke network to improve the cost performance of container barge transport in the hinterland: the case of the port of Rotterdam. *Journal of Transport Geography*, **29**, 63-73.
[13] Zondag, B., Bucci, P., Gützkow, P., & de Jong, G. (2010). Port competition modeling including maritime, port, and hinterland characteristics. *Maritime Policy & Management*, **37(3)**, 179-194.
[14] Aronietis, R., Markianidou, P., Meersman, H., Pauwels, T., Pirenne, M., Van de Voorde, E., & Verhetsel, A. (2010). Some effects of hinterland infrastructure pricing on port competitiveness: case of Antwerp. *In 12th World Conference on Transport Research*. Lisboa.
[15] Thai, V. V. (2016). The impact of port service quality on customer satisfaction: The case of Singapore. *Maritime Economics & Logistics*, **18(4)**, 458-475.
[16] Yeo, G. T., Ng, A. K., Lee, P. T. W., & Yang, Z. (2014). Modelling port choice in an uncertain environment. *Maritime Policy & Management*, **41(3)**, 251-267.
[17] Tang, L. C., Low, J. M., & Lam, S. W. (2011). Understanding port choice behavior—a network perspective. *Networks and Spatial Economics*, **11(1)**, 65-82.
[18] Wu, Y. C. J., & Goh, M. (2010). Container port efficiency in emerging and more advanced markets. *Transportation Research Part E: Logistics and Transportation Review*, **46(6)**, 1030-1042.
[19] Parise, G., Parise, L., Martirano, L., Chavdarian, P. B., Su, C. L., & Ferrante, A. (2016). Wise port and business energy management: port facilities, electrical power distribution. *IEEE Transactions on Industry Applications*, **52(1)**, 18-24.
[20] Esmer, S., Nguyen, H. O., Bandara, Y. M., & Yeni, K. (2016). Non-price competition in the port sector: A case study of ports in Turkey. *The Asian Journal of Shipping and Logistics*, **32(1)**, 3-11.
[21] Van den Berg, H. (2016). Economic growth and development. World Scientific Publishing Company.
[22] Rodrigue, J. P., Comtois, C., & Slack, B. (2016). The geography of transport systems. *Taylor & Francis*.
[23] Pavlic, B., Cepak, F., Sucic, B., Peckaj, M., & Kandus, B. (2014). Sustainable port infrastructure, practical implementation of the green port concept. *Thermal Science*, **18(3)**, 935-948.
[24] Haralambides, H. E. (2002). Competition, excess capacity, and the pricing of port infrastructure. *International journal of maritime economics*, **4(4)**, 323-347.
[25] Alderton, P., & Saieva, G. (2013). Port management and operations. Taylor & Francis.
[26] Mondragon, A. E. C., Lalwani, C. S., Mondragon, E. S. C., Mondragon, C. E. C., & Pawar, K. S. (2012). Intelligent transport systems in multimodal logistics: A case of role and contribution through wireless vehicular networks in a sea port location. *International Journal of Production Economics*, **137(1)**, 165-175.
[27] Nicolae F., Rosen I., Popa C., Nistor F., Cotorcea A., (2017). The relations between the port business framework and the qualified manpower compentencies – Literature review and proposed guidelines. *Mircea cel Batran Naval Academy Scientific Bulletin*, **18(1)**, ISSN 1843-6749, pp. 83-88.
[28] McConnell, M. L. (2011). The Maritime Labour Convention, 2006 - Reflections on challenges for flag State implementation. *WMU Journal of Maritime Affairs*, **10(2)**, 127.
[29] Taneja, P. (2013). The flexible port. Doctoral thesis
[30] Pantouvakis, A., & Dimas, A. (2013). The role of corporate agility and perceived price on the service quality–customer satisfaction link: some preliminary evidence from the port industry. *International Journal of Shipping and Transport Logistics*, **5(4-5)**, 412-431.
[31] Makris, S., Zoupas, P., & Chryssolouris, G. (2011). Supply chain control logic for enabling adaptability under uncertainty. *International Journal of Production Research*, **49(1)**, 121-137.

Using Gravity Potential Field and Inertial Navigation System in Real Time Submarine Positioning

Marius ROGOBETE[1], Octavian TĂRĂBUŢĂ[2], Ana Delia ROGOBETE[3], Sergiu EFTIMIE[4]

[1]Senior Software Eng. PhD. GE Power Romania
[2]Assistant Professor PhD. "Mircea cel Batran" Naval Academy
[3]Graduate teaching assistant PhD Candidate, Johns Hopkins University, Baltimore, USA
[4]PhD Candidate, Military Technical Academy
marius.rogobete@gmx.de

Abstract. Military submarines need specific skills when they operate in stealth mode, being not able to use their sonar system and need to travel in a total darkness environment. Therefore, a dead reckoning is used to evaluate the position, usually based on gyrocompass, measured speed and Inertial Navigation System (INS). This method has an acceptable distribution of errors for submarine position estimation, when it navigates in stealth mode for relative short time, but with serious errors for longer course. Our proposed method uses the Second Gradient of the Gravity (SGG) measured in real-time that is compared with a digital gravity gradient map, using an original method based on Kalman Filter. Finally, the INS estimation is improved based on correlation between the gradient map and the measured gradient value.

1. Introduction

Any military submersible embeds an inertial navigation system (INS), able to measure the boat's motion and constantly updates position, computing data from acceleration/deceleration, pitch and roll.

The main advantages of INS are:

- It does not rely on celestial sightings, GPS or radio signals;
- It allows the boat to navigate without sonar while remaining hidden under the surface.

Moreover, the estimation of position is doing with dead reckoning course information, processed up from gyrocompass information, measured speed and estimation of local ocean currents. The submarine is updating its reference position periodically, e.g. using outside navigational radio signals.

For deeper navigation is important the underwater terrain relief, usually detected with fathometer, used only in friendly water. Therefore, there is a challenging regarding positioning in stealth mode, when the sonar is not allowed. For this case, a silent measure replaces the active sonar, measuring a geophysical component that corrects the current point position by an indirect method. Matching the measured values with a previous obtained map, permits topographical coordinates extraction of the identified point on the map, which is the measuring point.

Our approach yields significant improvement, minimizing the drift tendency of the INS over time by matching the SGG (Second Gradient of the Gravity field) with gravity gradient map to determine the submarine position. Moreover, the method (especially SGG sensor) could be used on submersible drones, where the most specific characteristic is the extremely reduced size of the ship body.

2. Inertial Navigation System

A typical Inertial Navigation Systems performs navigation information from accelerometers and gyroscopes data but, for submarines, is added at least a pressure sensor for depth. The INS computer produces navigational state estimation using transformation matrix, Extended Kalman Filter and Unscented Kalman Filter. When possible, the position data is collected from GPS and INS set it as the reference position. These instruments are commonly referred as Inertial Measurement Units (IMUs).

The source of INS errors can be defined knowing the magnitude and direction of the submarine's accelerations, noted as vector $\ddot{\vec{s}}$, that is the second gradient of the submarine position $\vec{s}$ in respect to time. Therefore, the acceleration of the vessel $\ddot{\vec{s}}$ is equal with gravitational filed $\vec{g}(\vec{s})$ plus $\vec{a}$, the measured acceleration.

$$\ddot{\vec{s}} = \vec{g}(\vec{s}) + \vec{a} \qquad (1)$$

The position of the navigation equation (1) could produce the first order approximation of the position error in submarine's acceleration $\delta\ddot{\vec{s}}$,

$$\delta\ddot{\vec{s}} = \frac{\delta\vec{g}}{\delta\vec{s}}\delta\vec{s} + \delta\vec{g} + \delta\vec{a} \qquad (2)$$

Where the error in the gravitational field is $\delta\vec{g}$, $\delta\vec{a}$ is the error of measured acceleration and $\frac{\delta\vec{g}}{\delta\vec{s}}$ is the gravity gradient.

For modeling purpose, INS errors break down into three categories: inertial sensor error, alignment error and computational error. The inertial sensor error is accumulated over time into system and it isn't eliminated until the next GPS references. The other errors remain constant or oscillate over time between predictable limits.

The first order derivate equation estimates the sources of error at a given time, based on the initial sources of error and include the tilt error, heading error, velocity error, position error, gyroscope bias, and accelerometer bias. For any given time, the error may be found by multiplying a transition matrix by the assumed initial error source:

$$[\boldsymbol{\delta s}(t)] = [\boldsymbol{\Psi}(t - t_0)][\boldsymbol{\delta s}(t_0)] \qquad (3)$$

To improve the accuracy of the INS it should integrate the gained information (from gravity gradient instruments and map matching this aid) with the INS.

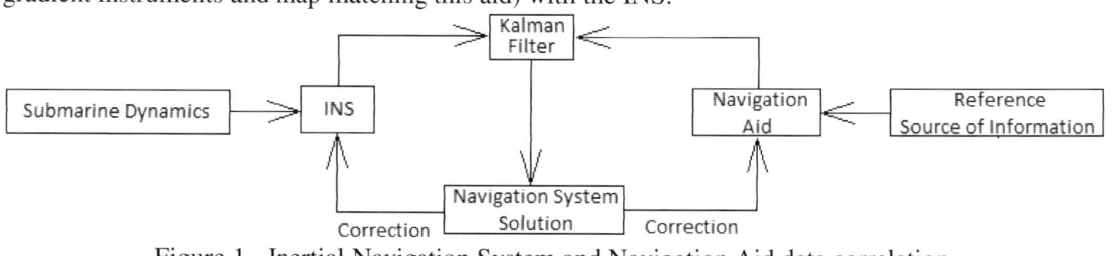

Figure 1 - Inertial Navigation System and Navigation Aid data correlation.

3. Kalman Filter

The information from navigation aids must be integrated using computer algorithms which maximize the accuracy of the navigation solution (e.g. a new position based on information from aid is updating the INS).

However, the Kalman filter is often choose as algorithm for integrating information from aids and INS. In literature it is described as a combination of two estimates of a variable to form a weighted mean, the weighting factors chosen to yield the most probable estimate. One estimate derives from INS information and the equations of motion, while the second comes from an aid, e.g. GPS.

The Kalman Filter equations (4), (5) are modeling the signal values evaluated with a linear stochastic equation (4), where $\boldsymbol{x}_i$ is the state vector at the moment $\boldsymbol{i}$, $\boldsymbol{w}_i$ is the white noise process with known covariance for Φ , and Φ represents the state transition process from the state of time $\boldsymbol{i}$ to $\boldsymbol{i+1}$.

$$\boldsymbol{x}_{i+1} = \Phi\boldsymbol{x}_i + \boldsymbol{w}_i \qquad (4)$$

To compute z_i that is actual measurement of $\boldsymbol{x}$ at time $\boldsymbol{i}$, the equation of the model for the observations (5) uses a state vector and measurement vector connection $\boldsymbol{M}$ without noise and specific measurement errors $\boldsymbol{v}_i$:

$$\boldsymbol{z}_i = \boldsymbol{M}\boldsymbol{x}_i + \boldsymbol{v}_i \quad (5)$$

Figure 2 presents the Kalman filter recursive algorithmic loop, where the fundamental matrix equations of a discrete, linear Kalman filter is Kalman Gain Equation (6).

$$\boldsymbol{\Lambda}_i = \frac{\mathbf{E}_{i|i-1}\mathbf{M}_i^T}{\mathbf{M}_i\mathbf{E}_{i|i-1}\mathbf{M}_i^T + \mathbf{N}_i} \quad (6)$$

The Update Estimate expression is:

$$\widehat{\boldsymbol{x}}_{i|i} = \widehat{\boldsymbol{x}}_{i|i-1} + \boldsymbol{\Lambda}_i\left(\mathbf{z}_{i|i} - \mathbf{M}_i^T\widehat{\boldsymbol{x}}_{i|i-1}\right) \quad (7)$$

Error of the Update Covariance equation:

$$\mathbf{E}_{i|i} = (\mathbf{I} - \boldsymbol{\Lambda}_i\mathbf{M}_i)\mathbf{E}_{i|i-1}(\mathbf{I} - \boldsymbol{\Lambda}_i\mathbf{M}_i)^T - \boldsymbol{\Lambda}_i\mathbf{M}_i\boldsymbol{\Lambda}_i^T \quad (8)$$

The state (9) and error covariance (10) projected in ***i+1*** form:

$$\widehat{\boldsymbol{x}}_{i+1|i} = \Phi\widehat{\boldsymbol{x}}_{i|i} \quad (9)$$

$$\mathbf{E}_{i+1|i} = \Phi\mathbf{E}_{i|i}\Phi^T + \mathbf{Q} \quad (10)$$

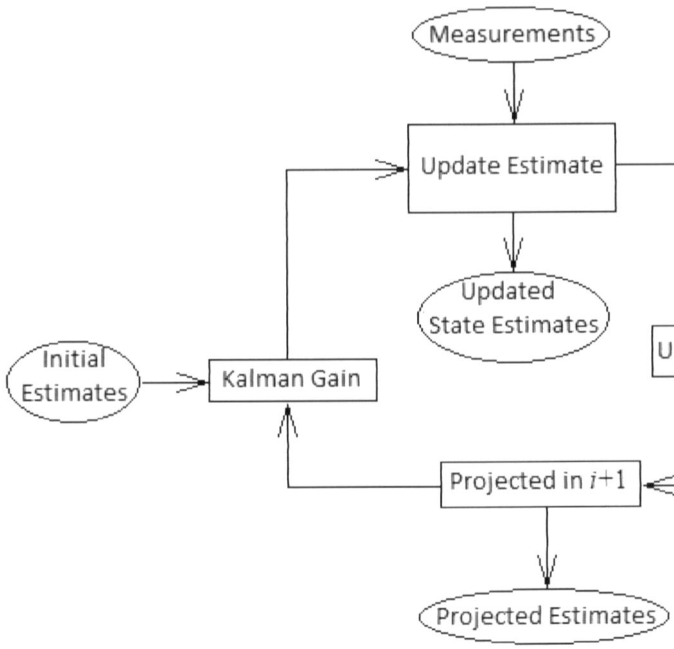

Figure 2 - Kalman Filter Algorithm. Squares represent matrix functions and ellipses - inputs/outputs.

Where $\widehat{\boldsymbol{x}}_{i|j}$ represents the estimate of $\boldsymbol{x}$ at observation time $\boldsymbol{i}$ up to but including at time $\boldsymbol{j}|\boldsymbol{j} \leq \boldsymbol{i}$ and $\boldsymbol{Q}$ represents a matrix of covariance that defines the system's noise, while the measurement noise is $\boldsymbol{N}$.

These equations assumes no correlation between the system and measurement noise and that both possess a zero mean and Gaussian distribution. In other words, the prediction process is defined by (9) and (10) and correction by (6), (7) and (8).

The most famous application of the Kalman filter occurred on Apollo mission moon flight, where it has provided corrections of the midcourse navigation.

4. Second Gradient of the Gravity Field

The vertical gravity gradient in geophysics is the rate of change of vertical gravity with high. However, in mathematical analyse it indicates the convexity/concavity (the sign of the first gradient) and inflexion points (the solutions of the second gradient) of the gravity graphic. The geopotential expression is:

$$\boldsymbol{W} = \boldsymbol{V} + \boldsymbol{V_c} \tag{11}$$

The potential derived from centrifugal force is:

$$\boldsymbol{V_c} = \frac{\omega^2}{2}(x^2 + y^2) \tag{12}$$

And the gravitational potential expression (which is the main part of geopotential field value) is:

$$\boldsymbol{V}(x,y,z) = \boldsymbol{G} \iiint_v \frac{\delta(\xi,\eta,\zeta)d\xi\, d\eta\, d\zeta}{[(x-\xi)^2 + [(z-\eta)^2 + [(x-\zeta)^2]^{1/2}} \tag{13}$$

With $\boldsymbol{P}(x,y,z)$ the observation point, $\Pi(\xi,\eta,\zeta)$ the source effect, $\boldsymbol{G}$ the gravitational constant and $\boldsymbol{\delta}$ the density of the source.

The gravity $\boldsymbol{g}$ is the first gradient of geopotential field, in 3D form equation:

$$\boldsymbol{g}(x,y,z) = \sqrt{\left(\frac{\partial W}{\partial x}\right)^2 + \left(\frac{\partial W}{\partial y}\right)^2 + \left(\frac{\partial W}{\partial z}\right)^2} = \sqrt{g_x^2 + g_y^2 + g_z^2} \tag{14}$$

As the horizontal components are practically null, (14) becomes:

$$\boldsymbol{g}(x,y,z) = \sqrt{g_z^2} \cong g_z = \frac{\partial g}{\partial z} \tag{15}$$

The vertical gradient of gravity is:

$$\boldsymbol{W}_{zz} = \frac{\partial^2 W}{\partial z^2} = \frac{\partial g_z}{\partial z} = \frac{g_z\left(z+\frac{l}{2}\right) - g_z(z-\frac{l}{2})}{l} \tag{16}$$

Moreover, in General Relativity the Riemann curvature tensor ($\boldsymbol{\Gamma}$) uniquely characterizes the gravity field, with corresponding gravity gradient tensor quantity in Newtonian field (*V*).
The gravity gradient tensor is:

$$\boldsymbol{\Gamma}_{ij} = \frac{\partial^2 V}{\partial x_i \partial x_j} \tag{17}$$

Where $\boldsymbol{\Gamma}_{ij}|(\boldsymbol{i} = \boldsymbol{x},\boldsymbol{y},\boldsymbol{z}),(\boldsymbol{j} = \boldsymbol{x},\boldsymbol{y},\boldsymbol{z})$ are the elements of the symmetric gravity gradient tensor.

4.1. *Gradiometer Principle*

The gradiometer is an instrument for vertical gravity gradient measurement. It measures the gravity in two points (using accelerometers) on the same vertical and divides the measured difference by distance.

An inline component gradiometer makes the linear acceleration difference between two aligned masses, A and B or C and D, figure 3.

The cross-component gradiometer could combine linear accelerations on four test masses with their aligned sensitive axes. It can be also constructed by differencing signals between two concentric angular accelerometers with their arms orthogonal (figure 4).

The combination between three in line and three cross component gradiometers constitutes a tensor gradiometer. In a rotating reference frame, the gravity gradient $\mathbf{\Gamma}'$ is measured in relation with the gradient in the inertial frame by (18), where $\mathbf{\Psi} = (\Psi_1, \Psi_2, \Psi_3)$ is the angular vector in gradiometer coordinate system.

$$\mathbf{\Gamma}' = \begin{pmatrix} \Gamma_{11} + (\Psi_2^2 + \Psi_3^2) & \Gamma_{12} + \Psi_1\Psi_2 & \Gamma_{13} + \Psi_1\Psi_3 \\ \Gamma_{21} + \Psi_2\Psi_1 & \Gamma_{22} + (\Psi_3^2 + \Psi_1^2) & \Gamma_{23} + \Psi_2\Psi_3 \\ \Gamma_{31} + \Psi_3\Psi_1 & \Gamma_{32} + \Psi_3\Psi_2 & \Gamma_{33} + (\Psi_1^2 + \Psi_2^2) \end{pmatrix} \tag{18}$$

4.2. *Gravity Gradient Linearization*

The function of gravity gradient is nonlinear. The linearization process replaces the surface of gravity gradient $h(x,y)$ with the plane equation $f(x,y)$ defined as linear expression of gravity gradient in the nearby INS referenced position $(\hat{x},\hat{y})$ using the parameter of gravity gradient linearization α, h_x and h_y:

$$f(x,y) = \alpha + h_x(x - \hat{x}) + h_y(y - \hat{y}) \tag{19}$$

Where h_x and h_y are the vertical and horizontal slope of gravity gradient and α is the referenced gravity gradient at position $(\hat{x}, \hat{y})$.

There are several gravity gradient linearization algorithms: First Order Taylor Series, Nine Point Fit, Full Plane Fit and Two Subgroup Fit.

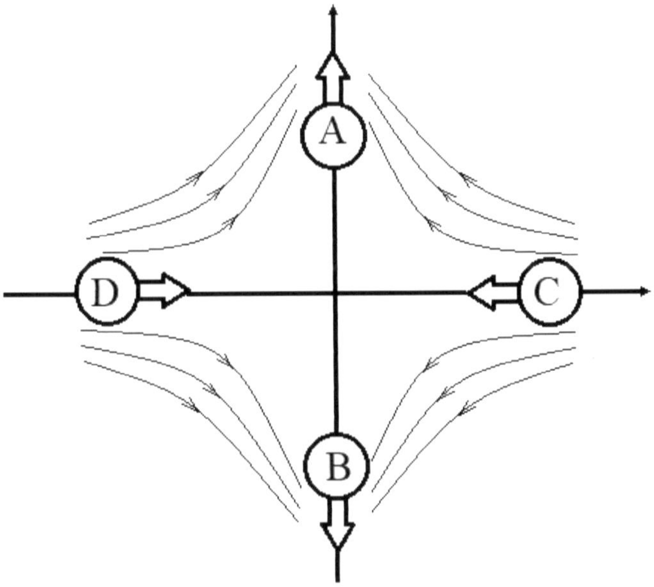

Figure 3 – In line component gradiometer

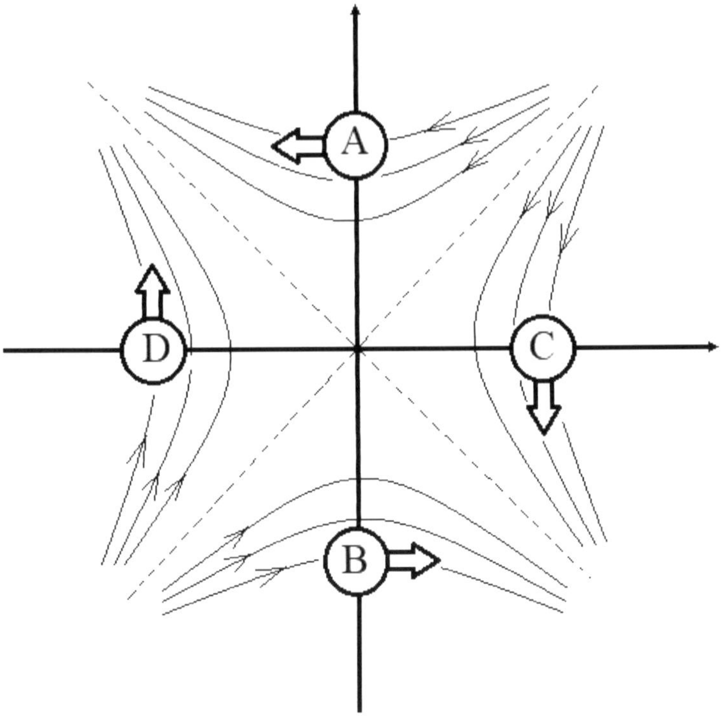

Figure 4 – Cross component type gradiometer

In the context of this research we propose to locate a submarine's position using an algorithm by matching the gravity gradient measurements profile and the digital gravity gradient map.

First are extracted the characteristic parameters of the series of measured values, then it is processed the corresponding gradient gravity map and finally is applied the map matching algorithm.

4.3. Method Description

(a) Gradiometer Measured Values

Each value measured by gradiometer is paired with a distance from the previous measurement point, approximated by Integration Navigation System along with $\boldsymbol{\delta}$, the submarine navigation direction. A model of these values is graphically represented into figure 8 a.

The graphic characteristics are defined using analytical parameters which full identify the graphical representation of the measurements series. In this work the parameters are minimum, maximum and inflexion points of the signal, where $\boldsymbol{M}$, $\boldsymbol{m}$ and $\boldsymbol{I}$ are maximum, minimum and inflexion point:

$$\boldsymbol{M} = \{p(x_i, y_i) \in \Gamma |\ y_i > y_{i+1}, y_i > y_{i-1}\} \qquad (20)$$

$$\boldsymbol{m} = \{p(x_i, y_i) \in \Gamma |\ y_i < y_{i+1}, y_i < y_{i-1}\} \qquad (21)$$

And $\boldsymbol{I}$ is the inflexion points set:

$$\boldsymbol{I} = \boldsymbol{I}_{up} \cup \boldsymbol{I}_{dwn}$$

Where (22)

$$\boldsymbol{I}_{up} = \{p_i(x_i, y_i), p_{i+1}, p_{i-1} \in \Gamma;\ y_i = a_i\, x_i + b_i \mid [y_{i-1} > a_i x_{i-1} + b_i\ AND\ y_{i+1} < a_i x_{i+1} + b_i]\}$$

$$\boldsymbol{I}_{dwn} = \{p_i(x_i, y_i), p_{i+1}, p_{i-1} \in \Gamma;\ y_i = a_i\, x_i + b_i \mid [y_{i-1} < a_i x_{i-1} + b_i\ AND\ y_{i+1} > a_i x_{i+1} + b_i]\}$$

As the equations (22) of the inflexion point show, the tangent parameter is enough difficult to be computed, comparing with minimum and maximum detection numerical expression, thus it is computed the second gradient of the numerical signal. As $\boldsymbol{I}$ points are solutions of the second gradient of the function that modeling the graphical representation, we apply the numerical gradient function (∇F):

$$\nabla F = \frac{\partial F}{\partial x}\hat{\imath} + \frac{\partial F}{\partial y}\hat{\jmath} \qquad (23)$$

Each specific point of one of type $\boldsymbol{M}$, $\boldsymbol{m}$ or $\boldsymbol{I}$, detected into the gradiometer measure series, is defined by type (typ) and evaluated distance ($dist$) from the reference point, $\boldsymbol{P_0}$ to the measuring point, and it is denoted $\boldsymbol{p_i(type_i, distance_i)}$ where the index $\boldsymbol{i}$ is increasing by the unity counter.

The specific MATLAB code segment that applies two times numerical gradient function is presented in figure 5.

```
x = [ 0 1 2 3 4 5 6 7 8 9 10 11 12 13 14 15 15 17 18 19 20 21 22 23 24 25 26 27 28];
y = [ 124 131 132 133 132 129 127 125 123 120 118 117 115 115 114 116 118 126 125 120 116 111 109 106 105 108 114 123 131];
grd1=gradient(y,x)
grd2=gradient(grd1,x)
id=sign(grd2)
idx=strfind(id,[-1 1])
inflexionP=x(idx+1)
```

Figure 5 – part of MATLAB code to detect inflexion points (noted *inflexionP*).

For matching process, the parameters $\boldsymbol{p_i(typ_i, dist_i)}$ are gathered into a vector of pairs, $\boldsymbol{G}$:

$$\boldsymbol{G(typ_i, dist_i)} = \begin{cases} \boldsymbol{typ_i = type_i} \\ \boldsymbol{dist_i = distance_i} \end{cases} \left|\begin{array}{c} \boldsymbol{i\ is\ index\ of\ p_i} \\ (\exists)\ \boldsymbol{p_i(type_i, distance_i)} \in \Gamma \end{array}\right. \qquad (24)$$

IOP Conf. Series: Earth and Environmental Science **172** (2018) 012005 doi:10.1088/1755-1315/172/1/012005

(b) Digital Gradient Map

The gravity gradient map data must offer specific information in terms of $\boldsymbol{M}$, $\boldsymbol{m}$ and $\boldsymbol{I}$ parameters for the geographical area where the route of submarine is. Then, based on data collected by submarine on board gradiometer that matching the digital map, the submarine position is approximated with a minimal error.

There are several pre-processing steps:

1. Computing the first gradient of the digital map – it will give the geometrical places for minimum and maximum of the map signal (represented as isolines of zero value of first gradient map), noted $\boldsymbol{Mg}$, $\boldsymbol{mg}$;
2. Computing the second gradient map that define geometrical place of the inflexion points (for isolines of zero value of second gradient map) noted $\boldsymbol{Ig}$;
3. Defining N directions around the INS evaluated direction, $\boldsymbol{\delta}$, using a predefined threshold to generate **Point extreme (+)** and **Point extreme (-)** around evaluated point $\boldsymbol{P_e}$;
4. Creates pairs $\boldsymbol{p_{k,i}(type_i, distance_i)}$ where $\boldsymbol{k} = \boldsymbol{1}$ to $\boldsymbol{N}$ is direction index and $\boldsymbol{i}$ is the point index of type $\boldsymbol{M}$, $\boldsymbol{m}$ or $\boldsymbol{I}$ on the profile/direction $\boldsymbol{k}$ of the map; $\boldsymbol{distance_i}$ is the distance between the reference point $\boldsymbol{P_0}$ and the $\boldsymbol{i} -$ th point $\boldsymbol{p_{k,i}}$ of measure by type $\boldsymbol{M}$, $\boldsymbol{m}$ or $\boldsymbol{I}$ along $\boldsymbol{k}$ direction (figure 6);

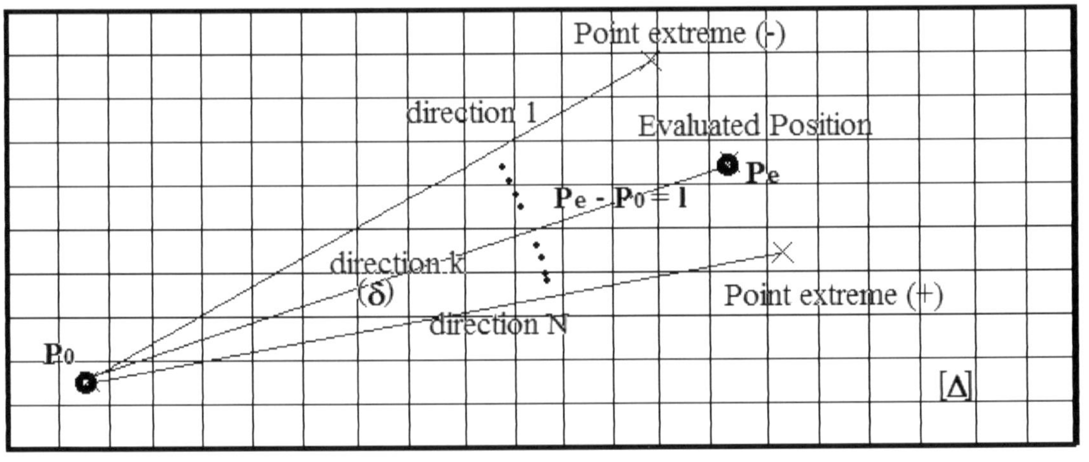

Figure 6 – Directions defined on Gravity Gradient Map, using evaluated point position and a configured threshold to define **Point extreme (+)** and **Point extreme (-)** for first and last direction.

5. Creates $\boldsymbol{N}$ vectors $\boldsymbol{H_k}$ of $\boldsymbol{p_{k,i}(type_i, distance_i)}$ where $\boldsymbol{k=1}$ to $\boldsymbol{N}$, N is the maximum number of directions to be tested on the map:

$$\boldsymbol{H}_k(type_i, distance_i) = \left\{ \begin{array}{c} type_i = t_i \\ distance_i = d_i \end{array} \middle| \begin{array}{c} \boldsymbol{k} = 1 \; to \; N \\ \boldsymbol{i} < N \\ (\exists)\boldsymbol{p}_{k,i}(t_i, d_i) \in direction\; k \end{array} \right\} \qquad (25)$$

(c) Matching Algorithm

The matching algorithm searches the $\boldsymbol{H_k}$ vectors, previously generated for the specific sequence defined by $\boldsymbol{typ_i}$ of measured vector values $\boldsymbol{G}$.

The algorithm is:

IOP Conf. Series: Earth and Environmental Science **172** (2018) 012005 doi:10.1088/1755-1315/172/1/012005

- Searching in $\boldsymbol{H_k}(type_i, distance_i)$ for $type_i == \boldsymbol{typ_i}$ of $\boldsymbol{G}(\boldsymbol{typ_i}, \boldsymbol{dist_i})$ and select the sequence number which has all $type_i == \boldsymbol{typ_i}$ and memorize it as an element of the vector ***A***:

$$\boldsymbol{A} = \left\{ k_j = k \;\middle|\; \begin{array}{l} \boldsymbol{k} = index\ \boldsymbol{k}\ of\ \boldsymbol{H_k} \\ \boldsymbol{type_i} == \boldsymbol{typ_i} \text{ for } \boldsymbol{H_k}(\boldsymbol{type_i},\ distance_i) \text{ and } \boldsymbol{G}(\boldsymbol{typ_i},\ dist_i) \\ \boldsymbol{j} = unity\ counter \end{array} \right\} \quad (26)$$

In other words, ***A*** will gather all ***k*** indexes of map directions that have the same ordering series of point types as the measured vector ***G***;

- Generating the vector of errors $\boldsymbol{E_j}$, as distances differences of identified type points:

$$\boldsymbol{E_j} = \left\{ e_i = |\boldsymbol{distance_i} - \boldsymbol{dist_i}|\ where\ \boldsymbol{H_{k_j}}(type_i, \boldsymbol{distance_i})\ and\ \boldsymbol{G}(typ_i, \boldsymbol{dist_i}) \right\} \quad (27)$$

- Choosing the minimum deviation error vector ***D***:

$$\boldsymbol{D} = \left\{ p = k_j | j = \min_j \boldsymbol{E_j}\ for\ \boldsymbol{H_{k_j}} \right\} \quad (28)$$

5. Model and Analysis

The matching results are improving the system position accuracy by integrating the new information via Kalman filter and Navigation System Solution with INS (figure 7).

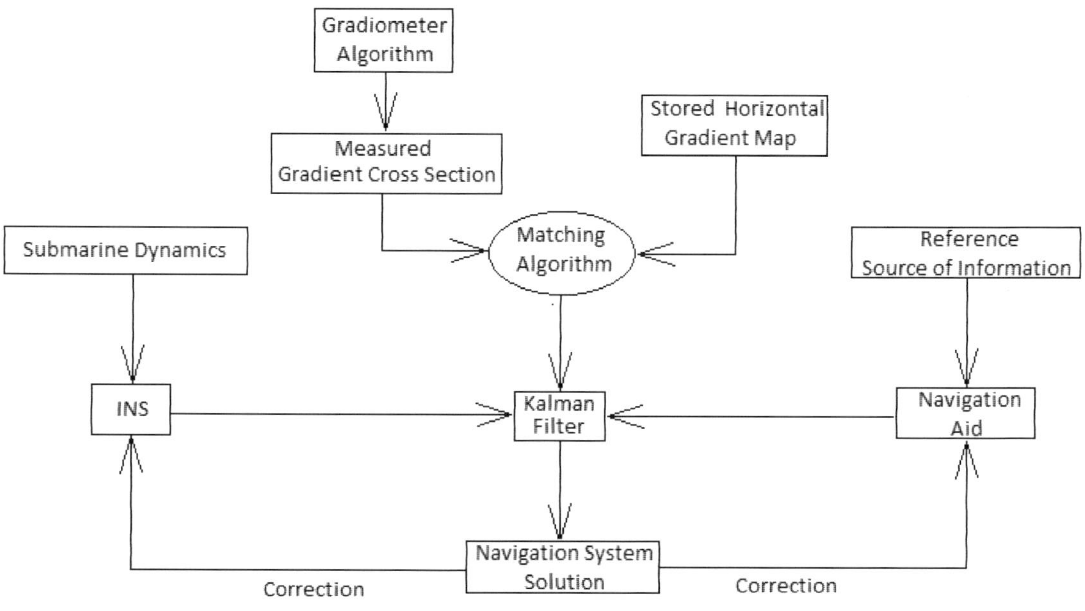

Figure 7 – INS correlated through Kalman filter with matching algorithm; squares represent matrix functions and ellipses inputs/outputs.

In equation (28) the ***D*** vector has the closest $distance_i$ (from $\boldsymbol{H_k}(type_i, distance_i)$) of $dist_i$ (from$\boldsymbol{G}(typ_i, dist_i)$). It means the map signal of the ***k*** direction is matching with the measured signal and the ***k*** direction is the computed navigation route closest of the real one, for our submarine.

A model representation for INS evaluation route (figure 8) and INS with Gradient information integrated (figure 9) demonstrate a reduced error of 65-80%.

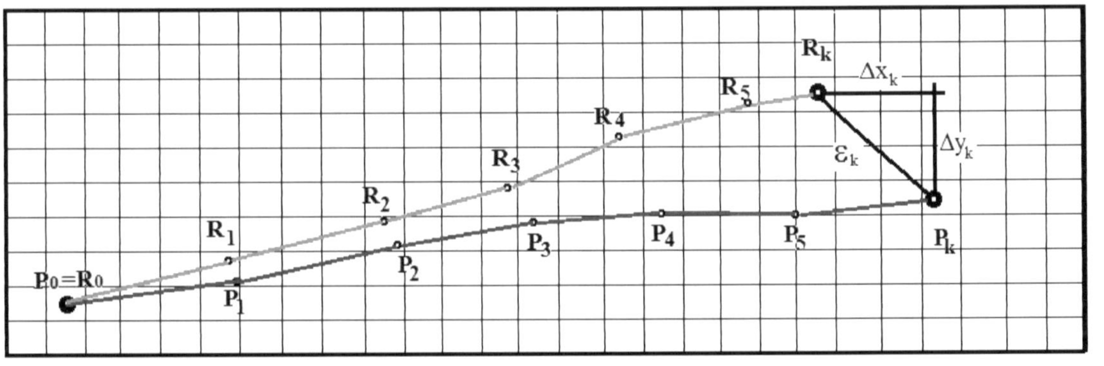

∘ **Intermediar Points** **R** **Real position**

● **Start/End points** **P** **Evaluated position (INS only)**

Figure 8 Error model deviations for several points for an INS evaluation without gradient matching system

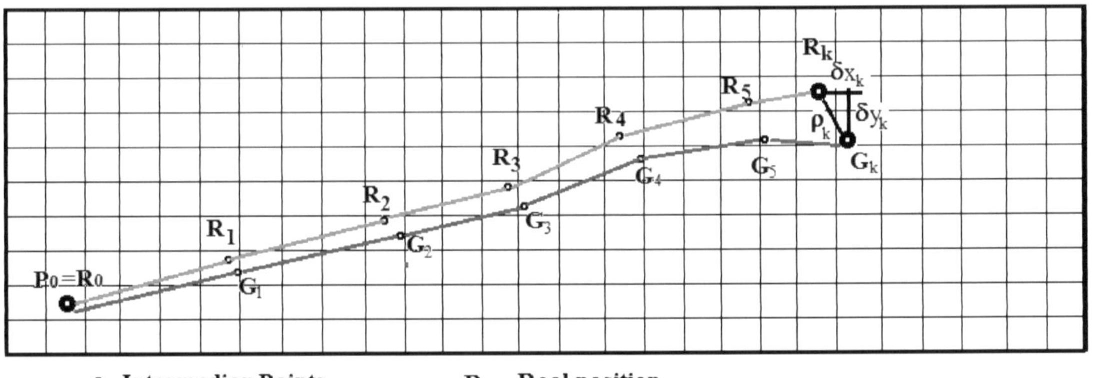

∘ **Intermediar Points** **R** **Real position**

● **Start/End points** **G** **Evaluated position (INS with Gradient integrated)**

Figure 9 Deviations of intermediary navigations points for INS with gradient matching system

In figure 10 b is presented a Vertical Cross Section of Gravity Gradient Map (Δ) where the map's isolines (orange) are correlated with the cross-section representation on 29 verticals for $\delta = k$ direction. ***Mg***, ***mg*** and ***Ig*** represents maximum, minimum and respectively inflexion points of the gravity gradient cross section function.

The alignment between figure 10 (a) and figure 10 (b) is performed on correlation between ***M***, ***m*** and ***I*** and ***Mg***, ***mg*** and ***Ig***.

IOP Conf. Series: Earth and Environmental Science **172** (2018) 012005 doi:10.1088/1755-1315/172/1/012005

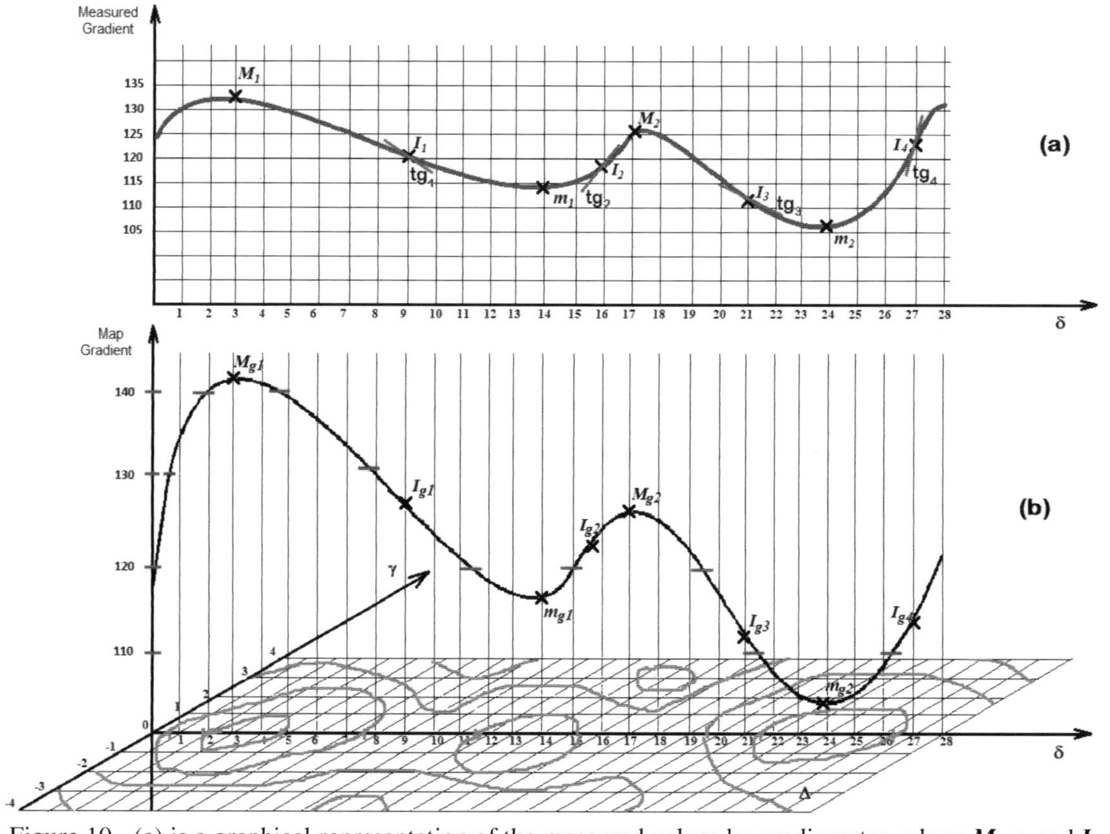

Figure 10 - (a) is a graphical representation of the measured values by gradiometer, where ***M***, ***m*** and ***I*** represents maximum, minimum and inflexion points respectively; - (b) shows the profile on the gradient map along the $\delta = k$ direction (vector of submarine navigation), where ***Mg***, ***mg*** and ***Ig*** are the parameters of the profile on the gradient map.

6. Conclusions

The presented method compensates the INS deviation using a digital gravity gradient map for comparison with the measured gradient using a spatial algorithm.

The minimal analyze processing method extracts the first and second gradients from both signals and compares several alignments from the map with the measured signal – using an original method able to be easy implemented even into embedded systems.

The models analyze for errors shows minimal qualitative deviations.

However, as the parameters of the used signals are qualitative, the dependency of any measurement correction is minimal (e.g. Bouguer correction, etc.) and thus any dependency of the gradiometer type is eliminated (figure 8).

While there is still some room for improvement in both the INS and measured geophysical parameters stage, we have demonstrated promising results especially on alignment process that are unlikely to be correctly aligned by earlier approaches usually based on frequency analyze, due to lack of analytical references positioning features: minimum, maximum and inflexion points.

References

[1] M. Rogobete, C. Racuciu, An Automatic Method for Digitizing Bathymetric Maps, "Mircea cel Batran" Naval Academy Scientific Bulletin, Volume XX – 2017 – Issue 1, DOI: 10.21279/1454-864X-17-I1-083, ISSN: 2392-8956

[2] Lee J., Kwon J.H., Yu M.,Performance Evaluation and Requirements Assessment for Gravity Gradient Referenced Navigation, Sensors 2015, 15, 16833-16847; doi:10.3390/s150716833

[3] Difrancesco D., (Lockheed Martin), Advances and Challenges in the Development and Deployment of Gravity Gradiometer Systems, EGM 2007 International Workshop Innovation in EM, Grav and Mag Methods:a new Perspective for Exploration Capri, Italy, April 15 – 18, 2007

[4] Marius Rogobete, First and second order image statistics in specific image artifact detection, International Conference of Innovative Technologies, September 26 - 29, 2012, Croatia

[5] Marius Rogobete, Ciprian Răcuciu – Using Potential Field Analysis into Image Artifact Detection Field, Indian Journal of Research, Volume III, Issue V, May 2014

[6] Lucian Besutiu, Marius Rogobete, "ON THE NORMAL GEOMAGNETIC FIELDS WITHIN CONFINED AREAS. THEIR MODELLING AND UNDERSTANDING", ‚Revue roumaine de géophysique, Volume 39,ISSN 1220-5303, Editura Academiei Romane, 1995, Bucuresti

[7] Lin Wu, Jiaqi Gong, et al., New method of underwater passive navigation based on gravity gradient, , MIPPR 2007, Proc. of SPIE Vol. 6790, 67901V, (2007) 0277-786X/07/$18 · doi: 10.1117/12.749408

[8] M. Ivan, 1994, Line integrals of potential fields data, Geophysical Prospecting, 42, p. 735-743.

[9] Zuo B., Andrew K., Hu X., Geng M., Full-tensor gravity gradient eigenvector analysis for locating complex geological source positions, Nonlin. Processes Geophys. Discuss., doi:10.5194/npg-2016-75, 2017

[10] Burlacu P., Dobref V., Bădără N.,Tărăbuţă O.; "A LQR controller for an AUV depth control"; ANNALS OF DAAM for 2007 & PROCEEDINGS of the 17 th International DAAM Symposium "INTELLIGENT MANUFACTURING & AUTOMATION" 24-26th 2007, Zadar CROATIA;Published:2007.IDSNumber: BHB34, ISSN:1726-9679, ISBN:3-901509-58-5,pp125-127.

IOP Conf. Series: Earth and Environmental Science **172** (2018) 012006 doi:10.1088/1755-1315/172/1/012006

Mathematical model of turbocharger as part of intelligent system component designed for technical operation of diesel engines and for training

Julian Moskov

Naval Academy, Varna, Bulgaria
y.moskov@naval-acad.bg

Abstract. The paper includes a basic module of a slow speed ship diesel engine turbocharger. Our experience in modelling and simulation makes it possible to produce flexible models. Next, a control system needs to have operational modules in real time, with sufficient range of precision. This is a light algorithm under Delphi area. Module makes up a part of smart intelligent pack. Pack is a software that combines and solves the task of exploitation and design of modern slow speed diesel engines. The module allows to construct the working mode line and to examine the behaviour of the operating point quasi-statically. The model is based on a quasi-static approach for real-time operation in a software package including a flexible input simulator and modules for solving constructive and operational tasks with graphical retrofit based on image recognition

1. Introduction.

Whether the current mathematical models of modern low-frequency marine diesel engines are up to date can determined by the following facts:

- The solution of systemic tasks through mathematical modeling replaces multiple ground experiments and reduces costs.
- Mathematical modeling gives clarity to processes and interrelationships (target functions, relationship between sub-themes) which is difficult, and in some cases impossible, for physical models.
- At this stage, sufficiently precise, flexible, adaptive and dynamic models are needed for management, technical diagnosis and training purposes.
- Physical modeling is still too expensive and time consuming, there is no unified memory of data in modern management systems.
- There is no known practical realization of trainable mathematical models, the requirements for real-time functionality of mathematical models are steadily increasing for management purposes.

2. A complex flexible mathematical model of a marine diesel engine for the purposes of technical diagnostics and management.

An efficient mathematical model has been developed taking into account the influence of actual processes in the exhaust and air supply systems, the laws of fuel supply, gas exchange and movement of the exhaust valve. Afterwards, accuracy, sensitivity, relevance and applicability are assessed and adapted by training with knowledge gained from the processing of experimental and model data for

IOP Conf. Series: Earth and Environmental Science **172** (2018) 012006 doi:10.1088/1755-1315/172/1/012006

specific engines. Based on this, a software called Flexible Intelligent Package has been developed [2]. The publication presents a sophisticated, functional, adaptive mathematical model of a marine diesel engine with an isobar system of overfill with specifying the sub-models of the main components; reporting the dynamics of the turbocharger aggregates and their joint work with the piston part and the specified gas exchange model.

A system has been developed for training the universal mathematical model for the management of marine engines, which includes an optimization procedure for adapting the universal model to a specific engine with specified accuracy and adequacy.

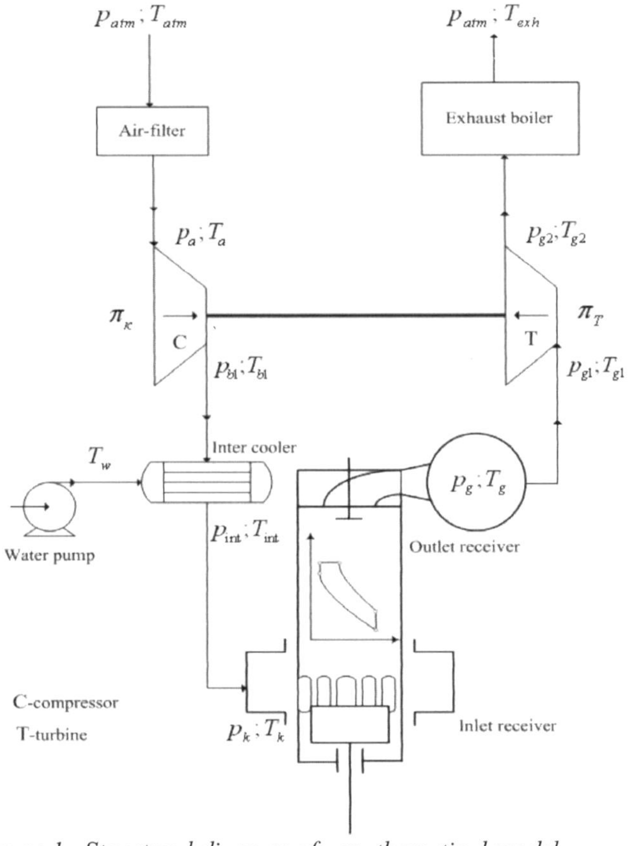

Figure 1. Structural diagram of a mathematical model.

The actual cycle is based on the concept of a physical-mathematical model of the components of the working process. This allows the elementary processes to be described by means of systems of private and ordinary differential equations expressing the laws not only of thermodynamics but also of aero-gas dynamics and chemical kinetics [1]. I'm going to discuss in more detail the processes in the turbocharger unit. For this purpose I will review the linear graphic model of the TA while analyzing separately the processes of transformation of energy in the compressor and turbine stage (Figure 2).

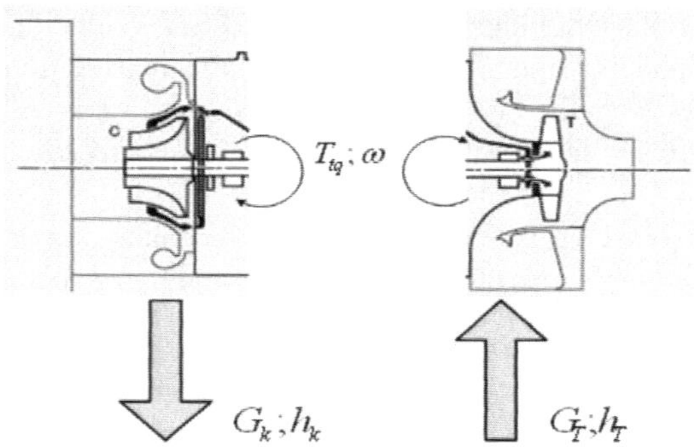

Figure 2. Processes of transformation of energy in the compressor and turbine stage

In linear graphical modeling, elements that transform or transmit energy are called converters. The main process of energy transformation in this type of unit is presented in Figure 2 by a two-input unit.

Figure 3. Two input element representing transformation of energy.

The definition of the two-input units is given by H. M. Paynter. There are two types of such units - a transforming unit and a "spinning" unit, which are presented in Figure 4.

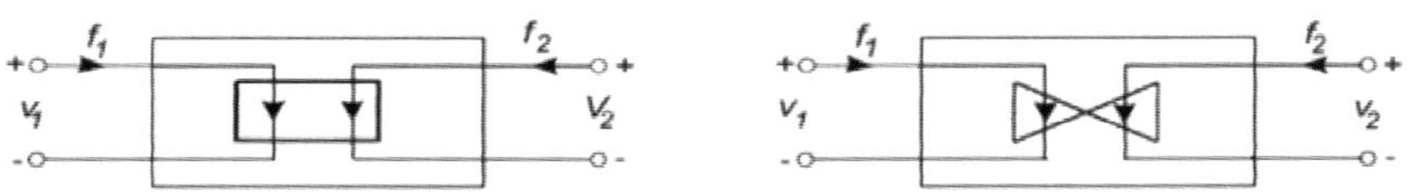

Figure 4. Transforming and spinning

The basic linear relation between the parameters of the two inputs is:

$$\begin{bmatrix} v_1 \\ f_1 \end{bmatrix} = \begin{bmatrix} c_{11} & c_{12} \\ c_{21} & c_{22} \end{bmatrix} \begin{bmatrix} v_2 \\ f_2 \end{bmatrix}$$

With transforming converters:

$c_{12} = c_{21} = 0 \quad c_{22} = -1/c_{11}$

And with 'spinning' converters:
$c_{21} = -1/c_{12} \quad c_{11} = c_{22} = 0$

And the basic relation now is:

$$\begin{bmatrix} v_1 \\ f_1 \end{bmatrix} = \begin{bmatrix} TF & 0 \\ 0 & -1/TF \end{bmatrix} \begin{bmatrix} v_2 \\ f_2 \end{bmatrix}$$ for transforming converters and

$$\begin{bmatrix} v_1 \\ f_1 \end{bmatrix} = \begin{bmatrix} 0 & GY \\ -1/GY & 0 \end{bmatrix} \begin{bmatrix} v_2 \\ f_2 \end{bmatrix}$$ for 'spinning' converters, where TF is the degree of conversion and GY is the 'spinning' module.

I consider the compressor and the turbine stage as two two-input power converters while looking for the degree of transformation between the fluid power and the mechanical power. I assume that power is transmitted without loss. The attitude of the variables is algebraic and independent of time, and is represented by coefficients that are constants and are linear.

The relationship between torque and flow is Euler's equation:

$$T_{tq} = -G_T\left(c_{u_{2T}} - c_{u_{1T}}\right)r_T \tag{1}$$

torque is a linear dependence on the consumption.

$$T_{tq} = -K_T G_T \tag{2}$$

therefore $K_T = \left(c_{u_{2T}} - c_{u_{1T}}\right)r_T$ (3)

If the energy is transformed without loss, the mechanical power is transformed as entirely fluid

the basic equations that express the relationship between

$$\boxed{P_m = P_{XT}} \quad \omega T_{tq} = G_T h_T \quad \omega K_T G_T = G_T h_T \quad \omega = \frac{h_T}{K_T} \tag{4}$$

The parameters are:

$$\begin{cases} T_{tq} = K_T G_T \\ \omega = \dfrac{1}{K_T} h_T \end{cases} \tag{5}$$

Similarly, we will consider the connection between the parameters in the compressor stage (Fig. 5)

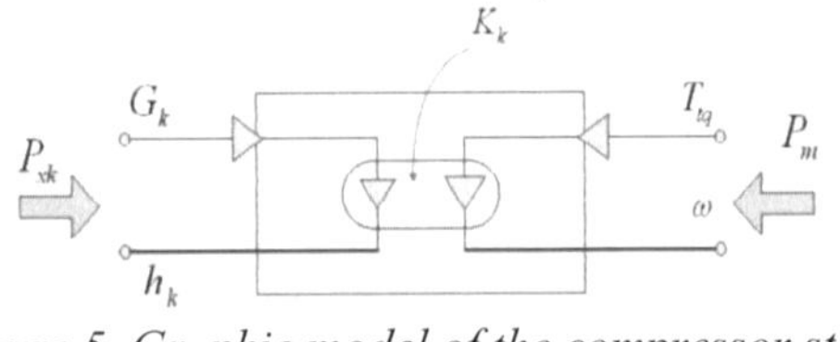

Figure 5. Graphic model of the compressor stage

And again I use the Euler equation and the assumption that the energy is converted without a loss

$$\boxed{P_m = P_{XK}} \quad \omega T_{tq} = G_K h_k \quad \omega T_{tq} = G_K h_K \tag{6}$$

$$h_K = \frac{\omega T_{tq}}{G_K} = \frac{\left(c_{u_{2K}} r_{2K} - c_{u_{1K}} r_{1K}\right)}{T_{tq}} \omega T_{tq} = \left(c_{u_{2K}} r_{2K} - c_{u_{1K}} r_{1K}\right)\omega$$

Therefore the equations that express the relationship between the variables are:

$$\begin{cases} G_K = \dfrac{1}{K_K} T_{tq} \\ h_K = K_K \omega \end{cases} \tag{7}$$

For the turbine and the compressor consumption relationships, and their heat fluctuations can be presented in the following form:

$$G_K = \frac{K_T}{K_K} G_T; \qquad h_K = \frac{K_K}{K_T} h_T \tag{8}$$

By replacing the K_k and K_T coefficients in the equations, we obtain dependencies between the main components of the compressor and the turbine stage, which do not specify the losses.

IOP Conf. Series: Earth and Environmental Science **172** (2018) 012006 doi:10.1088/1755-1315/172/1/012006

$$\frac{G_K}{G_T} = \frac{\left(c_{u_{2T}} - c_{u_{1T}}\right) r_T}{\left(c_{u_{2K}} r_{2K} - c_{u_{1K}} r_{1K}\right)} \tag{9}$$

$$\frac{h_K}{h_T} = \frac{\left(c_{u_{2K}} r_{2K} - c_{u_{1K}} r_{1K}\right)}{\left(c_{u_{2T}} - c_{u_{1T}}\right) r_T} \tag{10}$$

By analogy with the analysis of the ideal cycles of gas turbine engines (Brighton cycle) I make a correlation between the thermodynamic parameters of the engine and the gas dynamics parameters of the turbocharger. With sufficient precision for practice, we can accept a relative difference between the power consumed by the compressor and the power generated by the turbine within 1-2%, i.e.

$$\Delta P = \left| \frac{P_T - P_K}{P_K} \right| .100 \leq 2\% \tag{11}$$

The economical and reliable operation of ship diesel engine is only possible when the characteristics of all its elements - the compressor, the internal combustion engine and the turbine - are in concordance [4]. By concordance of the characteristics we mean the interconnection of all the main technical, operational, economical, gas-dynamical and constructive parameters of the internal combustion engine and turbocharger.

An algorithm for calculating the turbocharger has been developed. The input database for the algorithm is related to the mathematical model of the processes in the piston part of the slow speed diesel engine.

The equilibrium turbocharger mode includes the known equation of dynamics and equilibrium conditions in the non-compressor and turbine moments [3]. The primary equations are non-linear, and by Taylor's decomposition they are linearized, and then the system is obtained:

$$\left| \begin{array}{l} dT_{tqk} = \frac{\partial T_{tqk}}{\partial p_k} dp_k + \frac{\partial T_{tqk}}{\partial \omega_k} d\omega_k \\ dT_{tqk} = \frac{\partial T_{tqk}}{\partial p_k} dp_k + \frac{\partial T_{tqk}}{\partial \omega_k} d\omega_k \\ dG_k = \frac{\partial G_k}{\partial p_k} .dp_k + \frac{\partial G_k}{\partial \omega_k} d\omega_k \\ dT_{tqT} = \frac{\partial T_{tqT}}{\partial \omega_k} .d\omega_k + \frac{\partial T_{tqT}}{\partial p_{g_1}} .dp_{g_1} + \frac{\partial T_{tqT}}{\partial T_{g_1}} dT_{g_1} \\ J_K.\frac{d\omega_k}{dt} + F_k.d\omega_k = \frac{\partial T_{tqT}}{\partial p_{g_1}} .dp_{g_1} + \frac{\partial T_{tqT}}{\partial T_{g_1}} .dT_{g_1} - \frac{\partial T_{tqk}}{\partial p_k} dp_k \end{array} \right. \tag{12}$$

The result is the equation $T\frac{dx}{dt} + x = -k_n.\mu_n - b_1\frac{d\mu_n}{dt} - k_b.\mu_b - b_2\frac{d\mu_b}{dt}$ (13)

where $x = \frac{\Delta p_k}{p_{k_0}}$, $\mu_n = \frac{\Delta n_{\partial e}}{n_{\partial e_0}}$, $\mu_b = \frac{\Delta B_e}{B_{e_0}}$.

After the corresponding assignments and substitutions, a linear differential equation is obtained:

$$\left(T_0 + \tilde{T}t\right)\frac{dx}{dt} + x = C_0 + C_1 t + C_2 t^2 \ . \tag{14}$$

This deductive approach is to determine the turbocharger performance of an undetermined slow speed diesel engine regime. In new constructions and availability of manufacturer data for the dependencies used, MM can be updated. To find the functional relationship between $\mathbf{p_k}$ and $\mathbf{n_д}$, the similarity theory is used.

3. Program product FLEX Turbo.

The package was developed in DELFI environment [5,6]. The easy access for programmers non-specialized in the software environment of this type is the main reason for choosing it. The software complex is built on the modular principle. The modules are arranged hierarchically according to their importance, location and communication between them. A special place is assigned to the main sources of energy of the ship - the main engines. A simplified diagram of ship propulsion complex is presented in Figure 6.

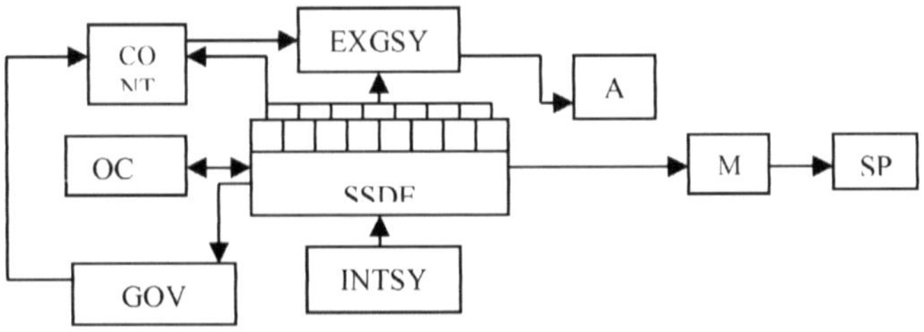

Figure 6. Structural diagram of MM.

The choice of the Delphi environment for the implementation of the model is mainly due to the fact that the Delphi programming language is a traditional continuation and refinement of the Object Pascal language. Variables, as in any other programming language, are of several types which allow partial combination of the environment with the imposed world-class multimedia software. Here we can recommend an application even such as Flash, which was previously believed to be incompatible with Object Pascal. The main argument for the use of Delphi is the modular principle in the realization of the final FLEX TURBO software. Although Object-Oriented Programming also uses other advanced languages such as C / C ++ Visual Basic, Delphi has been introduced into simulators by world-known manufacturers such as Turbo Diesel 2.0 and Virtual Engine Room by Dr. Stefan Cluj; Kongsberg; Maritime Software, and others. Unlike in Visual Basic and C ++, in Delphi variables cannot be assigned randomly, which is an advantage in our case of application. The database and the constraints associated with the process simulation are embedded in the software and can be retrieved by default. The wide range of components and modules developed by Delphi makes it extremely easy to develop the product. This allows to focus on the complex model and its improvement.

The software is designed to visualize the joint work of the piston and the turbocharger unit of modern low-frequency marine diesel engines [3]. After starting the main panel, an image is obtained as shown in Figure 7.

The next step is entering the mass dimensions of the particular ship engine to set up the model. The input data panel has default values.

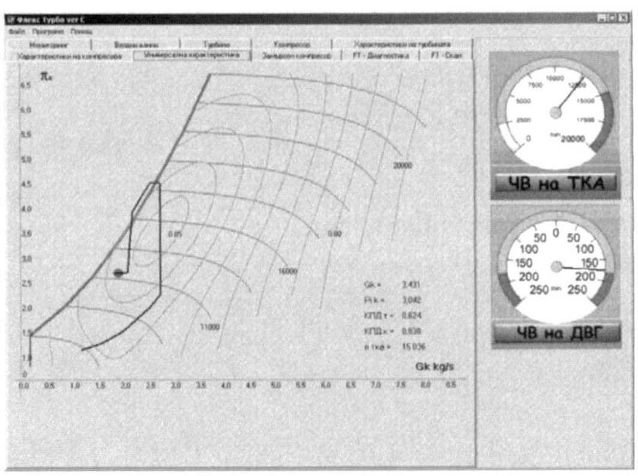

Figure 7. Control panel of the software product and preview of cost performance of a centrifugal compressor

The software dynamically visualizes the turbocharger unit's operating point on the universal feature of the centrifugal compressor. The mathematical model allows to observe the trajectory of the point when changing the ship engine mode factors. The character of the trajectory indicates a process that does not coincide with the hysteresis cycle discussed in previous versions of the product for loading and unloading the engine. Figure 7 depicts the graphics panel of cost performance of the input data for the mathematical model. The latter allows the pumping limit to be altered as a function of the operating factors and the contamination of the flow path of the compressor. Similar is the way of using the software product for simulated contamination of the flow part of the compressor.

The software is designed to meet the standards and requirements of ISO 9000-2000 and has the potential of further adjusting from a European to a North American metric standard. This makes it universally applicable to the turbocharger research by leading manufacturers from around the world.

4. Conclusion

The complex approach to using mathematical models in multi-purpose software is limited at this stage by the capabilities of the manufactured hardware. The realization of such a multi-purpose complex is useful for assisting the experts in day-to-day solving of the two basic engineering tasks - design and exploitation, and also for maintaining the connection between the two by means of feedback forms. When analyzing the behavior of the turbocharger, it should be taken in consideration that any change in the cost performance results in a corresponding change in the line of work modes. For example, if there is a decrease in the area of the windows as a result of contamination during the operation of the ship diesel engine, the air flow through the internal combustion engine also decreases and the cost performance is moved to the left of the initial position. At the same time, the airflow through the compressor is also reduced, and therefore the turbocharger operating mode line is also moved to the left, i.e. closer to the pump area. Analogous changes occur in the sailing characteristics of ships in the tropics as a result of an increase in the temperature of the air sucked up by the compressor. This possible move of the turbocharger line closer to the pump area must be taken into account when conforming the characteristics and operating ship diesel engine.

The airborne contamination of the ship diesel engine during operation results in an increase in the resistance of the treadmill, a reduction in air consumption, and discrepancies in ship diesel engine characteristics. Successful operation of ship diesel engine with poorly conformed characteristics of the compressor-ship diesel engine-turbine is virtually impossible. It is therefore necessary to quickly find and eliminate the causes of discrepancies in the characteristics.

References

[1] Hakan Yilmaz, Anna Stefanopoulou, Ann Arbor, 2002 Control of Charge Dilution in Turbocharged Diesel Engines via Exhaust Valve Timing *University of Michigan*

[2] Velitchko Yanakiev 1996 Ship diesel engines *"Св. Георги Победоносец"*, 200-202, 222-225

[3] Turbochargers for supercharging of ship diesel engines "*Машиностроение*", 1975, 27-95

[4] Межерицкий, А. Д.; Йосифов, Й. Турбокомпресорни агрегати за принудително пълнене на корабни двигатели с вътрешно горене, *"Техника"*, 1982, 113 – 160

[5] Velitchko Yanakiev, Yuliyan Moskov, Flexible mathematical model of turbocharger of slow speed diesel engine constant pressure, *International Technical Conference "MOTAUTO-2003", Sofia, 01-03.10.2003.*

[6] К у л ь т и н Н.Б. Программирование в Turbo Pascal 7.0 и Delphi. – СПб.: BHV –Санкт – Петербург, 1998 – 240 с.

IOP Conf. Series: Earth and Environmental Science **172** (2018) 012007 doi:10.1088/1755-1315/172/1/012007

A security mechanism based on anonymous signature schemes as a method to reduce the consumption of consumable materials

N F Antonie[1,3], C Răcuciu[2], D Glăvan[1] and F Medeleanu[1]

[1]Faculty of Military Electronic and Information Systems, Military Technical Academy, Bulevardul George Coşbuc 81-83, Bucureşti 050141, Bucharest, Romania
[2]Department of Computer Science, Titu Maiorescu University, 22 Strada Dâmbovnicului Tineretului, Bucureşti 040441, Bucharest, Romania
[3]Author to whom any correspondence should be addressed

Abstract. In an effort to reduce bureaucracy and the consumption of consumable materials Romania introduced a few years ago the "Single Integrated Information System" which ensures the collection, consolidation and processing of data from the entire Romanian health insurance system. Like any IT system the "Single Integrated Information System", may also pose some concerns regarding the confidentiality of personal data and information. In this respect, the present paper explores the possibility of mitigating these concerns by introducing anonymous signature schemes into the Romanian "Single Integrated Information System". Following their research, the authors concluded that this change can improve security and increase citizens' confidence in the system, without significantly increasing the implementation costs of the existing system. The authors are also confident that if implementing this method will result in a further reduction of the quantity of consumable materials, like paper and printer ink, used in the healthcare system. This will have a positive impact on the environment and on Earth in general.

1. Introduction

Nowadays the process of automation and the switching from physical to electronic means of data storage encompasses almost all sectors of society. Besides the great advantage of lowering bureaucracy, this transition also leads to reduced usage of consumables, especially paper, mainly used to store and manage data in a physical form. This advantage is even more relevant when it comes to protecting the environment and reducing pollution.

In Romania, a relatively new sector that has made this transition is the healthcare sector which previously relied on handwritten paper documents to maintain an archive, to issue medical documents or prescriptions, etc.. By using a large amount of paper and toxic ink, the healthcare insurance system was generating a significant impact on the environment. This sector introduced a few years ago the "Single Integrated Information System" which ensures the collection, consolidation and processing of data from the entire Romanian health insurance system. This system's main purpose is to help the national health insurance providers to perform their specific tasks of managing the health insurance budget.

Like any IT system the "Single Integrated Information System", SIIS for short, may also pose some concerns regarding the confidentiality of personal data and information. There are also legal considerations that need to be taken into account when thinking about the protection of private data and information.

In this respect, the present paper explores the possibility of mitigating these concerns by introducing anonymous signature schemes into the SIIS. These types of schemes can provide the same security services as standard signature schemes, but in addition they can hide the identity of the signer until such time when that identity needs to be proven.

After considering this possibility, the authors believe that adding anonymous signature schemes to the SIIS can improve security and increase citizens' confidence in the system, without significantly increasing the implementation costs to the already existing system.

2. Standard signatures and anonymous signatures

Digital signatures have become a standard element for many suites of cryptographic protocols available worldwide. They are mostly used to detect and/or to prevent the forgery or manipulation of digital messages. The most common examples of digital signatures usage are in software distribution, contract management software, in financial transactions or wherever there is a need to prevent forgery or manipulation.

2.1. Standard signatures

Standard digital signature schemes are mathematical schemes which are used to prove the authenticity of a digital message or document. A valid digital signature gives motives to a recipient to trust that the message was created by a known sender, that the sender cannot deny having sent the message (authentication and non-repudiation) and that the message was not altered in transit (integrity).

RSA cryptosystem (Rivest-Shamir-Adleman) is one of the first practical public-key cryptosystems and is widely used for secure data transmission. In such a cryptosystem, the encryption key is public and differs from the decryption key which is kept secret. In RSA, this asymmetry is based on the practical difficulty of factoring the product of two large prime numbers, the factoring problem [8].

The users of RSA algorithm create and then publish a public key based on two large prime numbers, along with an auxiliary value. The prime numbers must be kept secret. Anyone can use the public key to encrypt a message, but with currently published methods, if the public key is large enough, only someone with knowledge of the prime numbers can feasibly decode the message. In order to verify the origin of a message, RSA can also be used to sign a message.

2.2. Anonymous signatures

The first formalization of anonymous signatures notion was presented by Yang et al. in [1]. It was further explored by Fischlin in [2] and revisited by Saraswat et al. in [3]. This paper tries to make another proposal of the possible practical usage of these signature schemes in the Romanian national health insurance system.

Anonymous signature schemes use standard signature algorithms, but instead of proving the signer identity of the message (m) at any moment, the signature σ conceals the identity of the signer up to a specific moment in time. Anonymous signatures are used in many applications where the identity of the signer needs to be protected, e.g. key exchange protocols, electronic auction systems, electronic paper review systems or electronic voting systems like in [4].

There are several anonymous signature schemes, but most important and prevalent schemes are Yang [1], [5] and Saraswat [3]. Yang's scheme for anonymous signature guarantees the anonymity of the signer when the adversary obtains only the signature but not the message, or when the message contains a random string, called security parameter, which is kept hidden until the verification phase. To keep secret a part of the message may be not appropriate in some applications, but there are also enough applications which require not revealing the complete message.

Saraswat's scheme for anonymous signature splits the digital signature σ^* in two segments of data, $\sigma^* = (\sigma, \tau)$. The first part of digital signature (σ) is called anonymous signature or simply signature, and the second part of the digital signature (τ) is called verification token or simply token. The generation of signature (σ) and token (τ) makes use of signature generation algorithm which uses as inputs the signer's secret key and the message m. Verification phase occurs when anonymous signer

IOP Conf. Series: Earth and Environmental Science **172** (2018) 012007 doi:10.1088/1755-1315/172/1/012007

decides to prove publicly the anonymous signed message m belongs to him. At this moment, the signer has to make public m, σ and τ, and then the validity of the signature can be verified by everybody making use of the public key of the signer. During that time, τ is hidden, nobody can determine who the signer is, so the anonymity of the signer cannot be broken only from the message m and the signature σ.

An anonymous signature scheme Σ is a triple of algorithms, $\Sigma = (Gen, Sig, Vf)$, where the key generation algorithm $Gen()$ produces a key pair $(pk, sk) \leftarrow Gen()$. The signature generation algorithm $Sig()$ produces a pair of a signature and a verification token $\sigma^* = (\sigma, \tau) \rightarrow Sig(sk, m)$ using the secret key sk and a message $m \in \{0,1\}^*$. And the deterministic signature verification algorithm $Vf(pk, m, \sigma, \tau)$ produces an output „true" or „false". In the case that the signature, token or the message was not tampered, the following relation holds:

$$Vf(pk, m, Sig(sk, m)) = true \tag{1}$$

for $(pk, sk) \leftarrow Gen()$, and for any $m \in \{0,1\}^*$.

3. The Romanian electronic health insurance system

After an executive ministry order (645/2007) the plan for an electronic health insurance system in Romania was approved. It took some years to reach a functional solution and in 2015 the system became operational and available to the public at large.

The system has the following main components, the "Single Integrated Information System" (SIIS), the "Electronic Prescription" (EP), the "Electronic Health Insurance Card" (EHIC), the health service providers (HSP) and the electronic reporting applications (RA).

The most important component of the electronic health insurance system is the SIIS which has the role of collecting, consolidating and efficient processing of data from the entire national health insurance system. Its main purpose is to help the national health insurance providers to perform their specific tasks of managing the health insurance budget. It has a hierarchical structure distributed on three main levels, national, county level and health service providers – HSP level (figure 1).

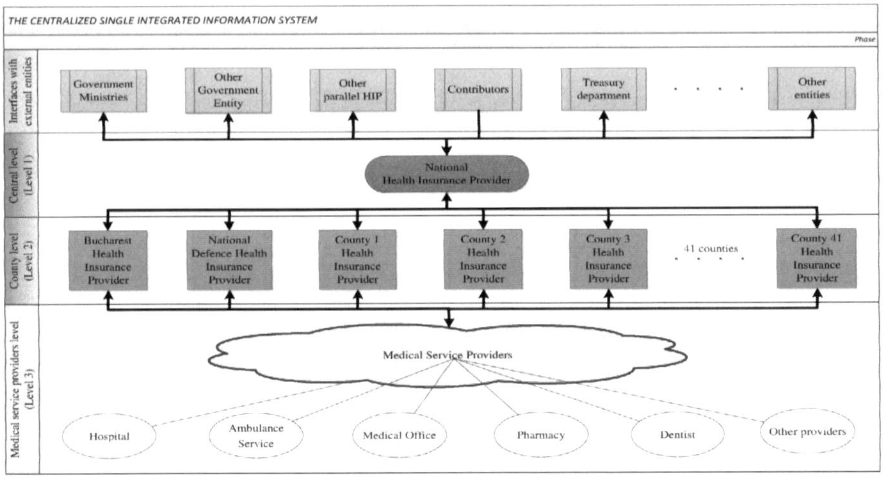

Figure 1: The hierarchical structure of the Single Integrated Information System.

The main functions of SIIS are, managing health insurance funds, managing the insured people, keeping an evidence of health service providers (medical and pharmaceutical), keeping the evidence of contributors and providing quality control for health services. The data from HSP is collected and analyzed locally by the level two of the hierarchical structure (figure 1) at the county level and afterwards is aggregated and maintained at the central level.

In the development of this system seven main objectives have been pursued, the collection and management of the economical and medical information required for an efficient operation of the Health Insurance System, transparency in the usage and management of health insurance budget by the National Health Insurance Provider, keeping a record of the insured people and of the health service providers by creating and administrating the National Registries for the insured people and for the health service providers, the simplification of data reporting by HSP, the standardization in the application of norms and laws at a national level, the highlight and the control of the costs for each insured person and the interfacing of other entities residing outside the system besides the HSPs, in online and/or offline modes.

The benefits of implementing such a system are obvious and intuitive.

3.1. The interfacing of the SIIS

The SIIS provides interfaces for the communication with the exterior through which electronic data transfer is being made. These interfaces are of two main categories:

- Interfaces with the health service providers (medical and pharmaceutical);
- Interfaces with other entities.

The main purpose of the interfaces with the HSPs is to report the provided services so that the health insurance provider can pay the HSPs for those services. On the other hand these interfaces are also used to exchange information related to the released medical prescriptions, sick leaves or other documents that a physician may issue. There are also other types of data exchanged on these interfaces, but that information depends on the specific contract each HSP has with the insurance provider.

The user (whoever it is, HSP, institution, etc.), does the reporting or receives details from the SIIS via the electronic reporting applications – the RAs. The RAs connect to the resources of the SIIS using https/SSL protocol (figure 2).

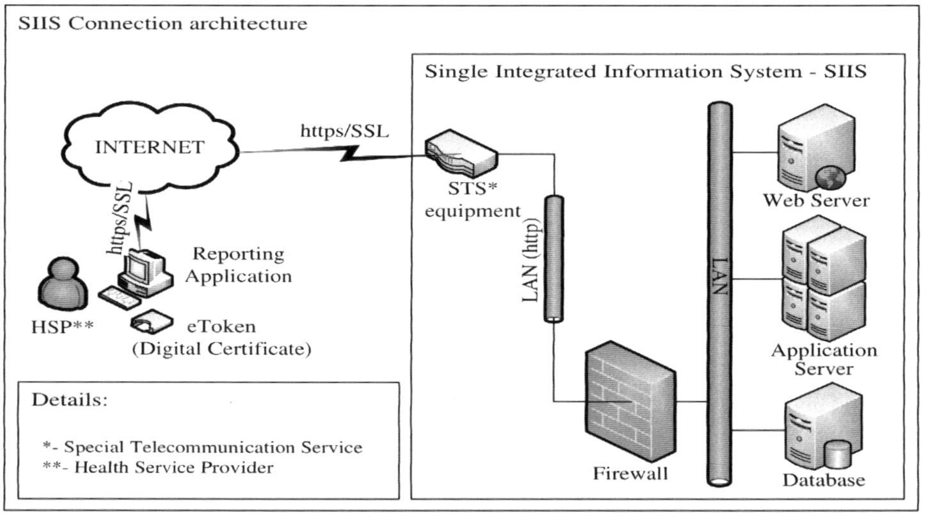

Figure 2: The connection of the reporting applications to the Single Integrated Information System.

The SIIS also has the option to validate the prescriptions online or to check in real time if a client is really insured, in order to avoid the possibility that a prescribed service, medical device or pharmaceutical product may not be covered under the insurance.

The interfaces of SIIS with other entities are defined according to some protocols between governmental institutions and are used to obtain or to provide data for the proper operation of the system. Some of these entities may be institutions that keep a population record, antifraud institutions, financial institutions, ministries, treasury departments, city halls, etc. (figure 1).

3.2. Information security in the Single Integrated Information System

To ensure data security for the exchange between the reporting applications and the System (i.e. SIIS), this project implements a public key infrastructure (PKI) designed to use digital certificates generated by well-known and approved certificate authorities.

The PKI infrastructure (figure 3), the technics, the practices and procedures that contributes collectively to the proper implementation and operation of the public key cryptosystems consists of hardware, software, databases, network resources, security procedures and legal obligations. These elements are linked together and are collaborating in order to provide and implement both certification services and other associated infrastructure services, like timestamps.

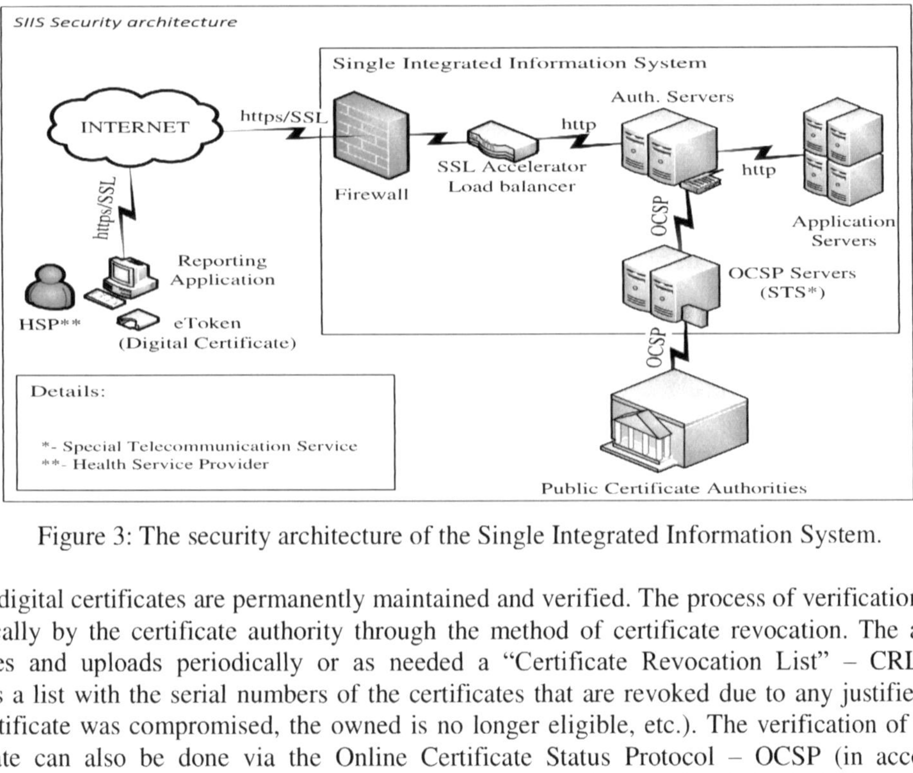

Figure 3: The security architecture of the Single Integrated Information System.

The digital certificates are permanently maintained and verified. The process of verification is done periodically by the certificate authority through the method of certificate revocation. The authority generates and uploads periodically or as needed a "Certificate Revocation List" – CRL, which contains a list with the serial numbers of the certificates that are revoked due to any justified reason (the certificate was compromised, the owned is no longer eligible, etc.). The verification of a digital certificate can also be done via the Online Certificate Status Protocol – OCSP (in accord with RFC2560). Using this method the applications do not have to download and check the long CRL file, but simply send a request to an OSCP service and check only the certificate of interest (figure 3).

The digital certificates will be handled using a dedicated database administered at the county level of the hierarchical structure of the SIIS (figure 1). The reporting applications will use the digital certificate to authenticate and to authorize its online requests to the SIIS by using the https/SSL suite of protocols (figures 2 and 3).

In order to access the resources of the SIIS, all HSPs will require the digital certificates generated by the certificate authorities mentioned earlier. Also those certificates must be registered in the SIIS in order for the HSP's reporting applications to work. For every operator of a specific HSP user accounts will be created which will give authorization to that operator. This will provide further accountability of the actions taken by each operator.

The access gateway into the SIIS is protected by a hardware firewall which also acts as a SSL Accelerator and load balancer. At the same time this equipment checks the integrity and validity of the digital certificates presented by the reporting applications when a new https/SSL session is initiated.

After integrity and validity verification the SSL accelerator forwards the request by http to the authorization and authentication servers (figure 3 – Auth. Servers). In the http header the SSL accelerator also includes data related to the digital certificate in question, information which will be used by the following servers for a further evaluation of the certificate revocation status by OCSP. The auth. servers are checking the dedicated database if the digital certificate was registered by an authorized user of the system. If it checks out the server forwards an OCSP request via the "Special Telecommunication Service" - STS, which provides this service, to the public certificate authorities in order to verify the revocation status of the certificate.

The STS offers as a part of the system a component that allows for simultaneous interrogation of all public certificate authorities in Romania. In this way the consequences when a certain certificate authority changes its structure or becomes unavailable are prevented.

If the OCSP request reveals that the digital certificate is not revoked, the auth. servers will check in a buffer database the status of the user's contract. If that will also check out the server will send a software token to the reporting application. The token is called session-id-hash and will be used by the application to announce that the session has already been approved. This will prevent next requests in the same SSL session to undergo the same process of authorization. This way will accelerate the operation of the application and will not overwhelm the auth. servers with verification requests which can result in slow operation and long waiting times.

3.2.1. Digital Signatures in the Single Integrated Information System. A digital signature represents additional electronic information attached (or logically attached) to an electronic document and has the same significance as the handwritten signature on a paper document. The signer is that person or entity that has the proper device or software to create the signature and is acting either in its own name or representing a third party.

Like the handwritten signature, the digital signature proves the ownership of an electronic document and in addition it offers the recipient the possibility to check if the document suffered alterations or tampering, either intentional or accidental.

For generating a digital signature the signer needs a digital certificate which follows the criteria presented earlier. The certificate links the digital signature to the person or entity that generated that signature.

In order to ensure non-repudiation of the data reported to the SIIS, all reporting files need to be signed with the signer's digital signature. Electronically signing these files leads to the possibility to fully automate the health insurance system and to simplify the procedures by eliminating paperwork.

The system keeps a record of all reported files, no matter if the digital signature was validated or not. The reasoning behind this behavior is to always have the proof for the revocation of a specific request, if the need to justify a decision made by the system arises.

4. Possible security concerns regarding the confidentiality of personal information in the Single Integrated Information System

Like any IT system, the SIIS is posing some concerns regarding information security. By using mainly proven security protocols and cryptographic means, the system can be considered to have a good security level when it comes to well-known methods of attack. The most obvious vulnerability of the system derives from the fact that the connection between the reporting applications and the SIIS is achieved via the Internet. Even though the employed security protocols provide good protection against attacks like packet sniffing, man-in-the middle, attempts to decrypt the traffic or to impersonate a legitimate user, the system is still vulnerable to attacks such as denial of service. The main reason behind this vulnerability is the fact that the Internet is opened to the whole world and any ill intended person or entity can perform a denial of service attack like the bandwidth attack for example. The attacker cannot penetrate the system or alter/read the traffic, but he or she can still be able to successfully prevent the reporting applications from establishing the SSL sessions with the system by occupying all available bandwidth.

The bandwidth attacks are usually succesfull attacks against most IT systems that run over the Internet and there are very few ways to protect against them. The most efficient method to mitigate these kinds of attacks is to have the system connected to more than one Internet Service Providers. This method is not an assurance that bandwidth attacks will fail, but doing so will substantially raise the cost of the attack, because the attacker will have to employ significantly more resources to perform the attack.

After analyzing the Single Integrated Information System, the authors of this paper manage to identify another possible vulnerability of this system. The data that is stored in the SIIS contains a lot of personal information regarding its users. This information might be medical information, general health insurance information, identification information and other type of information that may fall under the "Processing of personal data" laws.

When stored inside the system this sensitive data is well protected, but as it was shown in figure 1, the SIIS is connected to other entities through external interfaces. For instance one institution that requires a connection with the system is the "National Statistics Institute". This institute requires the data to compute detailed statistics regarding many characteristics of the national health system. These statistics may include statistics about the usage of different types of medication, about medical leaves, about types of diseases and their prevalence, etc.. In order to compute these statistics this institution requires the data generated by the reporting applications, like prescriptions, medical reports, etc.. These reports also contain data that may be used to identify the individual for which the reports were generated. And under the aforementioned laws this information is protected and access to it must be limited. Besides that, "The National Statistics Institute" doesn't need the identification information in order to achieve its objective of generating those statistics.

Due to security best practices any outside system connected to one's own system should be considered untrusted, as it is implied in [6]. And so, the identification data must be protected, even more so when it leaves the SIIS. But there must also be a way to retrieve the identification data when such need arises.

That is why the authors of this paper are proposing the introduction of anonymous signature schemes in the Single Integrated Information System.

5. The proposed solution

The proposed solution implies the implementation of anonymous signatures in the report generating process of the system.

In order to generate a report or to search a report in the system, the reporting applications use different identifications markers. These markers are the beneficiary's personal identification number, the serial and number of the report and the physician's stamp number. After the physician fills up a report, like a medical prescription, with the prescribed treatment and the aforementioned identification markers, he validates it in the system and digitally signs it. After that the system stores it. In order to access the user's health insurance information the physician is using the user's Electronic Health Insurance Card, the EHIC, which contains the user's keys and digital certificate. When the user goes to a pharmacy to get his prescription drugs, the pharmacist either scans the code on the paper prescription (which the doctor prints out after it was validated and stored in the system) or searches the prescription by introducing the same identification markers. The pharmacist also checks the user's health insurance status by using the data stored on the user's EHIC.

The possible vulnerability derives from the fact that the electronic prescription is stored with the name and personal identification number of the user alongside the prescribed drugs and diagnosis. The problem is even more noticeable when sending this kind of reports to other systems or institutions like the "National Statistics Institute".

It is the authors' opinion that the implementation of anonymous signature schemes is a way to eliminate this vulnerability.

The idea is to have an electronic dossier for each beneficiary of health insurance. This dossier will contain all documents released on that beneficiary's name and all its medical record. Each medical document released for that beneficiary will be added to this dossier. The dossier won't have any identification markers; instead it will be anonymously signed with a Saraswat anonymous signature generated using the data stored on the personal EHIC. This will also prove, if necessary, the identity of the owner.

At the reporting applications side (which are responsible for the generation of the documents that will be added to the dossier) all electronic documents generated (e.g. electronic prescriptions) will have the name and the personal identification number fields removed or encrypted (using the user's public key) by the application. Each document will be released and added to the person's dossier after

IOP Conf. Series: Earth and Environmental Science **172** (2018) 012007 doi:10.1088/1755-1315/172/1/012007

the validation from an insurance point of view. After adding the document to the dossier, the system will remove the previous signature and regenerate it including the latest modifications to the dossier (i.e. the addition of the newly generated documents). As it was expected the signature will be generated using the data from the personal EHIC.

The removal or the encryption of the name and the personal identification number from the generated documents is crucial because the other data on the document will be used to generate statistics and to be sent to other institutions or entities outside the system. And for that reason that data needs to be in the clear and the name and the personal identification number (i.e. the person that belongs to) needs to be protected.

Applying the anonymous signature scheme will result in the generation of the signature and the verification token, which will be used to verify the signature and to prove, if such need arises, the identity of the owner of the dossier. The verification token will be stored on the personal EHIC and it will be updated each time a modification to its dossier will be performed.

The dossiers for each insured person will be stored on the database servers (figure 2). The signature will be computed by a dedicated application server in conjunction with the reporting application.

By adding anonymous signatures the operation of the system will not change drastically. It will follow the operations presented in chapter 3 with the addition of this new functionality (figure 4).

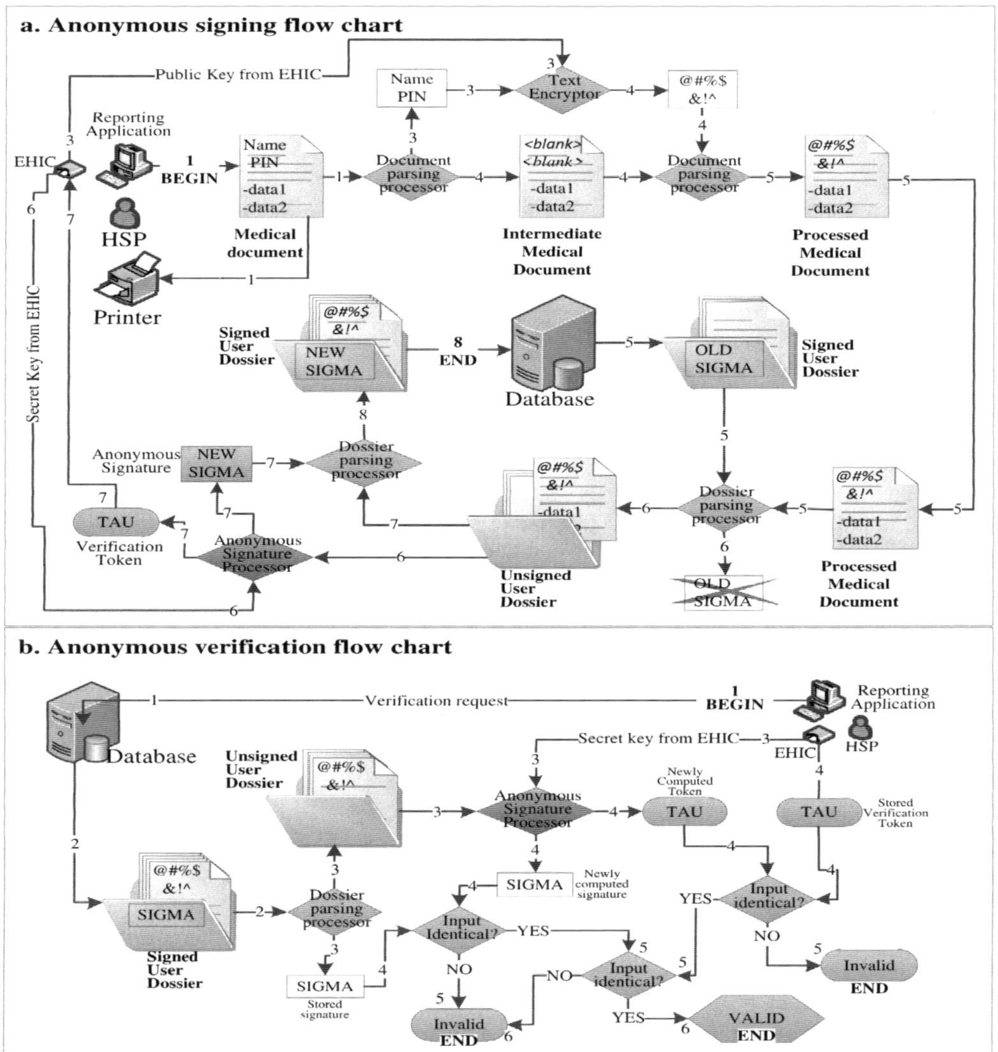

Figure 4: Anonymous signatures flow chart: (a) Signing flow chart; (b) Verification flow chart.

For example in order to release a prescription, the pharmacist (i.e. the reporting application of the pharmacist) will search the database for that prescription using the serial and number of the prescription. After finding it, the pharmacist app will check the validity of the dossier containing that prescription, by using the token from the user's personal EHIC. If the validation passes the pharmacist will be able to release the prescribed drugs to the client. The reporting app of the pharmacist will also have to make changes to the client's dossier in order to mark the prescription as being released. The resigning process will have to be repeated and the new token will replace the old one on the EHCI.

From a mathematical point of view we will denote $D \in \{0,1\}^*$ to be the user's dossier. Thus we can write $D = \sum_{i=1}^{n \in N^*} r_i$, where $r_i \in \{0,1\}^*$ is the report number i that was added to the dossier. We also have the key pair (pk, sk) generated by the key generation algorithm, where pk is the user's public key and sk is the secret key. The key pair is stored on the personal EHIC. Applying the anonymous signature will generate $\sigma^* = (\sigma, \tau) \rightarrow Sig(sk, D)$, where σ is the part of the signature that will be added to the dossier D and τ is the token stored on the EHIC used for verification. Thereby we will get $D' = \sigma + D = \sigma + \sum_{i=1}^{n \in N^*} r_i$ which will be the signed dossier stored on the system.

In the verification phase the system will have to compute $Vf(pk, D, \sigma, \tau)$ which is:

$$Vf(pk, D, Sig(sk, D)) \qquad (2)$$

If the result of (2) is true for any $D \in \{0,1\}^*$ and for (pk, sk) the initially generated key pair, we can judge that the dossier belongs to the claimer and has not been modified in the meantime.

A step by step flow chart example of the proposed solution is given in figure 4.

5.1. Security analysis

The anonymous signature scheme proposed by the authors uses a Saraswat scheme which can be combined with and asymmetric encryption scheme. The anonymous signature scheme is used to protect owner identity and to sign a user document dossier. It also provides non-repudiation and integrity protection for the dossier. The encryption scheme is not mandatory but it can provide encrypted identity markers (name and personal identification number) for each individual document in the user dossier thus allowing for more options and furthering security. It is also worth mention that for the anonymous signature verification there is no need for the asymmetric decryption of the documents contained in the user dossier (figure 4 - b).

In the signing phase (figure 4 - a), a dossier parser removes the old signature and adds the new document to the dossier. The new dossier is passed through the anonymous signature processor and a new signature is generated which is added to the dossier by a parser. The signature σ^* it is then split by the signature processor into two parts (σ, τ). σ will be used to sign the dossier and τ will be stored on the electronic health insurance card – EHIC and used in the verification process. In the final step the newly signed dossier is then stored on the database servers, which ends the signing process.

In the verification phase (figure 4 - b), the dossier parser will separate the signature σ from the user dossier. The dossier will be passed through the anonymous signing process again and the signature σ will be combined with the stored verification token τ from the user's EHIC in order to form the original full signature σ^*. The new signing process will generate a new $\sigma^{*\prime} = (\sigma', \tau')$ pair which will be compared to the original σ^*. If they are identical (i.e. $\sigma^* = \sigma^{*\prime}$), we can assess that the dossier belongs to the claimer and it was not tempered or modified.

The implementation of anonymous signature schemes into the Single Integrated Information System will also allow the user dossiers to be securely passed to other entities outside of the system. The outside recipients will not be able to uncover the identity of the dossier owner but can still read and process the medical data contained within.

6. Conclusions

Anonymous signature schemes can provide the functionalities of standard signature schemes but will keep the identity of the signer hidden until such time that the signer decides to reveal it.

This paper provides a practical example scenario where an anonymous signature scheme can be implemented and the benefits and improvements it adds to that scenario.

The Single Integrated Information System – SIIS, is the Romanian national health insurance system. It was recently implemented in Romania and provides electronic administration of citizens' medical data and health insurance information.

As any IT system, the SIIS, can be affected by security vulnerabilities, like, for instance, denial-of-service attacks or the illicit acquiring of user medical information. Proposing a solution to mitigate or eliminate the later form of attack is this paper's purpose because that kind of information is also protected by Romanian laws. Thus the authors proposed the implementation of anonymous signature schemes into the SIIS. This solution will allow the system to securely disseminate the medical and pharmaceutical data for statistics or other purposes without linking that data to specific individuals or entities. Beside anonymity and signing it also provides non-repudiation and integrity services. And if the need arises the solution provides a verification mechanism of the identity of the signer.

It is the authors' opinion that the implementation cost of this solution will be small in comparison with the whole implementation cost of the system. The reasoning behind this assumption is that the system hardware has already been deployed and paid for and the authors' solution will consist only in software development and implementation.

The implementation of such a security scheme to the health insurance system will also lead to the reduction of consumables like paper and the ink associated with the printing process of that paper. The reasoning behind this statement is that by properly securing electronic data and communications will no longer be required to generate and store hard paper copies of medical documents and records. This will lead to a better environmental footprint and a lower impact on planet Earth.

The authors of this paper also proposed an anonymity-based fair e-lottery system [7]. In that paper the proposal was made to use a hybrid anonymous signature scheme in order to restore people's faith in the lottery and increase the number of players thus increasing also the earnings for the lottery organizer. The downside of that proposal, compared with the proposal in this paper, was that the implementation cost would be quite large because a national-wide e-lottery system is not yet implemented.

7. References

[1] Yang G, Wong D S, Deng X and Wang H 2005 Anonymous signature schemes *PKC 2006* **3958** (Berlin, Heidelberg: Springer) pp 347 – 63

[2] Fischlin M Anonymous signatures made easy *PKC 2007* **4450**, ed T Okamoto and X Wang (Berlin, Heidelberg: Springer) pp 31 – 42

[3] Saraswat V and Yun A Anonymous Signatures Revisited *ProvSec 2009* **5848**, ed J Pieprzyk and P Zhang (Berlin, Heidelberg: Springer) pp 140 – 53

[4] Gang C An Electronic Voting Based on Multi-Party Computation *ISCSCT '08* 10.1109/ISCSCT.2008.80 (Shanghai: IEEE)

[5] Yang G, Wong D S and Deng X Efficient anonymous roaming and its security analysis *ACNS 2005* **3531**, ed J Ioannidis et al (Berlin, Heidelberg: Springer) pp 334 – 49

[6] Lehtinem R, Russell D and Gangemi G T Sr 2006 Computer security basics 2nd edition, ed T Apandi et al (Sebastopol CA: O'Reilly)

[7] Medeleanu F, Racuciu C and Antonie N F 2017 Anonymous signature schemes applied for a fair e-lottery system *The international conference "Education and creativity for a knowledge-based society" 11th edition* ISBN 978-3-9503145-5-7 (Vienna: Osterreichish Rumanischer Akademischer Verein) pp 32 – 7

[8] Stinson D R 1995 Cryptography: Theory and practice third edition, ed K Osen (Boca Raton: Chapman & Hall/CRC Press)

IOP Conf. Series: Earth and Environmental Science **172** (2018) 012008 doi:10.1088/1755-1315/172/1/012008

Synthesis of mathematical model of controlled electromotive device of pilotless floating vehicle

Ya Volyanskaya

Admiral Makarov National University of Shipbuilding, 9, Heroiv Ukrainy Ave., Mykolayiv, 54025, Ukraine

E-mail: yanavolyanskaya@gmail.com

Abstract. The features of mathematical modelling of the propulsion system of the electric motion of the autonomous floating vehicle are determined. Additional losses of mechanical energy of special-purpose asynchronous motors during their operation in a liquid dielectric medium are taken into account. As well as the correct modelling of the nonlinear characteristics of the screw-propeller throughout the range of angular velocity as essentially nonlinear hydrodynamic objects. The mathematical model is created based on the simplified functional scheme of the electromotive device of the autonomous floating vehicle as a control object. It contains an expression for the relative pitch of the screw-propeller. As well as the proposed method of clarifying the loss coefficient of braking hydrodynamic torque by taking into account the individual geometric characteristics of the electric motor and the characteristics of different types of filling the electric motor, dielectrics. It is shown with the use of experimental data that the mechanical characteristic of the engine depends significantly on the temperature and the type of the working fluid, so the influence of these parameters must be taken into account in the synthesis of the electric propulsion systems of the pilotless floating vehicle.

Currently, for the design of pilotless floating vehicle (PFV) with the necessary parameters, various approaches are used, which are often of an empirical nature [1]. To develop the perfect USW complex research facilities are built, a variety of measurements are conducted, data processing involves qualified personnel – all this requires significant financial costs. The mathematical approach to the solution of this problem is deprived of such shortcomings.

The method of mathematical modelling is the most common method of studying marine mobile objects, because it is characterized by high accuracy and almost unlimited possibilities for taking into account nonlinearities of any nature. The study carried out with the help of a mathematical model is often called an experiment on the model [2, 3].

Analysis of publications devoted to the development of mathematical models of controlled motion of PFV [3-6] shows that this problem has not lost its relevance until now. However, from the practical point of view, in modern conditions due to the development of computer technology, software, methods of mathematical description, it is advisable not to develop a new, and choose from existing models that meets the requirements of the problem.

Mathematical modelling of controlled motion PFV can be represented by a set of information, electromechanical and hydro-mechanical systems as components of the control object, interacting with each other:

– information system (IS) – sensor of spatial position of PFV and its dynamic characteristics (angles, velocity); further we neglect the errors of the sensors and consider them to be perfect;

– automatic control system (ACS) of movement, where is the processing of information coming from the sensors, and calculated control signals for actuators PFV (propulsion systems); a particular model of PFV should be built in accordance with the type of real ACS (analogue digital, fuzzy, neural network);

– electromotive device (EMD), to the peculiarities of mathematical modelling can be included the need to take into account additional losses of mechanical energy of engines due to their work in the liquid dielectric medium and correct modelling of nonlinear characteristics of screw-propellers over the entire range of their angular velocity as essentially nonlinear hydrodynamic objects;

– corps of PFV is a solid body moving in water under the action of stops of the movers, and is subjected to forces of hydrodynamic resistance considering the attached masses of water in dynamic conditions; the corps of PFV is also affected by forces of external disturbances; the corps of PFV can modify its operating parameters when working under water (the weight and the coefficients of hydrodynamic resistance).

When developing mathematical models of PFV elements, the following general assumptions are made that do not distort the physics of phenomena occurring in the object of study: water as a working medium is considered as an ideal unperturbed liquid; for each stage of freedom PFV has an appropriate engine; thrust vectors of movers and external perturbations are applied to the centre of mass of PFV and do not create a different and a roll; saturation of magnetic conductors of electric motors and nonlinear dependences of mechanical elements is neglected; PFV in water has zero buoyancy.

The behaviour of the PFV is considered in accordance with the laws of the mechanics of motion of a solid body in three-dimensional space. It has six degrees of freedom. Three of them characterize the linear displacement of the centre of mass and three to the rotation of the object about the centre of mass. Each degree of freedom corresponds to one velocity coordinate, which is the projection of the linear or angular velocity vector on the corresponding axis of the coordinate system.

Instantaneous values of the velocities of motion of a solid body are subject to the theorem on the change of the number of motion and the moment of the number of motion, which in a fixed coordinate system have the form

$$d\overline{K}/dt = \overline{R}; \qquad d\overline{L}/dt = \overline{M}, \tag{1}$$

where $\overline{K}$ – the main vector of the number of motion of PFV; $\overline{L}$ – the main vector of the moment of the amount of motion relative to the origin of the coordinate system; $\overline{R}$ – the main vector of external forces acting on PFV; $\overline{M}$ – the main moment of external forces relative to the origin.

However, the equations of PFV motion is written, as a rule, in the coordinate system, rigidly connected with the corps. This is explained by the fact that the forces and moments of interaction of the PFV body with the environment, as well as the forces and moments of external actions are easier expressed in the associated coordinate system. Since the point O – the origin of the associated coordinate system $Oxyz$ – moves at a certain linear speed ϑ and the PFV rotates around the origin at an angular velocity Ω, in the equations for changing the amount of motion and the moment of the amount of motion recorded in the mobile coordinate system, appear as additional components of the vector product

$$\begin{aligned} &\frac{d\overline{K}}{dt} + \overline{\Omega}\times\overline{K} = \overline{R}; \\ &\frac{d\overline{L}}{dt} + \overline{\Omega}\times\overline{L} + \overline{\vartheta}\times\overline{K} = \overline{M}. \end{aligned} \tag{2}$$

Two vector equations (2) correspond to six scalar differential equations written with respect to projections on the connected axes. According to the laws of mechanics, there is a connection between the kinetic energy of a moving body W, projections of linear velocities $\vartheta_x, \vartheta_y, \vartheta_z$, angular velocities $\omega_x, \omega_y, \omega_z$ and vectors $\overline{K}$ and $\overline{M}$:

IOP Conf. Series: Earth and Environmental Science **172** (2018) 012008 doi:10.1088/1755-1315/172/1/012008

$$\begin{cases} K_x = \dfrac{\partial W}{\partial \vartheta_x}; \quad K_y = \dfrac{\partial W}{\partial \vartheta_y}; \quad K_z = \dfrac{\partial W}{\partial \vartheta_z}; \\ L_x = \dfrac{\partial W}{\partial \omega_x}; \quad L_y = \dfrac{\partial W}{\partial \omega_y}; \quad L_z = \dfrac{\partial W}{\partial \omega_z}. \end{cases} \tag{3}$$

The pilotless floating vehicles differ by the fact that in the process of their own movement, they attract the adjacent layers of the environment (liquid). Part of the water near the hull of a ship moving with the viscosity of the fluid, begins to move, acquiring some kinetic energy due to the work of main propulsion of the vessel. The floating vehicle (FV) and the environment surrounding it form an interconnected, energy-related system. Therefore for calculation of parameters of movement of object it is necessary to proceed from the analysis of the total kinetic energy including together with energy of the most rigid body the attached kinetic energy of environment: $W = W_{\mathrm{FV}} + W_l$. In determining the magnitude of the kinematic energy, PFV W_{FV} is considered its constant mass, shape of the body and the distribution of masses inside the body. Under these conditions and the following assumptions: if the origin of the associated coordinate system is connected to the centre of mass of FV $x = y = z = 0$, and the coordinate axes are the main central axes of inertia $I_{xy} = I_{xz} = I_{yz} = 0$, the kinetic energy can be written in matrix form [4]:

$$W_{\mathrm{FV}} = 0{,}5\left[\vartheta^T \Omega^T\right] D \begin{bmatrix} \vartheta \\ \Omega \end{bmatrix}, \tag{4}$$

where $\vartheta = \left[\vartheta_x \vartheta_y \vartheta_z\right]^T$; $\Omega = \left[\omega_x \omega_y \omega_z\right]^T$ – matrix of the linear and angular velocities of PFV, to introduce the concept of matrix of inertia of a rigid body

$$D = \begin{bmatrix} m & 0 & 0 & 0 & mz & -my \\ 0 & m & 0 & -mz & 0 & mx \\ 0 & 0 & m & my & -mx & 0 \\ 0 & -mz & my & I_{xx} & -I_{xy} & -I_{xz} \\ mz & 0 & -mx & -I_{xy} & I_{yy} & -I_{yz} \\ -my & mx & 0 & -I_{xz} & -I_{yz} & I_{zz} \end{bmatrix},$$

where m – the mass of the rigid body.

The kinetic energy of the liquid can be represented as an expression [1]

$$W_l = 0{,}5\left[\vartheta^T \Omega^T\right] \Lambda \begin{bmatrix} \vartheta \\ \Omega \end{bmatrix}, \tag{5}$$

where Λ – the matrix of added masses and moments of inertia FV.

The total kinetic energy of the PFV moving in the liquid, taking into account (4) and (5) is determined by the expression

$$W = 0{,}5\left[\vartheta^T \Omega^T\right] (D + \Lambda) \begin{bmatrix} \vartheta \\ \Omega \end{bmatrix}. \tag{6}$$

To analyse these displacements, we need a complete model of the spatial motion of a solid body in a liquid medium. Simplifications are allowed only for certain stationary modes, which include movement and stabilization of the device at a given depth, ascent and immersion along an inclined trajectory or vertically.

In drawing up the equations of dynamics of PFV, it was taken into account that its shape does not correspond to the ellipsoid of rotation. Therefore, diagonal elements were taken into account in the matrix of the attached masses and moments of inertia. The symmetry of the apparatus with respect to the longitudinal axis except for diagonal elements $\lambda_{11}..\lambda_{66}$ ($\lambda_{11},\lambda_{22},\lambda_{33}$ – attached masses; $\lambda_{44},\lambda_{55},\lambda_{66}$ – attached moments of inertia) has a significant value $\lambda_{26}=\lambda_{62}$ and $\lambda_{35}=\lambda_{53}$ [4]. Therefore, the total kinetic energy of FV moving in the aqueous medium, according to (6) is

$$W=\frac{1}{2}\left[\vartheta_x \vartheta_y \vartheta_z \omega_x \omega_y \omega_z\right]\begin{bmatrix} m+\lambda_{11} & 0 & 0 & 0 & 0 & 0 \\ 0 & m+\lambda_{22} & 0 & 0 & 0 & \lambda_{26} \\ 0 & 0 & m+\lambda_{33} & 0 & \lambda_{35} & 0 \\ 0 & 0 & 0 & I_{xx}+\lambda_{44} & 0 & 0 \\ 0 & 0 & \lambda_{53} & 0 & I_{yy}+\lambda_{55} & 0 \\ 0 & \lambda_{62} & 0 & 0 & 0 & I_{zz}+\lambda_{66} \end{bmatrix}\begin{bmatrix} \vartheta_x \\ \vartheta_y \\ \vartheta_z \\ \omega_x \\ \omega_y \\ \omega_z \end{bmatrix}.$$

After multiplication of matrices we obtain

$$W=0{,}5(m\vartheta_x^2+m\vartheta_y^2+m\rho_z^2+I_{xx}\omega_x^2+I_{yy}\omega_y^2+I_{zz}\omega_z^2)+\lambda_{26}\vartheta_y\omega_z+\lambda_{35}\vartheta_z\omega_y. \tag{7}$$

Using the expression (7) we determine the projections of vectors and moments of the number of motion

$$\begin{cases} K_x=m_x\vartheta_x; \quad K_y=m_y\vartheta_y+\lambda_{26}\omega_z; \quad K_z=m_z\vartheta_z+\lambda_{35}\omega_y; \\ L_x=I_x\omega_x; \quad L_y=I_y\omega_y+\lambda_{35}\vartheta_z; \quad L_z=I_z\omega_z+\lambda_{26}\vartheta_y, \end{cases} \tag{8}$$

where $m_x=(1+k_{11})m$; $m_y=m_z=(1+k_{22})m$; $I_x=(1+k_{44})I_{xx}$, $I_y=I_z=(1+k_{66})I_{yy}$, I_{xx},I_{yy},I_{zz} – the moments of inertia relative to the associated axes; k_{11},k_{22},k_{44} – dimensionless coefficients depending on the geometric dimensions of the FV and vary in ranges $k_{11}=0{,}02...0{,}5$; $k_{44}=k_{66}=0...1{,}0$.

The equations of the dynamics of the spatial movement of FV in the generalized form are written as follows:

$$\begin{cases} m_x\frac{d\vartheta_x}{dt}+m_z\omega_y\vartheta_z-m_y\omega_z\vartheta_y+\lambda_{35}\omega_y^2-\lambda_{26}\omega_z^2=R_x; \\ m_y\frac{d\vartheta_y}{dt}+\lambda_{26}\frac{d\omega_z}{dt}+m_x\omega_z\vartheta_x-m_z\omega_x\vartheta_z-\lambda_{35}\omega_x\omega_y=R_y; \\ m_z\frac{d\vartheta_z}{dt}+\lambda_{35}\frac{d\omega_y}{dt}+m_y\omega_x\vartheta_y-m_x\omega_y\vartheta_x+\lambda_{26}\omega_x\omega_z=R_z; \\ I_x\frac{d\omega_x}{dt}+(\lambda_{26}+\lambda_{35})(\omega_y\vartheta_y-\omega_z\vartheta_z)=M_x; \\ I_y\frac{d\omega_y}{dt}+\lambda_{35}\frac{d\vartheta_z}{dt}+\omega_x\omega_z(I_x-I_z)+\vartheta_x\vartheta_z(m_x-m_z)-\lambda_{26}\omega_x\vartheta_y-\lambda_{35}\omega_y\vartheta_x=M_y; \\ I_z\frac{d\omega_z}{dt}+\lambda_{26}\frac{d\vartheta_y}{dt}+\omega_x\omega_y(I_y-I_x)+\vartheta_x\vartheta_y(m_y-m_x)+\lambda_{35}\omega_x\vartheta_z+\lambda_{26}\omega_z\vartheta_x=M_z. \end{cases} \tag{9}$$

The right-hand sides of equations (9) are formed by hydrodynamic forces and moments, gravity forces G, displacement A and screw-propellers F_{sp}. The specific features of the FV shape are reflected depending on the hydrodynamic characteristics of the kinematic parameters of motion. Many machines have split horizontal rudders (ailerons). At various turning left and right blades to the corners δ_l and δ_r differently the time of control is determined by their average value $\delta_h = 0{,}5(\delta_l + \delta_r)$ and heeling moment – difference $\Delta\delta_h = \delta_l - \delta_r$. Another design feature of the FV is that with a single motor installation, the axis of the mover shaft is combined with the longitudinal axis of the vehicle, and two screws with opposite rotation form the mover. Thanks to this, the thrust force is oriented along the axis Ox and the accompanying different and heeling moments are eliminated, which could play the role of additional disturbances.

Taking into account the above, it is possible to present the external forces and moments of equations (9) in the following form:

$$\begin{cases} R_x = F_{sp_x} - q\upsilon^{2/3}c_x(\alpha,\beta,\delta_h,\delta_v) + (A-G)\sin\psi; \\ R_y = F_{sp_y} - q\upsilon^{2/3}c_y(\alpha,\omega_z,\delta_h) + (A-G)\cos\psi\cos\theta; \\ R_z = F_{sp_z} - q\upsilon^{2/3}c_z(\beta,\omega_y,\delta_v) - (A-G)\cos\psi\sin\theta i \\ M_x = q\upsilon\xi_x(\beta,\omega_x,\Delta\delta_h) - Az_A\cos\psi\cos\theta - Ay_A\cos\psi\sin\theta; \\ M_y = q\upsilon\xi_y(\beta,\omega_x,\delta_v) - Ax_A\cos\psi\sin\theta - Az_A\cos\psi\cos\theta; \\ M_z = q\upsilon\xi_z(\alpha,\omega_z,\delta_h) - Ay_A\sin\psi - Ax_A\cos\psi\cos\theta, \end{cases} \tag{10}$$

where υ – volumetric water displacement of FV; c_x, c_y, c_z – dimensionless characteristics or coefficients of longitudinal, normal and transverse hydrodynamic forces; ξ_x, ξ_y, ξ_z – dimensionless characteristics of hydrodynamic moments of roll, yaw and trim respectively; α – angle of attack; β – the drift angle; ψ – the trim angle; θ – angle of heel; q – velocity head, $q = \frac{\rho\vartheta_w^2}{2}$; ρ – liquid density; δ_h and δ_v – the corners of the horizontal and vertical thrusters respectively.

In underwater applications are used bar unloaded motors (BUM), in which the inner space is filled with a liquid dielectric, in which the rotor spins [7]. To create a generalized mathematical model of the electric-motive device PFV with BUM of any kind it is necessary to use the system [8]

$$\begin{cases} M_{dyn_{x,y,z}} = M_m - Q_g - Q_{sp_{x,y,z}}; \\ Q_g = K_q(\tau,Д)\omega_{m_{x,y,z}}; \\ Q_{sp_{x,y,z}} = \begin{cases} \frac{k_Q\rho D_{sp}^5}{4\pi^2}\left(\frac{\omega_{m_{x,y,z}}}{i}\right)^2 & \text{при } J_{x,y,z} \in [-1;1]; \\ k_q\rho D_{sp}^3\vartheta_{x,y,z}^2 & \text{при } J_{x,y,z}^{-1} \in (-1;1); \end{cases} \\ F_{sp_{x,y,z}} = \begin{cases} \frac{k_F\rho D_{sp}^4}{4\pi^2}\left(\frac{\omega_{m_{x,y,z}}}{i}\right)^2 & \text{при } J_{x,y,z} \in [-1;1]; \\ k_f\rho D_{sp}^2\vartheta_{x,y,z}^2 & \text{при } J_{x,y,z}^{-1} \in (-1;1); \end{cases} \\ J_{x,y,z} = \frac{2\pi\vartheta_{x,y,z}}{\omega_{m_{x,y,z}}D_{sp}}. \end{cases} \tag{11}$$

where $\vartheta_{x,y,z}$ – projections of the PFV velocity vector relative to water on the corresponding axes of the PFV forward motion; D_{sp} – screw-propeller diameter (lag, vertical and running screws are the same); $Q_{sp_{x,y,z}}$ – the projection of the braking torque of the screw-propeller (SP) on the respective axes; $k_Q = f(J_{x,y,z})$ и $k_q = f(J^{-1}_{x,y,z})$ – dimensionless coefficients characterizing the torque of the screw-propeller; ρ – specific density of water; $F_{sp_{x,y,z}}$ – projection propulsion power (stop) of SP on the respective axes; $k_F = f(J_{x,y,z})$ и $k_f = f(J^{-1}_{x,y,z})$ – dimensionless coefficients characterizing the propulsion power of SP; $J_{x,y,z}$ – relative step lag, vertical and running SP; $J^{-1}_{x,y,z}$ – relative step lag, vertical, and running the SP in reverse [9, 10]. Coefficients k_Q, k_q, k_F, k_f non-linear, are determined from the graphic dependences given in [7, 8].

Equations (9–11) represent the generalized mathematical model of the electromotive device of the PFV.

Coefficients k_Q, k_q, k_F, k_f non-linear, their linearization is carried out by means of piecewise linear approximation of the action characteristics of SP. A set of graphs $k_Q = f(J_{x,y,z}), k_q = f(J^{-1}_{x,y,z}), k_F = f(J_{x,y,z}), k_f = f(J^{-1}_{x,y,z})$ [9] allows you to model the stop and torque SP in all modes. The hydrodynamic loss coefficient K_q is also non-linear. Its linearization is carried out by means of least-squares approximation of the experimental characteristics of the brake hydrodynamic torque loss BUM obtained for various dielectrics that fill it [7].

Inputs are changed in the ranges:

– the temperature of the working fluid (dielectric) BUM $\tau = -30...+70^{\circ}C$;

– type of working fluid (dielectric) BUM: ethanol, kerosene, oil T750, oil and I12A;

– the moment of resistance of the external environment $M_{nom} = Q_{sp_r} ... Q_{sp_d}$, where Q_{sp_r} – the maximum braking torque of the SP by the reverse rotation; Q_{sp_d} – the maximum braking torque of the SP by the direct rotation.

Source data of improved mathematical model are $\omega_{m_{x,y,z}}$ – angular velocity of the lag, vertical, and running SP; $F_{sp_{x,y,z}}$ – propulsion power lag, vertical and running SP; $\vartheta_{x,y,z}$ – projection of the velocity of the PFV during its spatial motion.

Thus, an improved mathematical model of electromotive device of pilotless floating vehicle with bar unloaded motor of any sort of current at spatial movement of the floating vehicle is linearized can be used for studies of complex sea mobile objects, including autonomous floating vehicle with significant nonlinearities.

The efficiency test and accuracy estimation of the improved mathematical model were performed by comparing the calculated data with the experimental data. The calculated data are obtained as a result of modelling with the help of the developed Simulink-model containing a system of equations of the improved mathematical model.

Experimental data were obtained in the study of the spatial motion of the hydrograph underwater vehicle in the test basing of the Admiral Makarov national University of shipbuilding. The Simulink model simulation conditions are identical to the experimental conditions in the test pool.

In the first part of tests were carried out at rectilinear movement of the floating device on each of axes. In the second part of tests, movement was carried out in two planes. In the third part – spatial movement: on simple trajectories (moving from point to point in a straight line); on a sphere.

Studies have shown that the error of modelling the spatial motion of PFV does not exceed 24 %, which is 6–11% lower than the known similar experiments [11, 12]. The resulting error allows us to talk about the reliability of the research.

At low speeds of PFV modelling accuracy reaches the maximum error and has alternating character. This is explained by the fact that the dimensionless dynamic characteristics of screw-propellers k_Q, k_q, k_F, k_f at low speeds have significant nonlinearity, which complicates their approximation with high accuracy.

IOP Conf. Series: Earth and Environmental Science **172** (2018) 012008 doi:10.1088/1755-1315/172/1/012008

References

[1] Ageev M D 2004 *Autonomous uninhabited underwater vehicles* (Vladivostok: Dal'nauka) p 272
[2] Zarubin V S, Krishchenko A P 2001 *Mathematical modeling in engineering* (Moscow: Izd-vo MSTU) p 496
[3] Chen-Wei Chen, Jen-Shiang Kouh, and Jing-Fa Tsai 2015 Modeling and Simulation of an AUV Simulator With Guidance System *IEEE Journal Of Oceanic Engineering* Vol. 38 2 p 211–226
[4] Lukomsky Yu A, Peshekhonov V G, Skorokhodov D A 2002 *Navigation and traffic control of ships* (Saint-Petersburg: Elmore) p 360
[5] Golding B, Ross A, Fossen T 2006 Identification of nonlinear viscous damping for marine vessels *14th IFAC Symposium on System Identification* (Newcastle, Australia) p 332–337
[6] Polyak B T, Shcherbakov P S 2003 *Robust stability and control* (Moscow: Nauka) p 303
[7] Blintsov V S, Kostenko D V, Shimchak P 2001 Determination of losses in a submerged electric submersible vehicle drive *Bulletin of the National Technical University "KhPI"* (Kharkov: National Technical University "KhPI") Issue 10 p 410–411.
[8] Volyansky S M 2014 Synthesis of the mathematical model of the propulsion-steering complex of the underwater vehicle as an object of control *News of the National University of Shipbuilding* (*Electronic Materials* vol 2) (Nikolaev: National University of Shipbuilding) Access mode: http://dx.doi.org/10.15589.evn20140305
[9] Korol Yu M 2005 Construction of a mathematical model for the automated design of thrusters of underwater vehicles *News of the National University of Shipbuilding* (Nikolaev: National University of Shipbuilding) 3 (402) p 20–30
[10] Korol Yu M 2004 Influence of hydrodynamic losses on the mechanical characteristics of encapsulated controlled asynchronous engines of underwater vehicles *News of the National University of Shipbuilding* (Nikolaev: National University of Shipbuilding) 6 p 15–24
[11] Lukomsky Yu A, Chugunov V S 1988 *Control systems for marine mobile objects* (Leningrad: Shipbuilding) p 272
[12] Voytkunsky I I 1985 *Handbook on the theory of the ship* (Leningrad: Shipbuilding) p 768

IOP Conf. Series: Earth and Environmental Science **172** (2018) 012009 doi:10.1088/1755-1315/172/1/012009

Diving Centre's Hyperbaric Complex Modernization for Medical Recovery and Treatment Throughout Hyperbaric Oxygen Therapy

Mihai Diaconu[1], Simona Rus[1,3], Cătălin Frățilă[1] and Edith-Hilde Kaiter[2]
[1] Diving Centre, Constanța, Romania
[2] "Mircea cel Bătrân" Naval Academy Constanța, Romania
[3] Author to whom any correspondence should be addressed, e-mail adress: simona_elena_rus@yahoo.com

Abstract: The main objective of the hereby research-development study is the modernization of the Diving Centre's hyperbaric complex facilities so as to carry out, in accgordance with the current European standards, medical recovery and treatment procedures by means of hyperbaric oxygen therapy (HBOT). The present study proposes setting up a hyperbaric chamber, pleasantly monitored and appropriately equipped, able to completely meet the provisions of the European Standard EN 14931-2006, adopted and confirmed by the Romanian Standardizing Society (ASRO), thus becoming SR EN 14931 of 2006. The research laboratory is presented in this work in order to build a consortium of institutions which are interested in launching a project on structural funds with regard to the compatibility of the hyperbaric complex, available within the base, with its modern HBOT centres established over the last twenty years in the developed states of the contemporary world.

1. Overview

The department of research, training and divers formation (SCAS) presents this work in order to develop a consortium of institutions interested in starting a cooperation with the Diving Center so as to improve the activity of the hyperbaric equipment existing in our unit. The Hyperbaric laboratory was built by the French firm COMEX in collaboration with Romanian specialists, and it can provide, if needed, good medical services of therapy with hyperbaric oxygen.

The theme supposes a monitored hyperbaric chamber able to meet the provisions of the European standard EN 14931- 2006, adopted by the method of confirmation of ASRO which became SR EN 14931 of 2006 (Hyperbaric Chambers for Human Use, Hyperbaric Chambers with More Places, for Therapeutic Use, Performance, Safety Requirements, Testing).

The complex will meet hyperbaric made compatible with good performance parameters and missions referred to in this document.

2. Analysis of system requirements

The increased number of possibilities which man have developed to work underwater revealed new physiological and pathological aspects of immersion and life in hyperbaric conditions, which have made us consider that the information contained in this paper is up-to-date, not only in terms of heuristic, but also of medical and engineering practice.

Hyperbaric medicine has come a long way, deriving from the history of diving activities, dating back to ancient times.

The history of hyperbaric therapeutics was closely connected with the development of technology and advanced underwater activities with the accumulation of knowledge about the physical and physiological mechanisms of breathing oxygen at pressures greater than atmospheric pressure. The use of hyperbaric oxygen therapy (HBO) has been addressed mainly in France as the practice of hyperbaric medicine on scientific basis. HBO effects on vessels and anaerobic bacteria can be used as a defense against infection and helps heal wounds.

The indications for HBO were established more precisely, especially for treatment of chronic diseases, and only for those that have been validated clinically based on the criteria (EBM).

During the recent years, the United States have drawn predominant indications for surgery, wound healing, traumatology and plastic surgery, mandibular osteonecrosis, and so on. In Italy, the most commonly used application was in the field of anaesthesia / intensive care. In Spain, within the domain of internal medicine, some specialists have initiated hyperbaric medicine.

The indications for HBO were agreed at the ECHM Consensus Conference in Lille in September 1994 and updated 10 years later, also in Lille in December 2004. In order to be considered acceptable, any indication had to be based on experiments and clinical trials conducted with strict methodology and demonstration of significant positive results.

The jury has issued recommendations using a three-tier scale of recommendations: Type 1 – recommended with demonstrated certainty, Type 2 - recommended with probable certainty, Type 3 - recommended with optional certainty.

Considering the basic missions of the Diving Centre hyperbaric complex, such as treating of diving accidents, carrying out oxygen and narcosis tests, training the divers within a dry and wet environment, as well as the mission covered by the present HBO study, the system will have to meet the following conditions cumulatively:

- the existence of equipment to ensure an optimal environment for the development of specific activities, according to current diving technologies and to divers training and testing programs,
- rigorous monitoring of all technical and medical parameters at the level of life and work,
- monitoring, analyzing, maintaining or regenerating the diving parameters provided within the unitary saturation technology and saturation, through modern equipments that provide:
  - measuring, adjusting and monitoring the following parameters:
    - The concentration of CO_2 in the respiratory gases (ppm);
    - The concentration of O_2 in the respiratory gases;
    - Humidity in the respiratory gases;
    - Pressure (m water column);
    - The temperature of water and air.
  - tracking medical parameters: pulse, voltage,
  - permanent monitoring of the activities in the Chamber and of the diving activity in the simulator,
  - recording and processing the parameters of the hyperbaric environment at the life and work level,
  - conditioning the atmosphere within the Chamber,
  - rapid analysis of the respiratory mix used during diving and the possibility of changing it as needed,
  - meeting the SR EN 14971 requirements regarding the automated fire extinguishing system or the fire extinguishing means which are to be used within hyperbaric environment.

The capabilities limitations of the existing equipment are listed below. These operational limitations are due to:

- the insufficient number of pressure transducers depending on the depth;
- the need for redundant measurements regarding temperature, dew point (humidity), CO_2 and O_2 concentrations. It is also necessary to equip all watertight compartments with independent measurement capability of all physical sizes;

- air conditioning of the hyperbaric areas throughout the diving process (compression – bottom time with life level – decompression);
- the lack of continuous monitoring possibility of essential physiological parameters (pulse, voltage, ppO_2).
- the oral-nasal masks and obsolete spills;
- the fact that there is no data acquisition and recording system;
- the fact that there is no centralized and managed monitoring through software tools;
- the fact that it is necessary to complete the video monitoring system with equipment that should be related to each hyperbaric compartment;
- the need for rigorous gas analysis.

The regulations applied to manufacturers of hyperbaric therapeutic equipment are severe.
A hyperbaric therapeutic chamber is considered to be a medical device in accordance with the European Council Directive 93/42 "medical products".
The design, test and safety requirements of new therapeutic systems must be in accordance with the new European standards EN 14931 CEN TF 127.
All new chambers will be marked with "CE" and those existing should reach the same safety level.
A risk assessment in accordance with the European standards EN 14971 must be carried out before bringing medical equipment into the chamber. To avoid treating a critically ill patient and a chronic outpatient in the same chamber, it is useful to have a series of chambers connected to each other with airlocks. A chamber is kept for emergencies and intensive care and another or others are scheduled for treatments of chronic illness. Ideal for a hyperbaric system would be to have separate compartments with specific equipment to perform the test function of divers, of workers in the hyperbaric environment or HBOT patients.

The jury strongly recommends that through the European Code of Good Practice for Hyperbaric Oxygen Therapy (ECGP) minimum requirements should be met by the European hyperbaric centres, as established by consensus among internationally recognized European experts.

3. Main Performances

3.1. Performance Parameters

Operational performance parameters are represented by environmental factors. These are the essential vectors of respirable atmosphere within the hyperbaric chambers.

Table 1 Operational performance parameters

No.	Performance parameter	Objective's Value	Threshold Value	Obs.
1.	Pressure measurement precision in hyperbaric compartments	±0.15% full scale	±0.25% full scale	
2.	Temperature measurement precision	1^0C	1^0C	
3.	Accuracy of ambient temperature control in stable pressure mode (level).	± 2°C	± 2°C	
4.	Value range for water temperature adjustment	0°C and 25°C	4°C and 25°C	With a pre-cision of ± 2°C
5.	Moisture regulation precision	± 5%	± 5%	For the air lock when not using the sanitary group

6.	The max. compression and decompression speed in the room and in the air lock is 25 m / min	25 m/min	20 m/min	
7.	O_2 measurement potential	• maximum domain of measurement: 0…100% oxygen; • minimum domain of measurement: 0…10% oxygen; • measurement precision (general use): ± 0,2% oxygen; • measurement precision (constant environment): ± 0,1% oxygen	• maximum domain of measurement: 0…100% oxygen; • minimum domain of measurement: 0…10% oxygen; • measurement precision (general use): ± 0,2% oxygen; • measurement precision (constant environment): ± 0,1% oxygen	
8.	CO_2 measurement potential	• area: 0-5.000ppm • resolution: 1 ppm • precision: ± 3% or ± 50 ppm	• area: 0-5.000ppm • resolution: 1 ppm • precision: ± 3% or ± 50 ppm	
9.	Measuring capacities for each airtight compartment	Monitoring potential of all parameters, independently for each compartment	A set of switchable sensors for a compartment plus the adjacent air lock.	

3.2. Functional Performances

The functional performances are provided by the brief technical environment in which the essential equipment, along with relevant technologies and trained responsible personnel, carry out missions during the planned time with minimal risk of accidents.

4. Current technical status. International Trends. The situation at internal level. Comparative analysis of main similar existing achievements.

4.1. Current technical status and international trends

The hyperbaric therapeutic equipment has developed and evolved a lot, being able to meet cumulative requirements for ergonomics, performance and efficiency. Specialized technical and medical training is well regulated and provided by the professional associations, materializing in skills and certifications regarding the exploitation of systems, technologies as well as the provision of specific medical assistance.

4.2. The internally situation

Situated in the centre of Constanta, “S.C. Eurohiperbar S.R.L.” possesses one of the most modern pressure rooms in Europe, equipped with the most modern computerized monitoring systems of treatment and patients at the level of the German therapy standards.

The “Hawar Starmed 2200” Chamber has the option of simultaneous treatment of 13 patients plus an attendant or up to 4 stretchers. Each treatment site has separate oxygen supply. The Chamber has the ability to perform hyperbaric treatments up to 50m deep / 6bar (absolute scale), 5bar (relative scale).

The creation of the Association of Romanian Hyperbaric Medicine is currently being undertaken at the initiative of the Hyperbaric Medicine Centre in Constanta.

IOP Conf. Series: Earth and Environmental Science **172** (2018) 012009 doi:10.1088/1755-1315/172/1/012009

The present configuration of the Diving Centre's Hyperbaric Laboratory, which covers the range of missions listed in Chapter 2, namely, treating of those involved in diving accidents, carrying out oxygen and narcosis tests, training the divers in a dry and wet environment, previously regarding unitary and saturation diving activities, comprises: a central hall, a compressor compartment, regeneration assembly and gas storage, a compartment for various groups of power and water supply.

- **The central Hall** contains (see figure 1 and 2):

The ASS 500 Hyperbaric Complex. This consists of:

- a humid chamber disposed as a symmetry between two identical decompression chambers. The wet chamber is positioned vertically, thus offering the possibility of flooding in the lower half (water column about 3000 mm). The dry side of the room is equipped with standard turret material. The compression and decompression speed in this room is 25 m/min (2,5 bar/min). In the dry part, the adjustment accuracy of the ambient temperature is ± 2°C. This precision is only valid under the constant pressure mode (level). Humidity is not constant. Water can be cooled for diving activities in which warm water suits are used. A refrigeration group allows the water temperature to be set between 0°C and 25°C with an accuracy of ± 2°C. When hot water diving suits are used, raising the water temperature in the Chamber will be around 4°C in one hour and the return time to the starting temperature between 0° C and 25° C will be around 4 hours. This data is calculated for 2 divers. The detachable bridging, between the wet and dry side allows only dry use when required. Permanent communications are possible between the divers from the dry side, the wet one and the assisted personnel due to an intercom system with a decoder. Visualization can be done through portholes or through the permanently functioning TV system.

- the decompression chambers. These are two chambers that allow for two separate dive programs. Each room has two parts: an enclosure with sanitary equipment (outer lock or toilet and shower room) and a chamber (room) for 4 divers. The maximum compression and decompression speed in the room and in the enclosure is 25 m/min. The accuracy of ambient temperature control is 1°C. This precision is only valid under a constant pressure regime (level). The accuracy of humidity adjustment is ±5%. This precision will not be met in the enclosure when using the sanitary group. Permanent communications are possible between divers and support staff due to an intercom system with a decoder. Visualization is done either through portholes or through a permanently functioning TV system.

- free climbing room. It is a vertical chamber, usually of 2300 mm in diameter and height, placed under the wet chamber. This room allows submarine staff training in the "free ascent" rescue technique. Training is done at simulated depths between 0 and 200 m. For pressure value convergence, the chamber is designed to allow homogeneity with other chambers in case of a different use (wet side to wet chamber extension, hydraulic tests of some equipment, etc.). The "free ascent" procedure is as follows: The wet room is half filled with water, and pressurized with air between 0 and 200 m meters. The training staff enters the "free ascent" room through the side gate. The room is filled with water. The staff breathes in the gas volume at the top of the room. This is rapidly pressurized at the pressure of the wet room. At the pressure equalization, the lower access door in the wet room is opened and the personnel climb to the top of the wet room that is immediately decompressed to simulate the surface climb. The compression speed of this room is 250m/min (25 bar/min) when free climbing. Water can be cooled if the wet room extension is used. Visualization is done through portholes and a permanently functioning TV system.

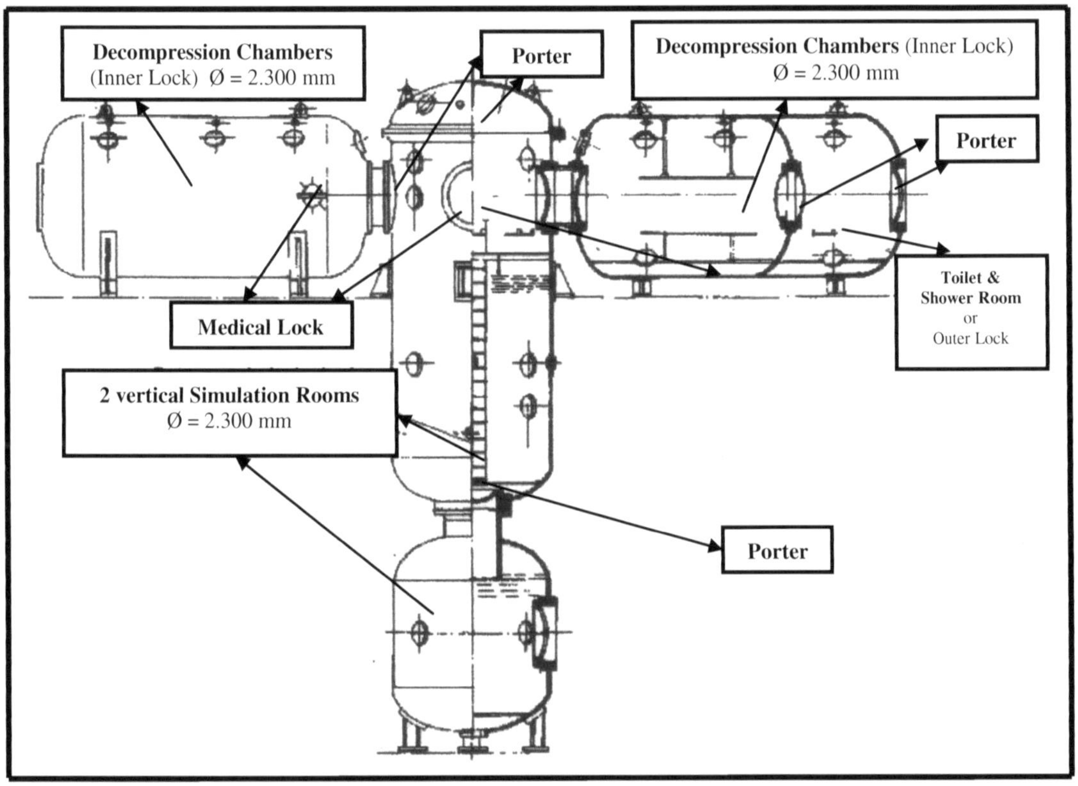

Figure 1. The Hyperbaric Complex

(consisting of 2 Decompression Chambers (Inner Lock), 2 vertical Simulation Rooms -one for diving training & one for the training of the people working on submarine, Medical Lock, Porter etc.)

The control panel:

- a four-section gas control panel, each section being the responsibility of a compartment, with pressure gauges and valves that allow compression and decompression management;
- a control panel for electrical circuits, communications, television;
- an analysis panel.

Figure 2. The Hyperbaric Complex. Side View

Figure 3. Gas storage racks

- **The gas conditioning and storage compartment** (see figure 3) contains:

Regeneration and storage of gases:
- racks or spheres for gas storage;
- regeneration units (soda lime, silica gel, reheater) to keep CO_2, humidity and temperature of the gas in the room at a certain level.

Compressor room:
- 1 compressor for gas transfer;
- 1 O_2 compressor;
- 1 vacuum pump (to remove the air from the Chamber before the He pressurization);
- flexible storage tank;
- a panel that allows distribution and control of the mixtures.

- **The compartment for different groups of electricity and water supply** contains:

Boiler (for warm water diving suits). It is used to heat water (up to 40°C) in warm water diving suits, and the circulation is carried out via a pressure changer (50 bar).

The cold group. It allows water to be held in the simulation room between 0°C and 25°C when divers use warm water suits (water heating will be of 4° in one hour for 2 divers).

Boiler (room heating). Each decompression room is heated by an individual electric boiler.

The water circulation pump in the rooms.

The emergency generator. It is an automatic generator of at least 40 KVA, which will automatically be switched on when electricity is disconnected from the national grid.

Figure 4. Monitor for Gas Monitoring and Management (one side of the control panel)

Figure 5. Station for atmospheric regeneration in compartments

5 Case study

We will further explore the current configuration of the system Pressure chamber ASS 500 Center divers in terms of compliance with SR EN 14931: 2006.

After the analysis that laid the foundation of the current R & D study, numerous conformities have been identified and the technical limitations of the pressure chamber owned by the hiperbar laboratory system are being presented, according to SR EN 14931 standard (annex 2). This standards applicable field is:hyperbaric chamber for general human use, hyperbaric chamber with several therapeutic uses and performance analysis, security requirements and testing mode of said requirements SR EN 14931:2006.

Identified Shortcomings

Presentation of the non-conformities according to the standard SR EN 14931:2006:
- the materials inside must be antistatic according to EN ISO 6941 or similar,
- the materials inside must be hardly combustible according to EN ISO 6941 or similar;
- each compartment must be equipped with fire extinguishing materials (ISO standard was developed after the adoption of this) or a special extinguishers pressure chamber,
- the existence on the outside of the pressure chamber of a breathable sources of air;
- control manometer of the inside pressure with 1% precision on the entire measuring scale;

- posting the no flammable or non-approved materials signs in visible spots ;
- an independent emergency illumination system from the main system with a illumination capacity of min. 90 lx,
- the inside electrical installation can not go over 42 V, fitted with a safety system;
- the existence of a calibrated safety decompression system allowing decompression of 2 bar. to Ambient in max. 2 min .;
- comprimated air purity according to STANAG 1458. The existence ballots analysis.
- conditioning system inside the hyperbaric atmosphere.
- monitoring and analysis data storage system with the possibility of analyzing at the last 3 h;
- visual and audible alarm system for when the O_2 % is exceeded inside the pressure chamber;
- at least two timers;
- additional power supply, source type UPS monitoring equipment.

As presented above, cutting-edge achievements do not reflect changes regarding the basic principles or structure. Built on a different scale, the systems retained the same functional framework.

The hereby study elaborated by the Centre's specialists presents also the theoretical training module for the specialized technical and medical personnel.

6. System / cost efficiency analysis. Comparison of possible solutions, conclusions on their efficiency and economy.

From the analysis carried out on the types of versions, it was established that the principle of efficiency and the technical-economic principle are optimized by using the II nd version above-studied.

7. Conclusions, proposals

The ASS 500 hyperbaric assembly provided by the Diving Centre is technically and constructively compliant with the SR EN 14931: 2006 standard which accounts for 90% of the endowment conditions necessary for its use according to the designation for which it was intended, in a safe mode.

Considering the fact that the Diving Centre, through the hyperbaric laboratory, broadly owns the necessary infrastructure for putting this kind of therapy into practice, it clearly shows it's potential to become one of the oxygen-specific areal entities.

In conclusion the idea of modernization revolves around "common sense" interventions.

Bibliography

1. Directives 1000.1, 1000.2, 1000.5 of the Armaments Department of the Ministry of Defense;
2. STP-M 40369-98: Quality assurance. Terms and definitions;
3. STP-M 40361-98: Technical requirements for the design and implementation of the quality assurance program by the contractor. Guide;
4. STP-M 40462-98: Military requirements for a quality assurance system in design, development, production, assembly and service;
5. STP-M 40463-98: Military requirements for a quality assurance system. Guide for application of STP-M 40462, STP-M 40464 and STP-M 40465;
6. STP-M 40464-98: Military requirements for a quality assurance system in production, assembly and service;
7. STP-M 40465-98: Military requirements for a quality assurance sys. in final inspections & tests;
8. STP-M 40466-98: Military requirements for a quality assurance system in final inspections;
9. STP-M 40469-98: Product Quality Assurance Program for Military Guide;
10. SR EN 14931:2006 Chambers for human use. Chambers with multiple seats for therapeutic use. Performance, security requirements and attempts.

A simple technique for identifying vessel model parameters

V. A. Golikov[1], V. V. Golikov[1], Ya. Volyanskaya [2], O. Mazur[1], O. Onishchenko[1]

[1]National University «Odessa Maritime Academy», 8, Didrihsona str., Odessa, 65029, Ukraine,
[2]Admiral Makarov National University of Shipbuilding, 9, Heroiv Ukrainy Ave., Mykolayiv, 54025, Ukraine

E-mail: oleganaton@gmail.com

Abstract. The marine autopilot system design is based on the mathematical model of steering dynamics. Ships maneuvering can be automated by using the autopilot system. A simple method for identifying the parameters of the ship's mathematical model is presented. In the present paper a study on Nomoto model has been undertaken for its selection for the ships steering dynamics. To identify the parameters of the ship model it is proposed to use: a) the results of «zigzag» tests; b) process the result of the tests using the Fourier transform. Next, a system of equations for 1, 3, 5 and so on harmonics of the Nomoto transfer function is compiled. The solution of the system allows us to find the time constants and the coefficients of the model. Verification of the results was carried out on the basis of «reverse» modeling.

The need to improve maritime safety and operational characteristics of marine vessels has led to the creation of different types of autopilots. Modern autopilots provide not only effective stabilization of the course of the vessel, but also self-tuning parameters of the course management system with changes in the condition of the vessel (draught, type of cargo) and sailing conditions (speed, weather, depth). The result is stabilization of the vessel at a predetermined rate with a minimum power loss of ship power plants (SPP) for handling (loss of yaw).

Such systems of autopilot steering control system (ASCS) allow to reduce the loss of running time by 3–4 %, increase the carrying capacity of the vessel and reduce fuel consumption for the flight by 3–6 %. The operation of any of the (ASCS) is based on a mathematical model for the dynamics of steering a ship [1–5]. Naturally, the more accurate the mathematical model will describe the real conditions of the vessel's motion, the more effectively it is possible to synthesize the (ASCS), the less resulting losses of the vessel's power plant [5–8].

Currently, the theory of control of the ship is fairly well developed and allows to build adequate physical processes of the movement of the ship mathematical model. The mathematical model of the vessel is usually based on six degrees of freedom of motion of a rigid body in space. However, the control system synthesized based on such model (control controllers) turns out to be cumbersome, in many respects - excessive and unreasonably sharply reacting to rapidly changing external influences. The compromise between the complexity of the mathematical model and the adequacy of the model to real physical processes led to the development of simplified mathematical descriptions of the vessel's motion.

The IMO (International Maritime Organization) has developed and adopted resolution A. 751(18), which regulates the need to use mathematical models of the vessel in solving practical problems in the field of navigation safety. As a result of the application of resolution A. 751(18), simplified mathematical models of Nomoto ship-models of the first and second orders of magnitude were recommended for practical use at several international conferences of experimental basins (e.g. the 14th conference) [9–11]. The first-order Nomoto model is extremely simplified and is often used for verification (evaluation, preliminary) calculations.

The second-order Nomoto model [10, 11] is more efficient in practice and is described by the differential equation:

$$T_1 \cdot T_2(d^2\omega/dt^2)+(T_1+T_2)(d\omega/dt)+\omega+H(\omega)=K\cdot\alpha_r+KT_3(d\alpha_r/dt), \qquad (1)$$

where ω – angular frequency (speed) of the vessel; $H(\omega) = \nu_1|\omega|\omega + \nu_2\omega^3$ – the nonlinear function of the angular frequency; T_1, T_2, T_3, K, ν_1, ν_2 – the parameters of the mathematical model; α_r – rudder angle.

The practical use of such a model (synthesis of control controllers for the course stabilization system) is extremely difficult due to the need to know for a particular vessel the basic parameters of the model (1): T_1, T_2, T_3, K, ν_1, ν_2.

One of the simple solutions to this problem of parameter identification of equation (1) is presented in the following material.

To identification of parameters of different mathematical models of the vessel, including-non-linear model (1) Nomoto, is devoted a very large number of publications, for example, [12–15]. This confirms the importance and relevance of studies related to the identification of Nomoto model parameters. Analysis of these [12–15] and other [16, 17] publications shows that the authors of the research use a variety of experimental data and a diverse mathematical apparatus of identification. These are statistical analysis, neuro-fuzzy identification and modelling, decomposition into periodic functions, approximation of various types and many other methods.

In resolutions IMO A. 160(ES.IV), A. 209(VII), A. 609(15) is mentioned the need to provide each of the vessels with objective data on their manoeuvrability characteristics. IMO resolution A. 751(18) specifically recommends the use to ensure safe navigation and build control systems for manoeuvring parameters such qualities as agility, risk level, stability on the course, braking performance [9]. Manoeuvrability qualities are determined for each vessel, mainly experimentally, and all the main test results are found in the vessel's manufacturer and in the ship's documentation.

It is this fact that the test results available on each vessel (or vessel of a similar design), such as the steady-state frequency of circulation and the «zigzag 10-10» or the «zigzag 20-20» [12–17], can be used.

Let us consider there are two experimental characteristics:

a) the static dependence of the steady-state angular frequency of circulations ω in the function of the rudder angle α_r (diagram of controllability, Fig. 1);

b) the results of the «zigzag» manoeuvre test (Fig. 2). Note that such features may be produced in the testing tank pool - with a small copy of the ship, or using a full-scale, taking into account the hydrodynamics of the vessel simulation.

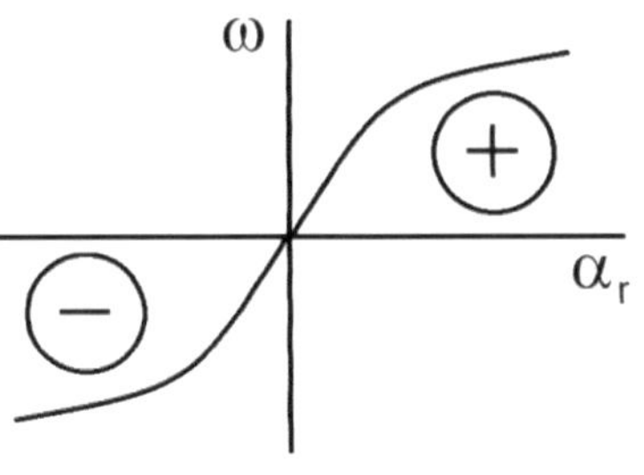

Figure 1. Manageability chart

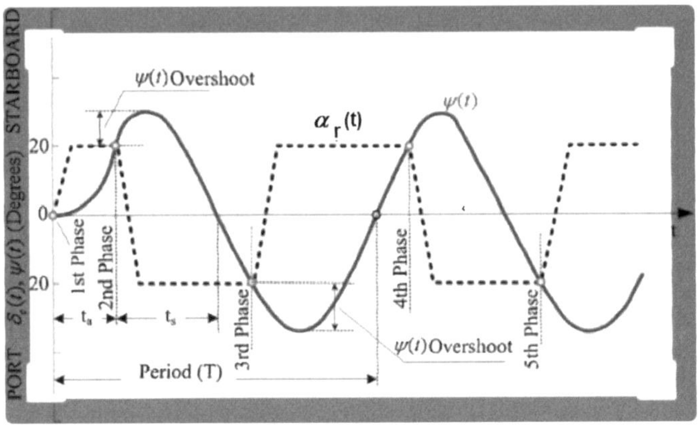

Figure 2. Figure «zigzag»

From the analysis of equation (1) shows that in the static mode ($d\omega/dt = 0$) we can write the equation $\nu_1|\omega_i|\omega_i + \nu_2\omega_i^3 + \omega_i = K\cdot\alpha_{ri}$, and for the positive circulation (Fig. 1) to obtain the expression:

$$\nu_2\omega_i^3+\nu_1\omega_i^2+\omega=K\alpha_{ri} \qquad (2)$$

Dividing the expression (2) to a work ($\omega_i \cdot K$), we get

$$\nu_2\omega_i^2 / K + \nu_1\omega_i / K + 1/K = \alpha_{ri} / \omega_i. \quad (3)$$

Substituting in (3) a few different values of the frequency ω_i for the respective angles of the rudder, for example, 2, 4, 6, ..., 30 degrees, we obtain a system of n-linear algebraic equations with three unknowns ν_2/K, ν_1/K and 1/K, and n > 3.

It is an overridden, redundant, system (RF – Redefined System or OS – Overdetermined System) and it does not have an exact classic solution. Known quasi-solution of the over determined systems of various kinds. For example, one of the possible solutions is based on the least squares method - Lagrange method [18, 19]. The result of such a decision may be some approximating expression. The graphical solution is a curve that does not pass through the experimental points of the steady-state angular frequency of circulations ω_i in the steering angle function α_r. It should be emphasized that at the same time the solution based on [18, 19] fully reflects the experimental dependence presented in Fig. 1, smoothing the errors (emissions, noises) of the experiment.

Thus, we find three unknown parameters of the equation (1) – K, ν_1, ν_2. It is necessary to find three more time constants T_1, T_2 and T_3. We will assess the required time constants in the following way.

We analyse the type of experimental dependence «zigzag» (Fig. 2) – dependence of the heading angle ψ function of the rudder angle α_r. It can be noted that for the steady state mode (excluding the first oscillations) both dependences are nonlinear, periodic and practically symmetrical with respect to the time axis.

We use the Fourier series decomposition method of the input α_r and output ψ signals to identify the three time constants. To do this, assign the main period T = 1/ω_1, based on the schedule shown in (Fig. 2). We limit ourselves to the first three odd harmonics of the decomposition: the first (main), the third and the fifth.

Because of series decomposition, we obtain three harmonics for the input signal. Each of the harmonics will be characterized by the amplitude $A_{1r}(\omega_1)$, $A_{3r}(\omega_3)$ и $A_{5r}(\omega_5)$ and phase $\varphi_{1r}(\omega_1)$, $\varphi_{3r}(\omega_3)$, and $\varphi_{5r}(\omega_5)$ for the respective frequencies ω_1, $\omega_3 = 3\omega_1$ и $\omega_5 = 5\omega_1$. By analogy, the decomposition obtained for the output signal values of the amplitude $B_{1\Psi}(\omega_1)$, $B_{3\Psi}(\omega_3)$ and $B_{5\Psi}(\omega_5)$, phases – $\varphi_{1\Psi}(\omega_1)$, $\varphi_{3\Psi}(\omega_3)$ and $\varphi_{5\Psi}(\omega_5)$.

Then the complex values of the transmission coefficients for each harmonic can be recorded (omitting to simplify the recording $\omega_{1,3,5}$) as the ratio of the output to the input signal:

$$\left.\begin{aligned} K_1 \cdot e^{j\alpha_1} &= \frac{B_{1\Psi}}{A_{1r}} \cdot e^{j(\varphi_{1\Psi} - \varphi_{1r})}; \\ K_3 \cdot e^{j\alpha_3} &= \frac{B_{3\Psi}}{A_{3r}} \cdot e^{j(\varphi_{3\Psi} - \varphi_{3r})}; \\ K_5 \cdot e^{j\alpha_5} &= \frac{B_{5\Psi}}{A_{5r}} \cdot e^{j(\varphi_{5\Psi} - \varphi_{5r})}. \end{aligned}\right\}. \quad (4)$$

For the regime of small deviations can be neglected in (1) the nonlinear term H(ω). Then get:

$$T_1 \cdot T_2(d^2\omega/dt^2) + (T_1 + T_2)(d\omega/dt) + \omega = K \cdot \alpha_r + KT_3(d\alpha_r/dt). \quad (5)$$

The transfer function corresponding to the expression (5) will look like:

$$W(s) = \frac{\omega(s)}{\alpha_r(s)} = \frac{(T_3 s + 1)}{(T_1 s + 1)(T_2 s + 1)}. \quad (6)$$

It is obvious that after the transition to the frequency domain (making a formal replacement $s = j\omega$), it is easy to write down from y (4) and (6) a system of three equations with three unknowns T_1, T_2 and T_3:

$$\left.\begin{aligned} W_1(j\omega) &= \frac{(T_3 j\omega_1 + 1)}{(T_1 j\omega_1 + 1)(T_2 j\omega_1 + 1)} = K_1 \cdot e^{j\alpha_1}; \\ W_3(j\omega) &= \frac{(T_3 j3\omega_1 + 1)}{(T_1 j3\omega_1 + 1)(T_2 j3\omega_1 + 1)} = K_3 \cdot e^{j\alpha_3}; \\ W_5(j\omega) &= \frac{(T_3 j5\omega_1 + 1)}{(T_1 j5\omega_1 + 1)(T_5 j5\omega_1 + 1)} = K_5 \cdot e^{j\alpha_5}. \end{aligned}\right\} \quad (7)$$

Such a system (7) is easily solved by modern computing means. For example, in Matlab the solution of the system of equations (7) is carried out by the command [T1, T2, T3] = *solve* (...) and as a result there are required values T_1, T_2 and T_3.

To build a mathematical model that allows verification of the results of identification of the parameters of the expression (1), use the tools Matlab/Simulink. Matlab/Simulink is a graphical simulation environment. This environment allows using flowcharts in the form of directed graphs, to build mathematical models of dynamic systems, including discrete, continuous, hybrid, nonlinear and discontinuous systems, to conduct Fourier analysis and other mathematical calculations and actions.

Verification of results of identification of parameters of (1) is feasible by mapping the results. Verification is carried out based on comparison of the results of the computer experiment «zigzag 10-10» (obtained on a full-scale DMI-model of the vessel) and the corresponding results of mathematical modelling in the environment Matlab/Simulink. The results of a computer experiment «zigzag 10-10» will be presented in the form of data tables as a function of time, approximating intermediate values (Fig. 3).

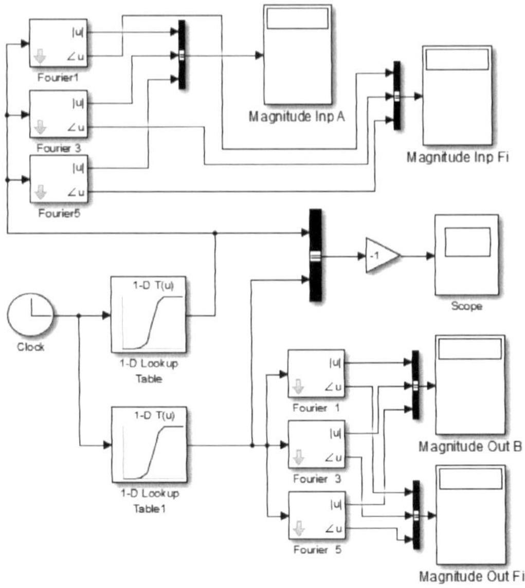

Figure 3. Fourier analysis of experimental results «zigzag»

Next, we will present (6) in the form of:

$$W(s) = \frac{\omega(s)}{\alpha_r(s)} = \frac{(T_3 s + 1)}{(T_1 s + 1)(T_2 s + 1)} = \frac{(T_3 s + 1)}{(T_1 \cdot T_2 s^2 + (T_1 + T_2)s + 1)}. \quad (8)$$

The theory of automatic control and regulation, for example, [20], allows us to present expressions like (8) in the form of block diagrams (block diagrams). Moreover, the structural scheme in the form of a Cauchy problem contains only the blocks of integration, and algebraic summation of the proportional gain.

It is obvious that with such a representation it is easy to take into account the (6). nonlinearity of H (ω), which was discarded earlier in addition, the mathematical model will take into account that:

a) the course angle is determined by the integration of the angular frequency ω;

b) steering gear α_r at the right time, according to the control algorithm «zigzag», modelled element «hysteresis loop» – «Relay block»;

c) the dynamics of the steering machine, limiting the speed of the steering, we present as a Subsystem.

Taking into account the said above, the block diagram of the mathematical model describing the «zigzag» experiment in the Matlab / Simulink environment is shown in (Fig. 4). The scheme takes into account the transition «rad/s – degrees». The (Fig. 5) shows the block diagram of the Subsystem, simulating the dynamics of the steering machine.

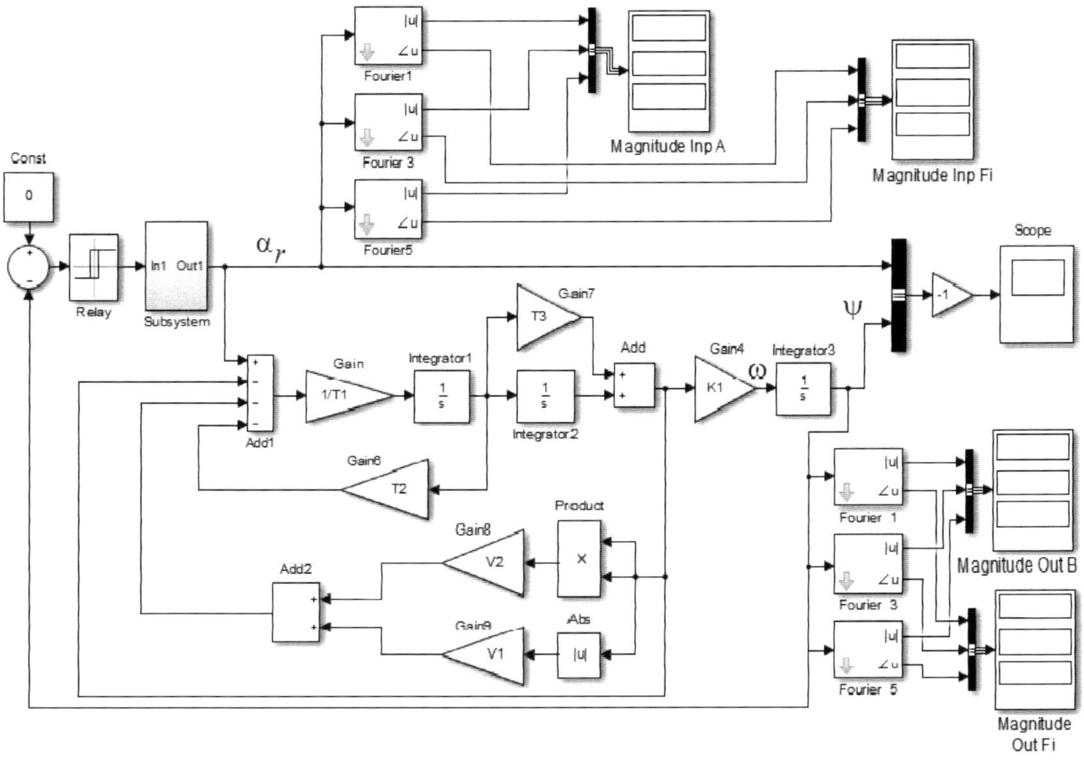

Figure 4. Block diagram, describing the experiment as «zigzag» and Fourier analysis

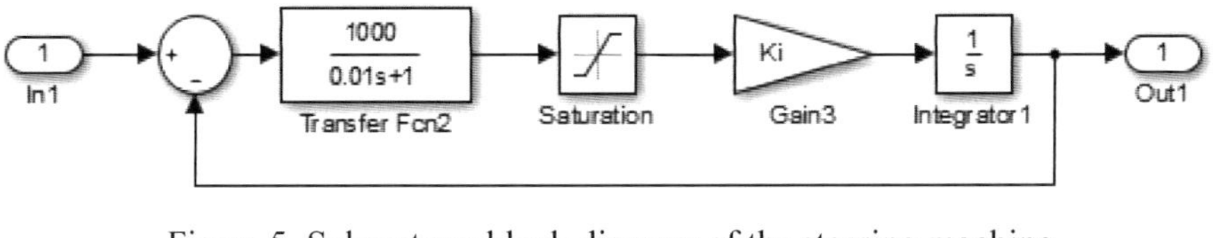

Figure 5. Subsystem: block diagram of the steering machine

The following parameters of a multi-purpose vessel of icebreaker type are taken as initial data.

1. Length of the vessel on design waterline, L_{dv} = 70,50 m.

2. Draught of ship at amidships, d_a = 3,50 m.
3. The ratio of the overall completeness, δ = 0,664.
4. The coefficient of completeness of amidships frame, β = 0,903.
5. Factor to the design load waterline, α = 0,858.
6. The wetted area of the vessel, S = 1240 m^2.
7. The area of the protruding parts, F_p = 5 m^2.
8. The speed interval of the vessel, V = 2...20 knots.
9. Volumetric displacement, W = 2864 m^3.
10. Total power SPP, N = 4600 kW.

With the help of a full-scale vessel movement simulator of the National University «Odessa Maritime Academy», for DMI-model of multi-purpose vessel icebreaker-type were conducted computer experiments and built:

a) static dependence of the steady angular frequency of circulations ω in the function of rudder angle α_r (controllability diagram;

b) manoeuvring test «zigzag 10-10».

The results of the manoeuvre test are presented in the form of appropriate tables to be inserted into the Lookup Table building blocks of the Matlab / Simulink simulation environment. The entrance table is the timer of model time Clock, outputs the rudder angle and heading angle.

Processing of the results according to the expression (3) and the subsequent solution of the excess system of equations allowed to identify the values K, ν_1, ν_2.

Simulation of the manoeuvre test «zigzag 10-10», based on the tables Lookup Table (Fig. 3) with treatment of results of Fourier analysis on (7) and the subsequent solution of this system, allowed to find the remaining members of the equation (1) – T_1, T_2 and T_3. On this basis, is constructed the model shown in (Fig. 4). The parameters of the model are given in (table 1).

Table 1. Identified parameters

K	ν_1	ν_2	T_1	T_2	T_3
0,031	$-1{,}7\cdot10^{-3}$	$-6{,}1\cdot10^{-4}$	31	15	5

The rate of increase in the change in the rotation of the steering machine was 3 degrees/s. In (Fig. 6) shows the results of verification of the experimental and identified model of the «zigzag 10-10» test. In (Fig. 6) the dotted line (index 1) shows the experimental data, solid line (index 2) – simulation data.

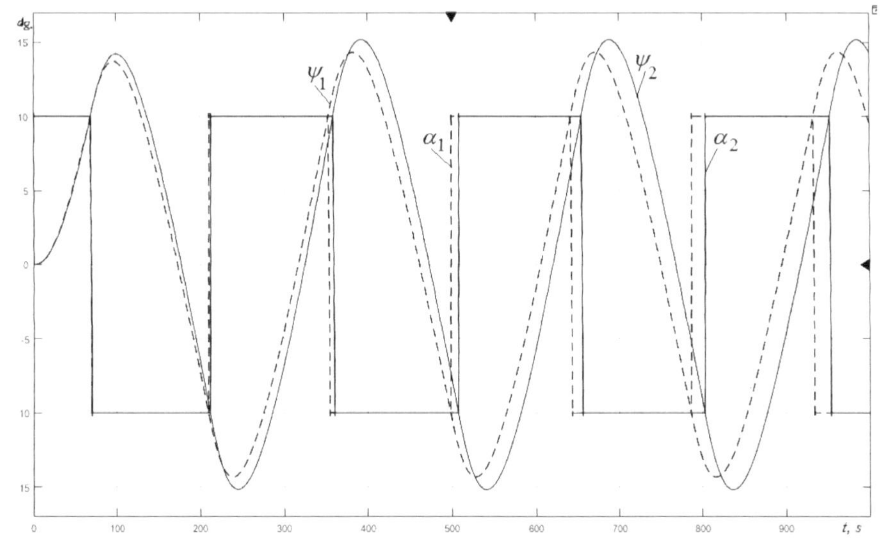

Figure 6. Verification of results of simulation of manoeuvre test «zigzag»: 1 – experiment, 2 – model

The proposed in the article calculation methodology and the set of end formulas such as (3), (4) and (7) make it possible easily to process the data on diagram of ship's steering quality as well as the zigzag maneuver. Subsequent processing consists of the following stages:

a) the method of solving an overdetermined equation system constructed on the basis of expression (3);

b) the use of the Fourier expansion of the data obtained from the maneuver tests «zigzag»;

c) the solution of equation system by standard mathematical means such as Matlab, Mathcad or others, allows us to quickly find the unknown coefficients of model (1) of ship's steering quality.

It should be noted that the use of calculation methodology requires careful maneuvering, reliable amplitude measurement, period and phase to obtain consistant results. The final result of identification of the parameters of Nomoto's equation (1) strongly depends on the quality of the actual experiments or the quality of similar experiments with ship's models (virtual or physical – in experimental basins).

It may be considered that it is sufficient to conduct one test of steady turning and one «zigzag» test, which reduces the required amount of initial data.

The proposed approach to solving the problems of identifying the parameters of the Nomoto equation is confirmed by modeling. When modeling, taking into account a very small number of parameters, this approach contributes to obtaining sufficiently accurate results. This accuracy is quite enough for practical application – in the synthesis and tuning of the regulator setting of ship's control system on course. In this regard, it should be taken into account that an area of using the considered technique has certain limitations, i.e. it can be used to study «weak» maneuvers.

In addition, it should be mentioned that parameters of the model (1) depend on the vessel's speed. The provided method of calculation methodology parameters of the mathematical model of the vessel using the results of full-scale experiment, enables to recommend it for parameters identification of mathematical models of different types of vessels. The further research is needed to carry out for the practical implementation of the methodology (controller synthesis) in the adaptive autopilot.

The authors are grateful to Professor V. F. Mirgorod (Odessa, Ukraine) for valuable comments and recommendations made during the writing of the article.

References

[1] Fossen T I and Smogeli O N 2004 Nonlinear Time-Domain Strip Theory Formulation for Low-Speed Maneuvering and Station-Keeping *Modelling, Identification and Control, MIC* 25(4) p 201–221

[2] Fossen T I 2005 A Nonlinear Unified State-Space Model for Ship Maneuvering and Control in a Seaway *Journal of Bifurcation and Chaos* (Plenary Talk ENOC'05, Eindhoven, The Netherlands) p 355–361

[3] Perez T and Fossen T I 2007 Kinematic Models for Sea-keeping and Maneuvering of Marine Vessels *Modelling, Identification and Control, MIC* 28(1) p 1–12

[4] Perez T and Fossen T I 2006 Time-Domain Models of Marine Surface Vessels for Simulation and Control Design Based on Sea-keeping Computations (Plenary Talk) *Proc. of the IFAC* (MCMC'06, Lisbon, Portugal) p 20–24

[5] Budashko V V, Onishchenko O A 2014 Mathematical foundations of simulation simulation of the control system of the power installation of a drilling vessel *Bulletin of the Kamchatka State Technical University* (Petropavlovsk-Kamchatsky: Kamchatka State Technical University) Issue 29 p. 6–13

[6] Budashko V V, Onishchenko O A 2014 Improvement of the control system of the propulsion device of the combined propulsive complex *Bulletin of the National Technical University "KhPI". Collection of scientific works. Series: Electric machines and electromechanical energy conversion* (Kharkov. NTU "KhPI") 38 (1081) p. 45–51

[7] Sutulo S and Soares Guedes 2011, Mathematical Models for Simulation of Maneuvering Performance of Ships *Marine Technology and Engineering* (Taylor & Francis Group, London) p 661–698

[8] Budashko V V, Onishchenko O A, Yushkov E A Physical modeling of a multifunctional propulsion complex *Zbirnik Naukovych prac Vijskovo Akademii (m. Odesa). Technical sciences* 2 p 88–92

[9] The Maneuvering Committee 2005 *Final Report and Recommendations to the 24th ITTC* (UK) 1, p 137–198.

[10] Nomoto K, Taguchi T and Hirano S 1957 On the steering qualities of ship *International Shipbuilding Progress* 4 35 p 56–64

[11] Yongbing Chen, Yexin Song, Mianyun Chen 2010 Parameters identification for ship motion model based on particle swarm optimization *Kybernetes* 39 Issue 6 p871–880 https://doi.org/10.1108/03684921011046636

[12] Ma F C and Tong S H 2003 Real-Time Parameters Identification of Ship Dynamic Using the Extended Kalman Filter and the Second Order Filter *IEEE Conference on Control Applications* Vol. 2 p 1245–1250

IOP Conf. Series: Earth and Environmental Science **172** (2018) 012010 doi:10.1088/1755-1315/172/1/012010

[13] Casado M H, Ferreiro R and Velasco F J 2007 Identification of Nonlinear Ship Model Parameters Based on the Turning Circle Test *J. Ship Res.* 51(2) p 174–181
[14] Sheng L, Jia S, Bing L and Gao-Yun L 2008 Identification of Ship Steering Dynamics Based on ACA-SVR *IEEE International Conference on Mechatronics and Automation* p 514–519
[15] Agarkov S A, Pashentsev S V 2015 Parametric identification of the generalized Nomoto model using the apparatus of the calculus of variations *Bulletin of MSTU* Vol. 18 1 p. 7–11.
[16] Pashentsev S V 2010 Parametric identification of maneuverability characteristics based on the results of full-scale tests of the Zigzag type in the non-linear model of vessel controllability *Vestnik MSTU* Vol. 13 4/1 p 730–735
[17] Journée J M J 1970 A simple method for determining the manoeuvring indices K and T from zigzag trial data *Delft University of Technology, Ship Hydromechanics Laboratory Report 267*
[18] Stevens Scott A 2017 System Analysis – Rank and Nullity *Math 220 – Matrices Handouts* (Pennsylvania State University) 3 p 115–119
[19] Anton Howard Rorres Chris 2005 Elementary Linear Algebra (9th ed.). *John Wiley and Sons* ISBN 978-0-471-66959-3
[20] Golnaraghi F, Kuo B C 2010 *Automatic Control Systems* (Wiley) p 786.

Digital control by the discrete systems of reactive power compensation in ship electrical power plants

Sergiy Dudko

Power Electronics & Energy Systems, Electrical Engineering and Radio Electronics Faculty, National University "Odessa Maritime Academy", Odessa – Ukraine

Corresponding author: sergiydudko.od@gmail.com

Abstract. The reactive power required to create the electromagnetic fields of electric motors, transformers and converters worsens the performance of the power system, the total current of the generator, transformers and cable lines must be significantly increased in relation to the required active load current by an amount that proportionately increases the installed capacity of the electrical equipment and the cross section cable connections. Instead of single work with full load, the parallel operation time of generating units with underload is increased to provide excess full power of electric consumers, including their reactive power. Solving the problem of energy saving, reducing the reactive power consumption allows capacitors Reactive Power Compensation Devices (RPCD). Varying the capacitance of the stages of the compensating capacitors, it is possible to fully compensate the reactive current of the actively inductive load of consumers. An analog (continuous) change in the capacitance of high-power capacitors is still an unsolved technical task. In most power plants, the capacity of capacitors varies stepwise, by connecting a different number of capacitor sections. This article is devoted to the investigation of RPCD, consisting of several blocks of compensating capacitors that are switched to the powernet independently of each other by means of high-speed thyristor switches. The change in the capacity of compensating capacitors in RPCD in power plants is accomplished by connecting a certain number of sections of capacitors. In this case, the capacitance value changes stepwise, i.e. discretely according to the adjustable value.
The capacitance of each capacitor section, the number of sections of capacitors and the algorithm of their switching to the network are the technical parameters of the discrete system of reactive power compensation. Analysis of these parameters and their optimization is the task of this article.

Keywords: reactive power compensation, compensating capacitors, high-speed thyristor switches, RPCD

1. Introduction

The main consumer of inductive reactive power on ships are asynchronous engines (60 ... 65% of its total consumption), transformers, including welding (20 ... 25%), valve converters, reactors and other consumers.

The inductive load in the process of operation is a source of reactive electricity, which oscillates between the load and the generator, is not connected with the performance of useful work, but is spent on creating electromagnetic fields and creates an additional load on the power lines. Accordingly, all network power, transmission and distribution equipment must be designed for large loads. In addition, as a result of heavy loads, the lifetime of this equipment can be reduced accordingly. A further factor in increasing costs is the heat transfer due to the increased value of the total current in cables and other switchgears, in transformers and generators.

Reactive power compensation improves the efficiency of generation, transportation, distribution and consumption of electricity. Effective compensation of inductive currents provides resource saving, improving the quality of electricity, [1,2].

The most common devices for reactive power compensation in ship distribution networks are static automated condenser installations. However, static RPCD provides compensation for the average value of reactive power, while rapid transient processes in the energy system are not taken into account. Their use is effective only for work in the established regime of loads of the grid system.

One of the leading electrotechnical firms in Europe Schneider Electric produces RPCD type Varset with capacitors of Varplus2 type with control from the functional modules Varlogic. As the switches that switch capacitors, specially designed contactors are used in the Varlogic modules. At the first moment of switching capacitor C is connected to the network through the resistor R and the reactor L, which allows avoiding dangerous throws of charging currents. Then the power contact of the key closes. However, the lack of synchronization of the moments of switching on the capacitors with the mains voltage leads to transient processes at the commutation moments.

The use of power semiconductor elements as the keys allows the capacitor to be switched in each phase at the time of equal network voltage and capacitor. Schneider Electric Company produces thyristor RPCD type Varset FAST with a key operating time less than 40 ms. Another example of a reactive power controller with contactor switching of capacitors is the MULTICOMP 96 Eco controller from MKS technology. The controller BR7000 of firm EPCOS is also designed for performing similar operations for measuring the electric power parameters and switching capacitor sections. Change of reactive power and set in contactor installations is performed with a certain clock frequency according to the specified control program. The required value of the capacitive current is achieved by connecting and disconnecting the respective stages of the capacitors of the compensation device. Regulatory programs take into account the power of the stages.

All the presented RPCD systems are static and do not provide dynamic (continuous) compensation of reactive power.

A more promising and completely unsolved problem is the use of dynamic reactive power compensation devices in the ship power system.

2. Model of the system

The mathematical model of a ship electric power plant includes differential, algebraic and logical equations describing physical processes in the elements of the system and the connection between aggregates. The model includes accepted assumptions in the mathematical description.

The composition and structure of the mathematical model is determined by the functional scheme of the ship's power plant. Each element of the installation under study corresponds to a specific model. The links between the elements are modeled by introducing common variables and combining the subsystems of the equations of the corresponding parts of the SEPP.

Simulation of a three-phase thyristor-capacitor block consisting of several switched sections requires a separate description of the connection of currents and voltages in each phase separately. An equivalent circuit of one phase of a four-digit thyristor-capacitor block is shown in Fig. 1.

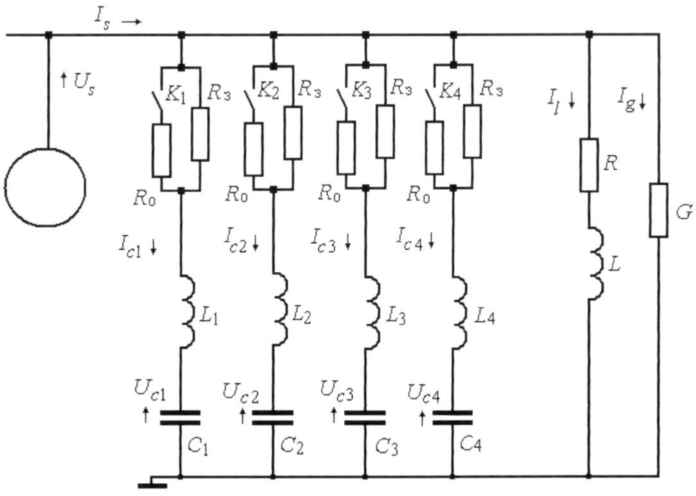

Figure 1 Equivalent circuit of one phase of a four-digit thyristor-capacitor block

The symmetrical thyristor switch is represented by two parallel-connected resistors with an open state resistance R_0 and a closed state $R_з$. When the switch is open K_i the resistance of the switch will be equal $R_з$, and when it is closed - $(R_0 \cdot R_з)\cdot(R_0 + R_з)^{-1} \approx R_0$, as $R_з >> R_0$. This representation of the thyristor switch in the mathematical model allows to write the same equations when state is closed and when it is opened, assuming the switch resistance is equal $R_к$. In the open state $R_к = R_0$, in the closed state - $R_к = R_з$.

The thyristor switch is simulated by software dy changing the resistance of the switch.

Each thyristor switch K_i switches the capacitor C_i of one of the sections through the safety throttle L_i.

The active-inductive load of the generator is modeled by the serial connection of the active conductivity G and the circuit RL.

In the mathematical model, stator chains are described for each phase separately. Equations in the Cauchy form that connect the voltages and currents in the equivalent circuit in Fig. 1 has the following form.

$$U_s = (I_s - I_l - I_{c1} - I_{c2} - I_{c3} - I_{c4})\cdot(G)^{-1}$$

$$\frac{dI_l}{dt} = (U_s - I_l R)\cdot(L)^{-1}$$

$$\frac{dI_{c1}}{dt} = (U_s - I_{c1}R_{k1} - U_{c1})\cdot(L_1)^{-1} \qquad \frac{dU_{c1}}{dt} = I_{c1}\cdot(C_1)^{-1}$$

$$\frac{dI_{c2}}{dt} = (U_s - I_{c2}R_{k2} - U_{c2})\cdot(L_2)^{-1} \qquad \frac{dU_{c2}}{dt} = I_{c2}\cdot(C_2)^{-1}$$

$$\frac{dI_{c3}}{dt} = (U_s - I_{c3}R_{k3} - U_{c3})\cdot(L_3)^{-1} \qquad \frac{dU_{c3}}{dt} = I_{c3}\cdot(C_3)^{-1}$$

$$\frac{dI_{c4}}{dt} = (U_s - I_{c4}R_{k4} - U_{c4})\cdot(L_4)^{-1} \qquad \frac{dU_{c4}}{dt} = I_{c4}\cdot(C_4)^{-1}$$

Equations of the stator load and compensating condenser sections of one phase of the generator are described by a system of ninth-order differential equations. For three phases, the order of the equations is 27.

Such a number of additional differential equations which describe each capacitor circuit in each phase substantially increases the order of the mathematical model of the electrical installation. Together with the equation of the diesel drive, the synchronous contactless generator, the voltage regulator and the frequency of rotation and the filters of the measured parameters, the order of the system is 38. If the capacitor bank is modeled without inclusion in each phase, the order of the system is 15 orders.[3]

An increase of the order of the system of equations substantially increases time of calculation of the transient process.

Accounting for the features of the operation of thyristor switches when switched on and off is modeled using logic functions. The thyristor is closed when the control signal is removed after the current is reduced to a value less than the thyristor holding current. The thyristor will close when the The current through the thyristor is calculated at each integration step and the instant of change of the current sign can be determined when the product of the capacitor currents at the current step $I_{ci}(t)$ and at the previous step of calculation $I_{ci}(t-T_{\text{p}})$, i.e.. $I_{ci}(t-T_{\text{p}}) \cdot I_{ci}(t) < 0$.

A shock-free thyristor unlocking commuting the capacitor is possible, if the voltage on the thyristor is zero. If the closed thyristor is modeled by a resistor with a large resistance $R_{з}$, then the voltage at the anode and the cathode of the thyristor will equal when the current is zero through the resistor $R_{з}$. When the voltage across the thyristor is zero, the current through the resistor $R_{з}$ will also change sign. Thus, condition for changing the state of the switching thyristor can be determined by the general condition $I_{ci}(t-T_{\text{p}}) \cdot I_{ci}(t) < 0$.

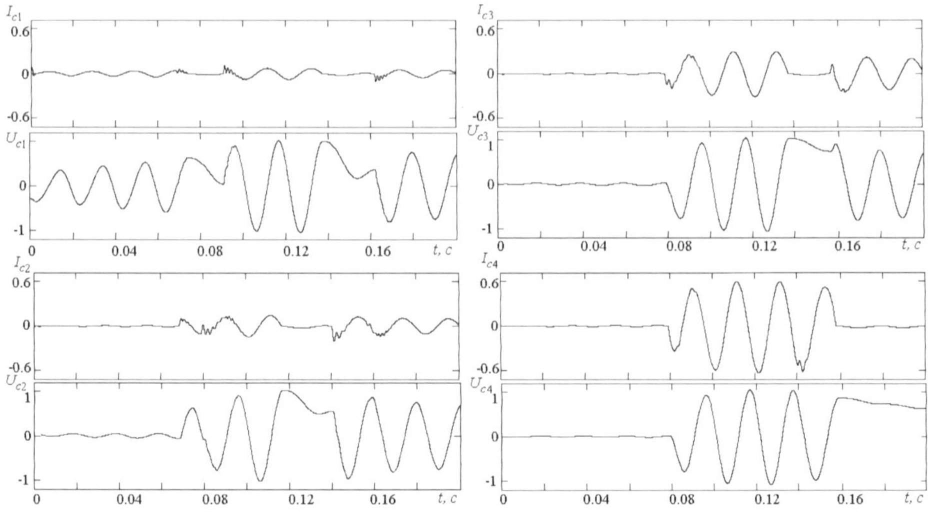

Figure 2 Processes in one phase of a four-digit capacitor unit

The logical condition for changing the resistance of the thyristor switch R_k will be:

If $I_{ci}(t-T_{\text{p}}) \cdot I_{ci}(t) < 0$ and $I_{ci}(t) < 0$ then $R_k = R_{\text{o}}$ or $R_k = R_{з}$.

Selection of resistance value of the switch $R_k = R_{\text{o}}$ or $R_k = R_{з}$ is determined by the control device. The condition $I_{ci}(t) < 0$ allows to clock the capacitor capacitance change once during the generated voltage period. Synchronization of control during simulation is determined by the phase transition of the phase voltage through zero: If $U_a(t-T_{\text{p}}) \cdot U_a(t) < 0$ and $U_a(t) > 0$ then $C_k = f(\Delta U, P_q)$. The calculated processes of changing currents $I_{ci}(t)$ and voltages $U_{ci}(t)$ in each section of capacitors of one phase are shown in Fig. 2.

3. Results and Discussion

3.1. Two-digit RPCD

The reactive energy compensation device with two sections of capacitors, whose capacitance is two times different, is the simplest adjustable system. It is possible to combine only three values of the total capacity: ΔC , $2\Delta C$ and $3\Delta C$.

Nevertheless, it allows to ensure sufficient accuracy of compensation of the load angle at the radian level $\Delta\phi_{\min} = \arcsin(\sin\phi_n \cdot (2^2-1)^{-1}) = \arcsin(\sin\phi_n \cdot 3^{-1}) = 0.2...0.27$ radian. At the same time, the power factor can be provided not less than

$\cos\phi_{\min} = \cos(\arcsin(\sin\phi_n \cdot (2^2-1)^{-1}) = \cos(\arcsin(\sin\phi_n \cdot 3^{-1}) = 0.96\ldots0.98$.

The capacitance of the smaller section of the capacitors is equal $\Delta C = C_k \cdot (2^2-1)^{-1} = \sin\phi_n \cdot 3^{-1} =$ 0.2…0.27 r.u.

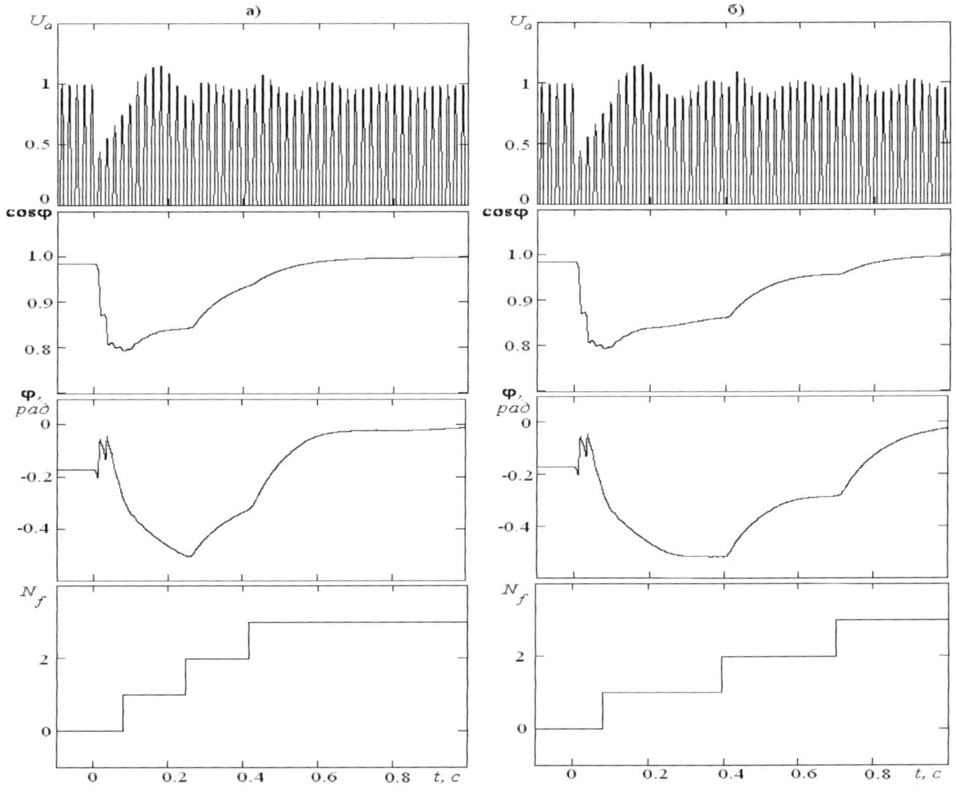

Figure 3 Transient processes in a two-digit discrete-pulse compensation system for the reactive energy of a synchronous generator when 50% of the active-inductive load is switched on a) - $T_0 = 8T_c$; b) - $T_0 = 15T_c$

Figure 3 shows the transient processes of switching on 50% of the active-inductive load of a synchronous generator with a two-digit regulator. The implementation of a complex control algorithm [4] in a two-digit controller is difficult and does not lead to the desired result. Improvements in the dynamic properties of the voltage stabilization system can not be achieved.

3.2. Three-digit RPCD

The increase in the number of switched sections and the number of control bits to three increases the accuracy of reactive energy compensation by more than 2.2 times. The accuracy of maintaining the load angle near zero is $\Delta\phi_{\min} = \arcsin(\sin\phi_n \cdot (2^3-1)^{-1}) = \arcsin(\sin\phi_n \cdot 7^{-1}) = 0.09...0.12$ radian, and the power factor equal to unity with a three-digit regulator is

$\Delta\cos\phi_{\min} = 1-\cos\Delta\phi_{\min} = 1-\cos(\arcsin(\sin\phi_n\cdot(2^3-1)^{-1})) = 1-\cos(\arcsin(\sin\phi_n\cdot 7^{-1}))$ =0.3…0.6 %.
Such accuracy of compensation is quite sufficient for ship and other technical electric networks.

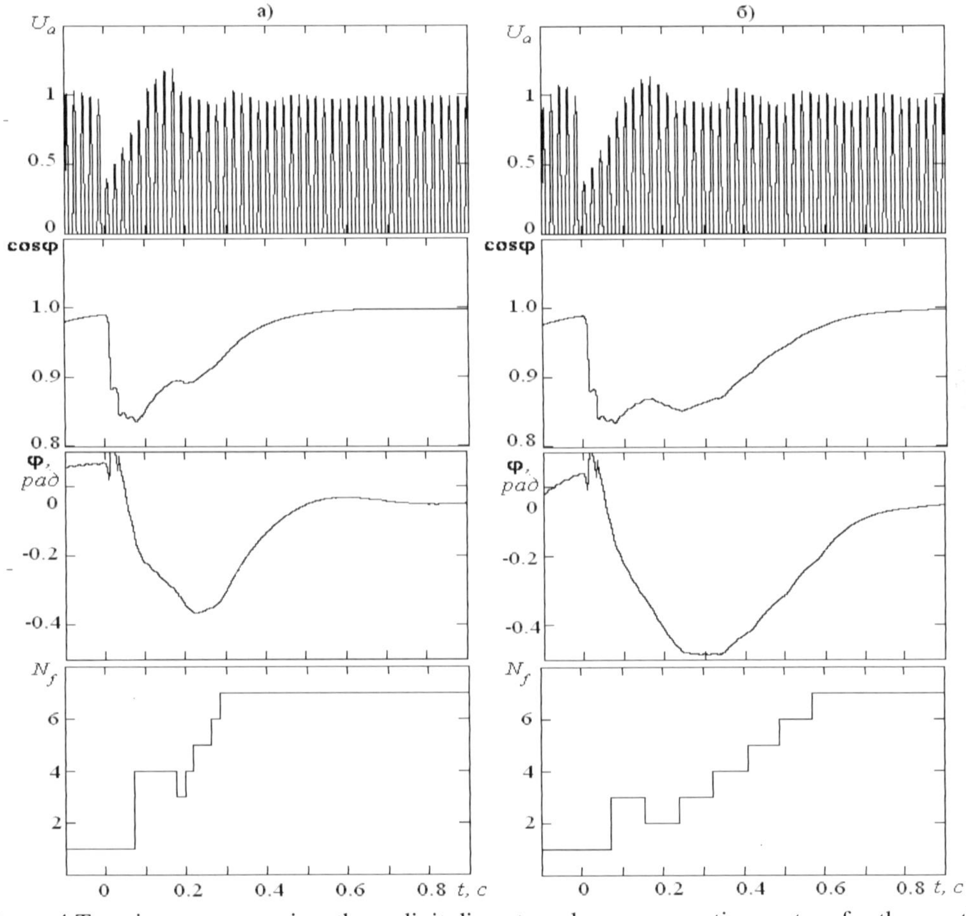

Figure 4 Transient processes in a three-digit discrete-pulse compensation system for the reactive energy of a synchronous generator when 50% of the active-inductive load is switched on a) - $T_0 = T_c$; b) - $T_0 = 5T_c$

An increasing of the minimum switching capacity of the compensating capacitors worsens the dynamic properties of the synchronous generator voltage stabilization system.
Figure 4 shows the calculated transient processes of switching on 50% of the active-inductive load of a synchronous generator with a three-digit regulator under the U-PI control law [4] with the optimum tuning parameters: $k_u = 1.5$, $k_q = 0.05$, $T_q = 0.003$ c.

The time-quantization period in Figure 4a was also selected $T_0 = T_c$ and $T_0 = 5T_c$ in Figure 4b. With a longer period T_0 compensation time increases from 0.5 to 0.8 s, the power factor failure remains the same, and the dynamic quality of the generator voltage is somewhat degraded[4,5].

3.3 Four-digit RPCD

The ratio of the capacitance values of the three-phase sections will be chosen proportionally to the weights of the bits of the binary number system: 1: 2: 4: 8. The functional diagram of the control device of the compensating device is shown in Figure 5

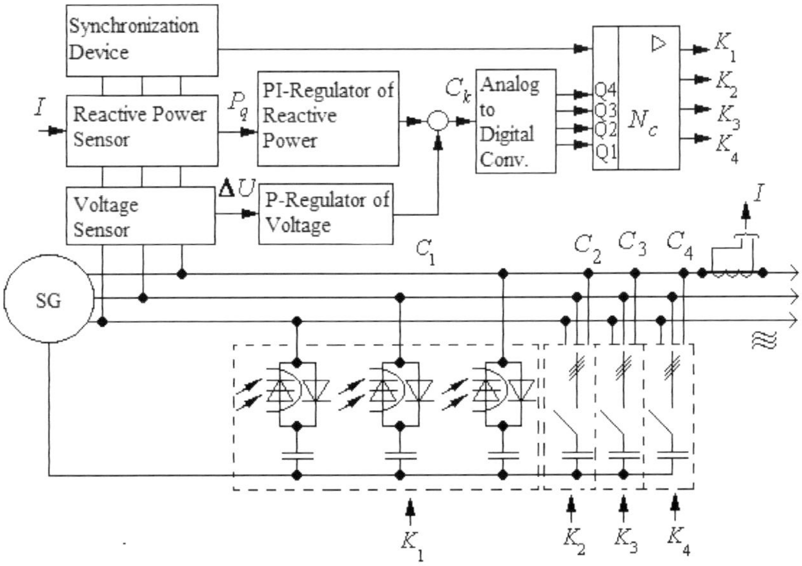

Figure 5 Functional diagram of the control device for four-digit RPCD

Signals from voltage sensors ΔU and the reactive energy parameter are fed to analog regulators that form the law of variation of the compensating capacitance $C_k(t)$. Further, the ADC converts it into a binary number $N_c = Q_4Q_3Q_2Q_1$, that controls the corresponding switches K_1, K_2, K_3, K_4, connecting the sections of the capacitor block C_1, C_2C_3, C_4. The synchronization device determines the moments of changing the binary number $N_c = Q_4Q_3Q_2Q_1$ accordingly to the period of the generated voltage. The minimum period of discrete switching of capacitors is equal to one period of the generated voltage $T_0 = T_c$. However, it can be increased multiples of the network period $T_0 = K_c \cdot T_c$. That is, capacitors can be switched once for one period, for two periods and more periods of the network. Calculate the transient processes in a system with a four-digit compensator under the U-PI control law with optimal adjustment parameters $C_k(t) = k_u \Delta U_m(t) + k_q \cdot P_q(t) + T_q^{-1} \int_0^t P_q(t)dt$.

Figure 6 shows the reactive power compensation processes in the digital system when 50% of the active-inductive load is switched on with the U-PI control law.

The time-quantized period is equal to one period of the network $T_0 = T_c$. Adjustment parameters of the controller are: $k_u = 1.5, k_q = 0.05, T_q = 0.003\ s$. The detailed simulation of each switched capacitor in each phase and section, taking into account the synchronization of the inclusion and the discreteness associated with a limited number of sections, shows that the splitting of the capacitor block into four three-phase sections allows to achieve the required parameters of the reactive energy compensation system. The accuracy of maintaining the load angle near zero is $\Delta\phi_{\min} = \arcsin(\sin\phi_n \cdot (2^4 - 1)^{-1}) = \arcsin(\sin\phi_n \cdot 15^{-1}) = 0.04...0.05$ radian, and the power factor equal to unity with a four-digit regulator is $\cos\phi_{\min} = \cos(\arcsin(\sin\phi_n \cdot (2^4 - 1)^{-1})) = \cos(\arcsin(\sin\phi_n \cdot 15^{-1}))$ =0.1…0.2 %. The dynamic failure of the power factor does not exceed 0.2, the recovery time is not more than 0.5 s.

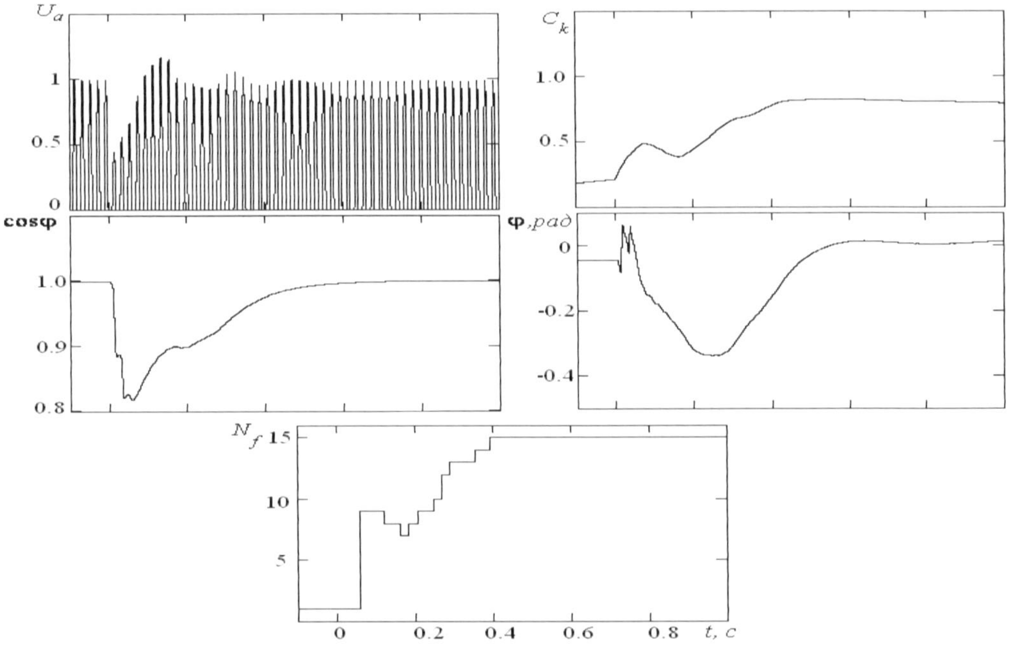

Figure 6 Processes in a four-digit discrete-pulse compensation system for reactive power

Figure 7 shows the calculated transients in a four-digit discrete-pulse compensation system for reactive energy when 50% of the active-inductive load is switched on. All processes are constructed with the same settings of the regulators and differ in the time of the time quantization period $T_0 = 2...10 \cdot T_c$. Comparison of these processes shows that the compensation of the power factor in all cases satisfies the requirements for accuracy and compensation time. Thus, the recovery time of the power factor does not exceed 0.6 s at. $T_0 = 2...10 \cdot T_c$.

Therefore, the main criterion for choosing the discrete period T_0 for a four-digit discrete-pulse compensation system for reactive energy will be the quality of the transient process along the voltage stabilization channel of the synchronous generator.

Analysis of voltage transient processes in Figures 5, 6 and 7 shows that the best is the recovery process with minimal discrete time $T_0 = T_c$, Fig. 5 b.

IOP Conf. Series: Earth and Environmental Science **172** (2018) 012011 doi:10.1088/1755-1315/172/1/012011

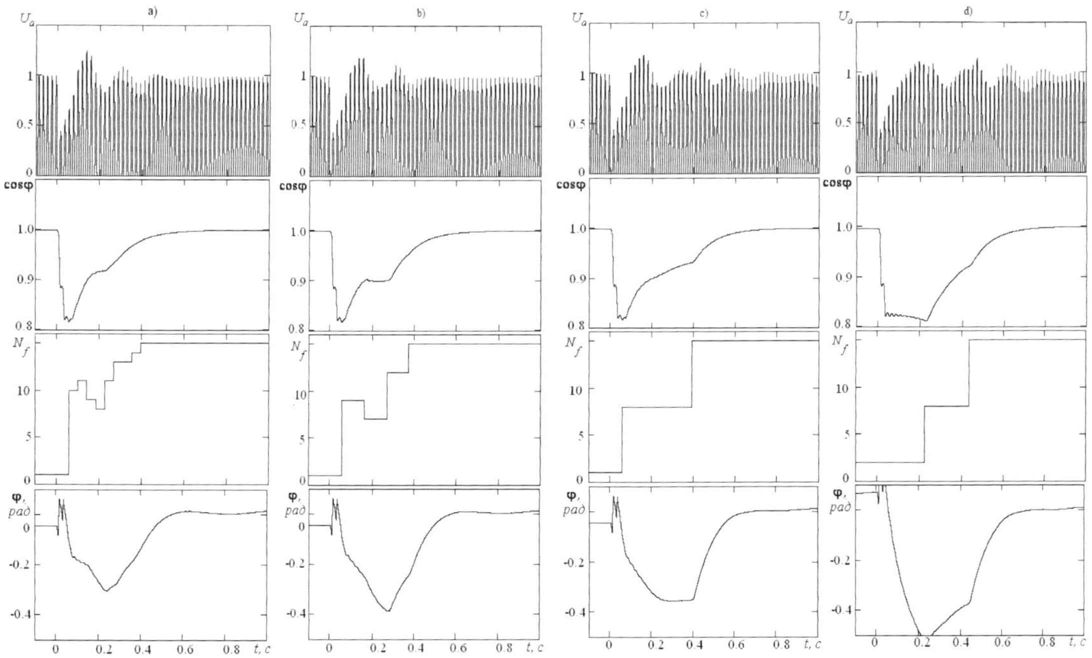

Figure 7 Transient processes in a four-digit discrete-pulse compensation system for reactive energy when 50% of the active-inductive load is switched on

a) - $T_0 = 2T_c$; b) - $T_0 = 5T_c$ c) - $T_0 = 8T_c$; d) - $T_0 = 10T_c$

4.Conclusion

The choice of a binary system for capacitor capacitance control is justified by the wide application of the binary system in computer devices. The capacitance values of the capacitor sections of the RPCD are referred to as the weights of the bits of the binary number system: 1, 2, 4, 8 .... The value of the control number will correspond to the value of the compensating capacity, provided that the corresponding bit of the binary number is controlled. From the switching period of capacitors T_0 the time of reactive load compensation t_q depends on the dynamic quality of the voltage stabilization system. For discrete level systems, the dependencies of the integral quality index of the voltage transient process I_{Du} from the period of discreteness in time T_0 are determined. These dependences for a two-, three- and four-digit compensator of a reactive load are of a different nature. The best processes for a two-digit compensator ($N=2$) will be for $n=7...8$. For a three-digit compensator ($N=3$) the best processes are achieved at $n=2...3$. In the four-digit compensator ($N=4$) the optimal value of the discrete period is equal to the network period, $T_0=T_c$.

5.References

[1] M. Bollen, Understanding Power Quality Problems. Piscataway, NJ, USA: IEEE, 2000, ch. 1, pp. 1–35.

[2] Agunov M.V., Agunov A.V., Verbova N.M. Opredelenie sostavlyayuschih polnoy moschnosti v elektricheskih tsepyah s nesinusoidalnyimi napryazheniyami i tokami metodami tsifrovoy obrabotki signalov // Elektrotehnika, 2005. — # 7. — P. 45 — 48.

[3] V M Juromskiy Engineering model of the electric drives of separation device for simulation of automatic control systems of reactive power compensation by means of serially connected capacitors 2016 J. Phys.: Conf. Ser. 751 012010

[4] Dudko S.A., Muha N.I. Analiz algoritmov upravleniya sudovyimi ustanovkami kompensatsii reaktivnoy energii [Tekst] //XIII – mIzhnarodna naukovo-tehnIchna konferentsIya. Problemi

energoresursozberezhennya v elektrotehnIchnih sistemah. Nauka, osvIta I praktika. – Kremenchuk: KNU, 2011. – Vip. 1. – S.96-97.
[5] Vishnevskiy L.V., Muha N.I., Dudko S.A. Upravlenie sudovyimi kondensatornyimi ustanovkami kompensatsii reaktivnoy energii [Tekst] - ElektrotehnIchnI ta komp'yuternI sistemi: nauchn.tehn. sb. – KiYiv: Vidavnitstvo «TehnIka». – 2011. – #3(79). – P.330-333
[6] Muha N.I., Dudko S.A. Realizatsiya diskretno-impulsnyih zakonov upravleniya kompensatsiey reaktivnoy energii v sudovyih elektroustanovkah [Tekst]/ - VIsnik SevNTU. ZbIrnik naukovih prats. SerIya: MehanIka, energetika, ekologIya. – Sevastopol: SevNTU, 2012. – Vip. 132/2012. – P.86-92.

IOP Conf. Series: Earth and Environmental Science **172** (2018) 012012 doi:10.1088/1755-1315/172/1/012012

Modelling and simulation of oil spills in coastal waters

A Pocora[1], A A Purcarea[2], F Nicolae[3] and A Cotorcea[4]

[1] PhDc eng., University Politehnica of Bucharest, RO
[2] PhD eng., University Politehnica of Bucharest, RO
[3] PhD eng., "Mircea cel Batran" Naval Academy of Constanta, RO
[4] PhD eng., "Mircea cel Batran" Naval Academy of Constanta, RO
E-mail: andrei.pocora@anmb.ro

Abstract. Over the years many navigation incidents took place in coastal areas or in ports. As a result of these incidents (collision between vessels, groundings, flooding, etc.) many wrecks remained. Due to the inability of authorities to manage these situations, or the lack of interest for the repercussions that may occur in such an incident, many problems may arise in medium and long term. The impact on the environment and local economy may prove to be important if no pollution measures are taken in order to prevent ecological disasters. In this paper the case of a sunken vessel in the Romanian Black Sea coastal area, in terms of management of a possible leakage of hydrocarbons, is analysed. The method used in the paper allows the modelling and simulation of oil products spreading, through specialized software solutions, thus providing the authorities with a starting point in managing the response to the emergency situation.

1. Introduction

The aim of this paper is to show how a simulation software can be used in monitoring a potential oil spill in Romanian Black Sea coast. Although no major spills were recorded in the Romanian waters and the general trend of oil spills worldwide has a downward trend, the limited equipment owned by the authorities responsible impedes that exercises and simulations should be conducted, as the latest incident dates February 2018 in Constanta port.

The main causes of large oil spills in coastal areas occur due to collision or grounding of vessels, hull or equipment failure, fire or explosions on board and illegal discharge of oil contaminated water. Most of the oil spills float on the sea surface and can be easily spread over a wider area due to wind, waves and currents. Usually, the impact of the spills can be found in the upper areas of the waters, but some type of oil can sink and affect the seabed.

Constanta port and the Black Sea are important regions, being the link, in term of trade, between Europe, Asia and Africa and the number of vessels passing through this region is increasing yearly. Also, this region is widely known for its underground resources which consists of large amounts of oil and natural gas. If we also take into consideration that all these resources are near the coastal area, an accidental oil spill can have a major economic, social and environmental impact.

The impact on the environment is portrayed as a long-lasting disaster, and it affects not only the wildlife, plankton, fish, mammals, seabirds, but also the human activities in the near vicinity of the spill. Also, the environmental regeneration will last for years to come after the incident and it involves human resources and large amounts of money.

2. Oil spill statistics and causes

The latest statistics in the field published by ITOPF show a tremendous downward trend in the last five decades for spills over 7 tonnes. While in the '70s the average number of incidents was 79, since 2010 this number decreased to no more than 7 incidents per year (Figure 1). This important decrease is the result of combined efforts of the governments and the shipping industry stakeholders which act in order do prevent pollution and increase safety at sea. Nonetheless, oil is still spilt each year, by accident or willingly, so the risk of pollution still remains.

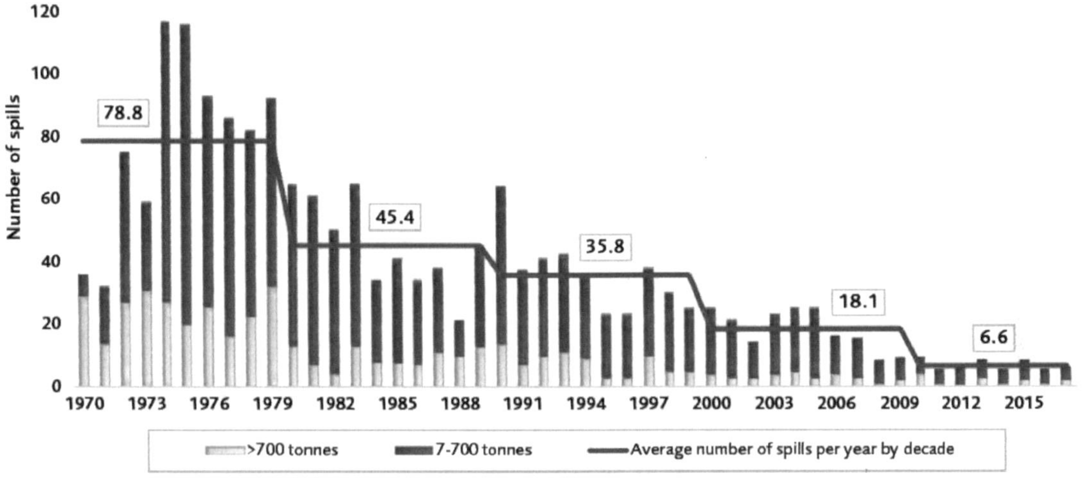

Figure 1 – Oil spills between 1970 – 2017

In the last five decades, most of the oil spills occur while vessels were underway in open or inland/restricted waters, 67%, while only 9% of the incidents occurred while the vessels were performing loading/discharging operations. Other noticeable spills occurred while the vessels were at anchor in open or inland/restricted waters, 6%, and the last 18% are unknown.

In terms of causes, the main causes are related to grounding, 32%, respectively collision, 29%. Other noticeable causes are related to hull or equipment failure, fire and explosions onboard or are of unknown causes (Figure 2).

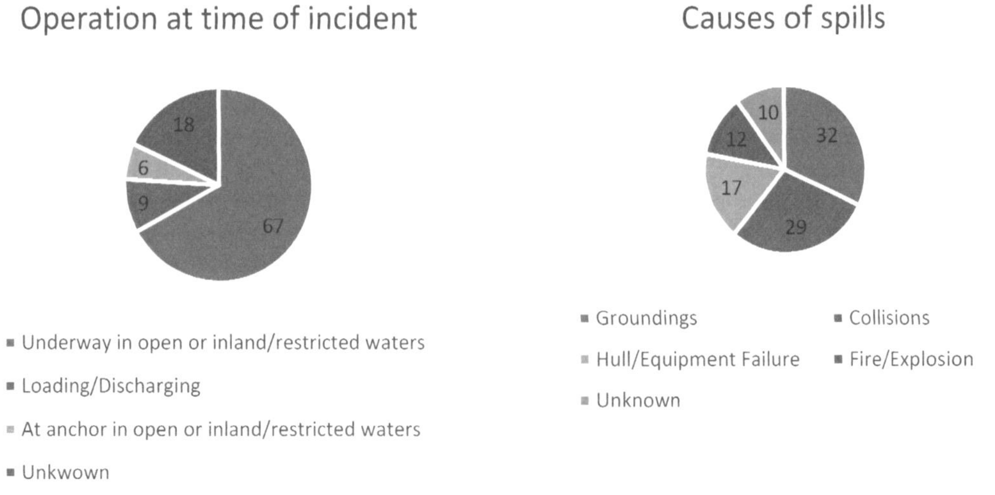

Figure 2 – Operation at time of incident and causes of spills

3. Case study – Medy wreck

In order to simulate the evolution in space and time of an oil spill, PISCES II (Potential Incident Simulation, Control and Evaluation System) simulation software was used. This system was developed by Transas Ltd. and it has the ability to simulate mathematical models of oil spills on water, calculate the trajectory and also model the interaction between the spill and response forces such as skimmers, booms or chemical dispersants. By simulation one incident, multiple plans for intervention can be developed and according to the results, the most effective one can be put in force. Thus, this system can provide not only simulations for training of response teams, but it also it has the capability to conduct real-time actions.

On the 1st of September 2010, M/V Medy (Figure 3), a 6358.9 tdw cargo loaded with 3427,4 of scrap metal, left Constanta South – Agigea at about 1 AM. Due to an incorrect ballasting procedure, little after 3 AM the cargo captain issued a distress call for sinking and after the successfully rescue of the 17 member of the crew the vessel sank at 12:51 AM. Her tanks were fully loaded with 22 tonnes of fuel and 2.33 tonnes of lubricating oil, making the vessel a potential pollution factor. In an effort to evaluate the seriousness of the situation, a team of divers was sent to wreck. Unfortunately, two of them died as the meteorological conditions were harsh and they could not get back to the surface. All other actions were stopped by the Romanian National Authority.

Figure 3 – M/V Medy before sinking

In our research, we simulated the possible leaks of fuel and oil from Medy wreck, based at 4 nautical miles from the shore, near Cape Tuzla, with the following coordinates: 44°00.567' N and 28° 45.471'E. For visualization capability, Transas TX-97 charts were used, similar to the ones used by ECDIS.

Regarding the environmental conditions, several databases were consulted in order to establish the most realistic meteorological characteristics for the western part of the Black Sea. One of the most influent factor in oil spills is the current. According to Mihailov et.all. the current for the westerns part of the Black Sea has the following characteristics: speed between 0.5 and 1.2 knots and direction between 200° and 220°. In our research we used a speed of 1.2 knots with a general direction of 210°.

Regarding the wind conditions, a mean value of 6,17 m/s with direction 67° from 01.04.2017 to 30.04.2017 was loaded into PISCES II system. The same period was considered for the water surface temperature of 6°C and 10°C as an average air temperature. In term of the sea state, Douglas scale was used with an average value of 4 – moderate sea, resulting in wave heights up to 2.5 m. Water density was extracted from nautical publications and the value used for this region was 1025 kg/m^3.

As mentioned above, the vessel had 22 tonnes of diesel fuel with a density of 837 kg/m^3 and 2,33 tonnes of lubricating oil, with a density of 874 kg/m^3. These exact products were used as pollutants and the simulation scenario was started on the 1st of April at 12.00. The next figures show the initial conditions and the general trend of the pollution path.

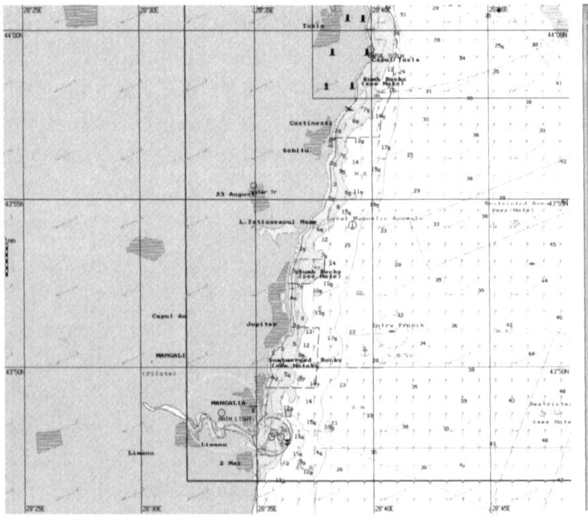

Figure 4 – Initial conditions H + 0.00 h

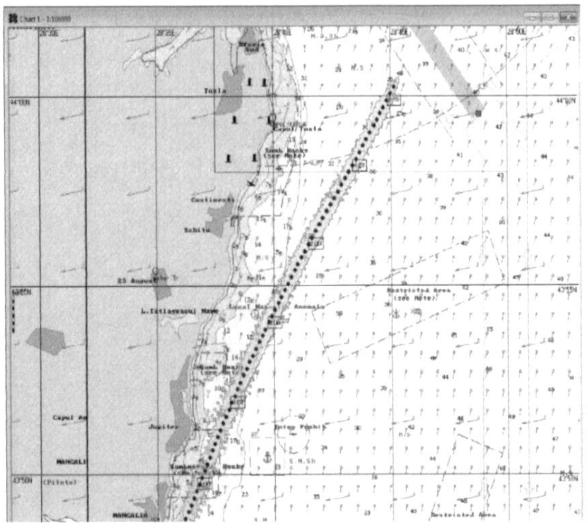

Figure 5 – Oil spill path at H + 5.5 h

When analyzing the simulation, it was found out the oil spill will reach the surface of the water after 10 minutes since the beginning and its general trend is following the current. After H + 1.30 minutes the oil spill starts to disperse and it starts evaporating, maintaining it direction towards SW.

After H + 6 hours the oil spill is near Mangalia harbour, but since the quantity is quite low it is dispersed in proportion of 95.5% while the other 4.5% has already evaporated. Simulation was runup to H + 60 hours and the parameters remained the same, taking into consideration the above mentioned hydro-meteorological parameters.

In this case, no part of the oil spill reaches the shoreline, thus no major environmental effects would occur on the shore (Figure 6).

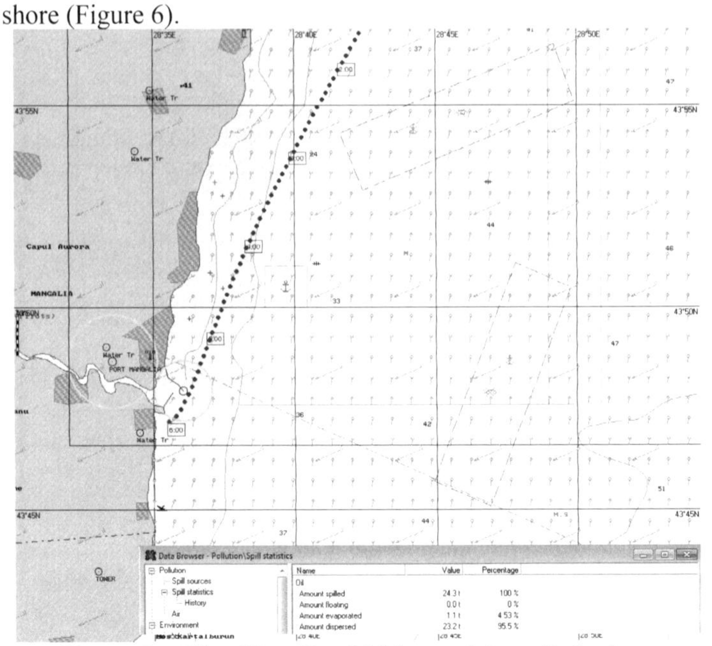

Name	Value	Percentage
Oil		
Amount spilled	24.3 t	100 %
Amount floating	0.0 t	0 %
Amount evaporated	1.1 t	4.53 %
Amount dispersed	23.2 t	95.5 %

Figure 6 – Oil spill at H + 6.00 h near Mangalia harbour

In order to underline the seriousness of an oil spill in this region of the western Black Sea coast, another simulation was conducted with the same hydro-meteorological parameters, but with a larger quantity of pollutant: 220 tonnes of diesel fuel with viscosity 837 kg/m^3.

Again, 10 minutes after the start of the simulation, the oil spill reaches the surface, but in this case, it covers a larger area and immediately starts dispersing. After H + 6 hours the oil spill will reach Mangalia harbour and has contact with the shoreline, northbound and southbound of the harbour. The simulation was kept for 60 hours and the oil spill kept going south having contact with both Romanian and Bulgarian shoreline, showing that a major environmental disaster can take place in this case (Figures 7 and 8).

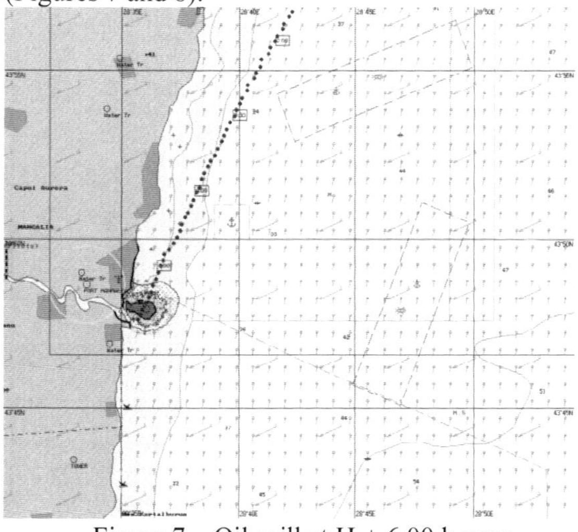

Figure 7 – Oil spill at H + 6.00 h near Mangalia harbour

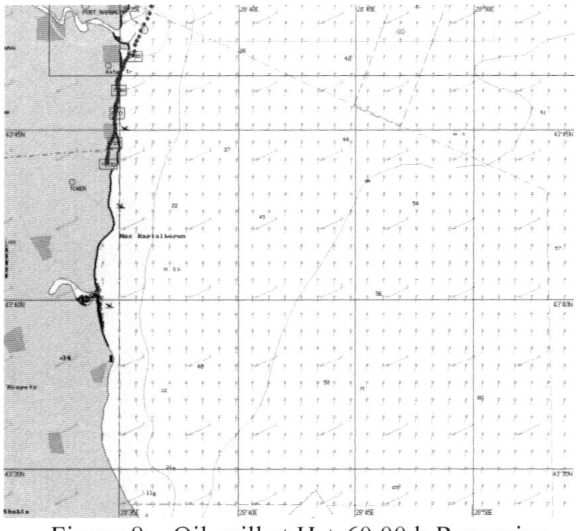

Figure 8 – Oil spill at H + 60.00 h Romanian and Bulgarian shorelines polluted

4. Conclusions

The simulations presented in this paper prove that PISCES II is a state of the art software that can be used for space – time prediction of oil spills, as well as providing the necessary tools for developing an intervention scenario. By being able to simulate multiple environmental situations for the same incident, this system may help the authorities in charge to chose the best way of intervention in order for the impact to be at a minimum. Nonetheless, by developing complex scenarios, PISCES II can also be used for conducting trainings for crisis management in case of oil spills. Thus, research and obtained results from PISCES II will increase the readiness for action.

References

[1] Bandoc G *et all* 2014 Systematic approach of the impact induced by climate changes on hydrothermic factors at the Romanian Black Sea Coast *J Environ Prot Ecol* **15** 455

[2] Datcu L and Nicolae F 2010 Fesasibility Study on the Implementation of Strategy and Modern Policy Environment Management in Naval and Port Activities of the Danube–Black Sea Canal *J Environ Prot Ecol* **11** 1085

[3] *** 2018 International tanker owners pollution federation handbook

[4] Delgado L et all 2006 Simulation of oil spill behaviour and response operations in PISCES WIT *Transactions on Ecology and the Environment* **88** 279

[5] Nicolae F *et al* 2016 Method for monitoring the space-time development of oil spilled in marine environment using PISCES II simulation software *J Environ Prot Ecol* **17** 136

[6] Rusu L *et al* 2014 Analysis of extreme storm events in the Black Sea considering the results of a ten-year wave hindcast *J Environ Prot Ecol* **15** 440

[7] Mihailov M E 2012 Black Sea Water Dynamics on the Romanian Litoral – Case Study: The Upwelling Phenomena, *Romanian Reports in Physics* **64** 232

[8] Medy wreck - http://www.divingromania.com/images/epave/wreck-epava-medy.jpg

International regulations related to naval mines and protection of navigation

Valentin S Vasilev*
* Nikola Vaptsarov Naval Academy, 73 Vasil Drumev Str., Varna, Republic of Bulgaria, valentin-vasilev@naval-acad.bg

Abstract. The free movement of people and goods is one of the fundamental values of the European Union. Nowadays, the security environment out of Territorial waters is changing its characteristics. The statute of any military action in the Exclusive Economic Zone (EEZ) becomes more sensitive than ever. The use of naval mines is demanding adequate responsibility so that to correspond to trends in maritime security. It is vital, the perception for laying off mines at sea to be placed impartially between international legislation and military morale inspired by inherent civilization values, adequate to requirements of Geneva Conventions and United Nations Convention on the Law of the Sea (UNCLOS-82) in respect with human rights. The Convention relative to the laying of automatic submarine contact mines (Hague VIII) is an example that there are some needs for regulation of the use of naval mines at sea. Moreover, the significance of the Hague VIII Convention released in 1907 is more or less melting. UNCLOS-82 doesn`t treat the use of naval mines in details, so there is certain vacuum in regulation of those security aspects. The existence of sound and transparent principles is extremely important for building firm trust and mutual understanding between coastal states.

1. Introduction

Safety and security are a required as a prior condition for the development of the maritime transportation system, offshore industry and prosperity for coastal states.The safety and security environment out of territorial waters (TTW) of Coastal states is changing its characteristics. The statute of any military action in Contiguous zone and in Exclusive Economic Zone (EEZ) becomes more sensitive than ever. Nowadays, the use of naval mines is demanding adequate reflection so that to correspond with trends in regional security. It is vital for the perception for laying mines at sea to be placed impartially between international legislation and military morale inspired by inherent civilization values. The determination of Rules of engagement (ROE) for the coastal State naval forces is another milestone that shapes out the use of mines at sea. The synthesis of proper ROE supports achieving mission goals in accordance with the requirements of the United Nations Convention of Law at sea (UNCLOS-82) and Law of armed conflict (LOAC) in respect with human rights.

The Black Sea is a crossing point of interests and spheres of influence. In order to maintain a stable environment of safety and security, it is necessary to balance the EU, NATO and UN security environment management norms, as well as the national interest. However, taking into account all the factors affecting freedom of navigation and the preservation of economic interests, it is a complex process and a serious challenge for researchers and security experts. In this study, some limitations have been placed in order to sharpen the focus and to achieve specific depth and details of the research work. Limitations are needed in spatial aspect. The projection of the sea mines usage on the security safety and environment in the Western Black Sea is reviewed. A wide range of risks and threats influence the security and safety environment, but among them only impacts of the mine threat has been explored. Significant factors are considered, such as inherent rights and values for free movement at sea, understanding of naval mines nature as well, as existing international legal norms.

Research work results are obtained by using methodology based on system analysis. Using methodology given by system approach, it is possible to assess maritime security elements related with legal aspects of naval mine threat. Undoubtedly, it affects the maritime transportation system as part of global seaborne activities. The thesis is formed using the author`s accumulated experience as mine warfare specialist influenced by a specialized legal training courses.

2. Diverse dimensions on freedom of navigation

Where is the Rubicon River when we speak about freedom of navigation, innocent passage of a vessel or need of temporal denial of passage through given maritime area? Actually, The Rubicon is a river that marked the ancient boundary. Julius Caesar led his army across it into Apennine Peninsula in 49 BC, breaking the law forbidding a general to lead an army out of his province, and so committing himself to war against the Senate and Pompey. Since then, 'crossing the Rubicon' means to break the Law. And after that, there is no way back.

Nowadays, international merchant vessel traffic constitutes more than 90 per cent of global transportation of goods worldwide. Transportation through the sea is the most cost-effective way of dealing with merchandise. It provides a reliable and secure commerce and helping to increase the well-being of nations. Maritime security is related to global food security and an economic security. As an essential part of the Maritime Transportation system Merchant shipping also 'produces' security and is a life-line for the energy sector and industry. Safety and security of sea lanes are compulsory factor for balanced economic upgrowth and continuous predictable development [1].

The sea has been a source of livelihood since time immemorial. Actually, the freedom of navigation has solely meaning concerning wellbeing of the society. The free movement of people and goods is one of the fundamental values of the European Union. The free movement of goods is the *'sui generis'* value of the European Union. The free movement by sea has its economic significance. It is a sound basis to build markets from which European citizens and businesses are gaining benefits and which is at the core of European policies. Safety and security cover a wide spectrum of maritime economic activities, especially in the littoral zone. Essential activities are transport of cargo, cruise and passenger transport, offshore energy industry, pipeline transport of gas, tourism, fishing and aquaculture.

According to the EUROSTAT data base, delivered results of maritime economic indicators for the EU members in Black sea are convincing. For instance, there is an indicative ascending trend showing loaded and unloaded goods in harbours of Republic of Bulgaria presented by year:
2012 - Total 26,012 million tons; 2013 - Total 28,841 million tons; 2014 - Total 27,235 million tons; 2015 - Total 27,166 million tons; 2016 - Total 28,685 million tons [2].

Further, the Maritime Transportation System is a part of global logistics system. It is international in nature and it becomes a subject to a complex array of policies and regulations that transcend national frameworks. Therefore, multiple states have intersections of interests because of connected markets. It settles a distinctive pillar for co-operation between governmental and non-governmental, regional and local organizations, industry stakeholders and other interested parties. In other words, contemporary life depends on the Maritime Transportation System and its services.

Accordingly, the maritime transport industry meets requirements of different legislations and jurisdictions. Seeing from the point of view given by the system analysis, the components of the Maritime Transportation System could be decomposed into subsystems and elements. It is system`s structural aspect. Desired system`s behavior and characteristics are obtained by forming a particular appropriate links between discrete elements. There are a wide spectrum of authorities and organizations involved in the legislation and policy-making activities. Furthermore, stakeholders of different nationalities are players in the routine business of the shipping industry. Participators are owners or are associated in specific way with the design, operation, management and manning of merchant vessels. Other elements are involved in seafarer education and training, as well as in the classification and insurance aspects of shipping. It could be stated that the subjects listed above are indirectly influenced by the factors of maritime safety and security. Physical objects at sea are the elements directly influenced by security risks and threats. These elements are consisted of ultimate contractors such as merchant vessels, auxiliary vessels, pilotage, vessel traffic services, towage and salvage [3].

Maritime security threats to shipping affect the predictability and efficiency of sea trade flows. The destructive outcome of security-related incidents has the potential to affect wide spectrum of maritime transportation elements. Naval mines pose inevitable risk to vessels because the world trade is becoming wider, extending to new sea lanes and new critical infrastructure in littoral zone. It makes shipping traffic more intensive in certain regions. Further, congested areas arise, shaping new peculiar shock points being a safety and security challenges.

Decomposing the Maritime Transportation System into elements makes possible to recognize the institutions involved in safety and security aspect. Governmental authorities and legitimate international organizations including naval and coastguard forces are responsible for maritime law enforcement. They are involved with the protection and security of sea lanes used for international trade as well as harbors and other crucial maritime infrastructure. Their responsibilities are related with the implementation and enforcement of security requirements under the ISPS Code. In this way, the ISPS Code as an amendment to the Safety of Life at Sea (SOLAS) Convention (1974/1988) covers the majority of issues concerning maritime security but in fact, it does not treat mine-laying and interdiction of navigation by naval mines.

The ISPS Code is required to be implemented on board all ships in all ports engaged in international maritime transport. In parallel with all benefits there are some imperfections because of existing gaps in the scope. Unfortunately, protection measures provided by ISPS have to respond to the threats posed to the seaborne traffic but does not cover mine risk. The arranged in advance objectives of the Code are to establish an international framework involving interaction and coordination between all competent institutions, such as contracting governments, governmental agencies, local administrations and port authorities to detect security threats and take proactive measures against security incidents affecting ships or port facilities. The structural and content analysis points out that mine threat is not considered at all. In other words, it describes scenarios with 'unauthorized weapons, incendiary devices or explosives' to ships but it does not cover a case study when a ship is struck by a naval mine en route to its next destination [4].

3. The nature of mine threat in littoral zone

3.1. Emerging of mines as a naval weapon

A naval mine is an explosive device typically used to create spatial structures into water called minefields. Naval mines can be used offensively in order to hamper enemy shipping or block vessels into a harbor. To use mines defensively means to protect own or friendly vessels and create 'safe' zones beyond designated areas. The emergence of this kind of capability of the Navy not only enables more effective and easier offensive warfare but also allows for defensive and reactive warfare.

The concept of sea mine came into being in medieval ages but naval mines have been used in armed conflicts at sea since the 19th century. Mine utilization is well documented for the first time in the American Revolution. The construction originated by David Bushnell was very common decades after that. Such devices were placed under the hulls of British ships in New York harbor. Moreover, during the American Civil War, the Confederate navy protected its infrastructure and shores using submerged contact explosive devices. The imperial Russian Navy was also a pathfinder in the field of mine warfare. Mines were used for a very first time as a tactical weapon during the Crimean war of 1853-1856 and during the Russo-Turkish War of 1877-1878.

The mine warfare tactics was further developed in the beginning of XX century. The modern shape of mine warfare is distinctive because of their vast use in the Russo-Japanese War in 1904-1905. However, that is not set as focus of this study. Tactical capabilities to be adapted easily from one to other naval tasks make mine preferred naval weapon in modern conflicts. For sure, cost-effectiveness is another reason for practically all navies to obtain mine-laying capabilities. Another factor in security aspect is the effectiveness of mine countermeasures. For instance, it would take much longer time to clear a minefield as to lay it. The total spending for manufacturing and laying a mine are normally up to 10% of the cost of mine countermeasures aiming at the removal.

Moreover, concerning legal regulation for the use of mines in terms of naval tactics, minefields could be classified as offensive, defensive and protective. Offensive mine laying can affect harbors and shipping channels, river mouths, canals and interior water lanes, choke points, and straits or coastal waters. It might even be possible to be determined that offensive minefields are reaction minefields designed to interdict surface traffic in TTW. In comparison with other naval weapons mines are easy to design and easy to camouflage once manufactured. The use is characterized by a significant psychological effect and a long period of impact being hard to neutralize.

3.2. Some illustrative examples of existence of a mine threat

Particular elements of World War II naval minefields still pose a threat because there are not sufficient detailed information concerning geographical coordinates and types of used mines. Discovered mines along the Bulgarian Black Sea coast in recent years enable research work and systematization, analysis and evaluation of data related to the mine threat in the region. Different types of naval mines were widely used in the Black Sea during the World Wars. Lost mines in the Black Sea create a number of hazards not only for shipping but for economic activities in general - offshore industry, tourism, etc. Influenced by currents, waves, seismic activities or other natural factors, they could change their position and could emerge suddenly on unexpected locations drifting along sea lanes or get washed up on shore. Nowadays, the mine threat causes fractal influence to safety of navigation. It has real dimensions and should not be underestimated.

Over the last decades, mines have undoubtedly been proven as relevant and effective naval weapon. In the summer of 1984, magnetic sea mines damaged at least 19 ships in the Red Sea. The US put the blame on Libya for the minelaying. In response, the US, the Great Britain, France, and three other nations launched Operation Intense Look, a minesweeping operation in the Red Sea and Gulf of Suez involving more than 46 ships. It acknowledges that mining of particular 'hot points' could jeopardize the shipping through this critical waterway harming the economy beyond national boundaries.

Another case study is mining Nicaragua's Sandino port in 1984 in support of the Contra guerrilla group. A Soviet tanker was among the ships damaged by these mines. In 1986, in the case of Nicaragua versus United States, the International Court of Justice ruled that this mining was a violation of international law. The Law aspects of Mine warfare are found during 'Guerrilla war' in Central America.

A lot of bottom and moored mines were prepared to be thrown out in the Persian Gulf. For example, during the Iranian-Iraqi War from 1980 to 1988, the belligerents mined several areas of the Persian Gulf and nearby waters. In 1988, Iran placed approximately 150 mines in the Strait of Hormuz On April 14, 1988 USS *Samuel B. Roberts* (FFG-58) struck an Iranian M-08/39 mine in the central Persian Gulf shipping lane, wounding 10 sailors. The ship had to be repaired in dry dock for six months. The cost of repair works amounted to over 89.5 million USD while the approximate cost of mine was up to 1500 USD.

During the Second Gulf War, Iraq places 1 157 mines along the Shat-el-Arab River in 1990 and 1991, making its waters extremely dangerous to navigate. The presence of mine fields terminated a planned landing operation.Other security incidents are related with the USS *Tripoli* (LPH-10) and USS *Princeton* (CG-59). The ships were severely damaged by Iraqi minefields consisting of mines

IOP Conf. Series: Earth and Environmental Science **172** (2018) 012013 doi:10.1088/1755-1315/172/1/012013

respectively contact type LUGM-145 and influence bottom type MANTA - MN103. A rough calculation of damage is 21.6 million USD corresponding to total cost of mines 11.500 USD.

During the Operation Iraqi Freedom in 2003, coalition forces pursued prolonged surveillance of suspicious activities by small craft to prevent Iraq from deploying sea mines. Thus, coalition forces detained three vessels refitted as minelayers. It was revealed after boarding that barge was designed to hide type LUGM-145 contact mines and MANTA influence mines. The mines could be placed through specially designed hatchways at the aft side of barge. A dozen LUGM-145 contact mines were found on the deck of another tug, under the masking structures made of empty drums. Thus, it was prevented from laying 137 contact and influence mines, which would seriously have jeopardized coalition ships in the area [5].

There are contemporary examples of use of improvised mines close to the studied area. A Ukrainian cutter was seriously damaged after the detonation of an 'improvised mine' near the stern. The security incident happened in Sea of Azov on the 8 of June 2015. As a result, a border patrol boat was lost after an explosion off the port city of Mariupol, Ukraine. Unfortunately there were casualties, with one dead, one injured and one declared missing. The security incident could be classified as violation of International Law, because it is unnoticed use in practice of uncontrolled armed floating sea mine. Ukrainian authorities blame the separatists for this security incident. But it is just like writing on water, because the perpetrator is still unknown.

So, it is necessary to regulate and coordinate activities at inter-institutional level, including a series of demands for far-reaching integrated approaches to reactive and proactive mine countermeasures at international level. EU and NATO membership create convenient operating conditions in the region to find such cooperation in the framework of international bilateral and multilateral projects and initiatives. This is not the case since yesterday. Illustrative examples of real actions towards improving the environment for overcoming the mining threat are conducted mine countermeasure (MCM) exercises 'Andromeda' and 'Poseidon' with the participation of MCM forces from the Republic of Bulgaria and the Republic of Romania. Moreover, the NATO Standing MCM Group in Mediterranean Sea participates regularly in opened exercises in Western Black Sea such as 'Poseidon', the naval exercise 'Breeze', etc. It is an undeniable fact that mines are used by belligerents as a means of compensating for their shortage of surface ships and submarines seeking the balance of capabilities with its opponents. Implementing system approach, the security in the Black Sea could be assumed to be a subsystem of wider Global maritime security system. Particular connections and relations could be found between system elements. Moreover, processes, interacting elements, the structure of subsystem and monitored relations are fractal to global ones. Further, the use of mines worldwide would have particular security reflection upon safety and security environment specifics in the Black Sea.

4. International regulations

Are there any International regulations concerning the laying of Naval Mine? That sounds as a tricky question. Actually, that area of Law in the second decade of XXI century still sounds as 'Terra Incognita'. Freedom of navigation is a right provided by UN conventions, European Union directives and bilateral and multilateral regional agreements.

4.1. Freedom of navigation guaranteed by United Nation Convention of Law at sea

The UNCLOS-82 is applicable to all activities at sea, including safety of navigation all over the world. It could be defined as the main contemporary source of law at sea. It shows how to determine sea in spatial aspect. Thereafter, the jurisdiction of coastal states over particular areas of sea could be determined. The coastal State has the right to adopt laws and regulations, in proportions with provisions of article 21 and other principles of international law, studied in the presented research work.

According to the UNCLOS-82, ships of all nations have the right of innocent passage through the territorial sea in terms of article 19. States seeking to lay off naval mines in peacetime are limited in how and what they may do by both treaty and customary international law obligations. The UNCLOS-82 is the most relevant treaty for determining the circumstances under which a state may lawfully lay mines in peacetime. It must not obstruct the appropriate passage of foreign ships. This rule is derived from both treaty and customary international law.

The coastal state have rights anytime to establish *Mine Danger Areas* in internal waters and TTW pursuing rights in terms of article 8 but obligation arises in terms of article 21 in order to provide the safety of navigation and the regulation of maritime traffic. According to article 24, the Coastal State shall give reasonable notice to any danger to navigation within its territorial sea, not excluding inevitable mine threat. In addition, it is entirely in the spirit of Article 25 concerning rights of protection of the coastal State.

If a belligerent decides to lay mines or arm previously laid mines in international waters, it must have 'due regard' for the legitimate interests of neutral states; for example, this may entail the provision of safe alternative routes. Belligerents may, if deemed necessary, lay mines in the EEZ and continental shelf areas of neutral states. It is important to be matter of significant concern that in the Black Sea there are no international waters according to the interpretations of UNCLOS-82 in articles 55 and 56. Moreover, the delimitation of EEZ according to the implementation of international and customary regional law is paramount as it is mentioned in article 57 and article 59. Article 74 covers at full extent the relations between Coastal States with adjacent or with opposite coastlines and it is applicable to the Western part of the Black Sea. Article 139 is applicable when a mine causes damage to a third party vessel. Then the Coastal State bears responsibility for ensuring compliance and liability for the damage.

4.2. The dawn of attempts to regulate the use of naval mines

The use of naval mines in armed conflict at first time is governed by the Hague Convention promulgating maritime international humanitarian law rules. The Hague VIII Convention aims to protect innocent passage both during and after conflict by limiting the indiscriminate effects of naval mines. To date, it remains the only treaty governing naval mines. There are essentially two reasons why the modern international humanitarian law and Law of Armed Conflict began its development in the middle of the nineteenth century, one idealistic and the other profoundly pragmatic. The first was about the altruistic concern for basic humanity in relation to the treatment of the victims of war. It resulted in the creation of the Red Cross movement and the series of Geneva conventions, the first of which was agreed in 1864. It also produced the first international agreement in modern times prohibiting the use of a weapon for reasons of humanity, the 1868 St Petersburg Declaration. The second reason was driven by national interests of the most powerful states, further known as the Great Powers. The essential purpose was to secure freedom of navigation on the high seas for commercial shipping, specifically to ensure the continuation of the extremely profitable maritime trade in time of war.

The 1907 conference came to be heavily focused on maritime concerns. Of the total of fourteen conventions produced at the conference, nine were devoted to aspects of war at sea [6]. The Number eighth was that dealing with automatic submarine contact mines, the type of mine that had caused substantial damage to both sides during the Russo-Japanese War. So, by 1907, sea mines had become a focus of serious concern and attention. Commercial trading interests tended to want them banned altogether, while naval interests wanted them regulated, but also saw the value of them as sea-denial weapons and as a means of disrupting enemy trade. It was inspired by the desire to keep sea routes open for some navigation, while going on to stress that it was about regulating rather than banning the use of automatic submarine contact mines. The 1907 Convention clearly needed to cover the possibility of moored mines floating to the surface, however, hence the operative articles omitting the word 'submarine'. Defining the type of mine covered by the 1907 Convention, makes it not applicable to sea mines in general. In the 'the spherical, horned' contact mine is triggered by the contact of a 'passing ship against one of the Hertz horns…' This is an important point to acknowledge, because, if the Convention agreed in 1907 had covered all devices described as sea mines, and all forms of sea mines subsequently developed.

Article 3 requires the laying of mines to be notified to ship owners and other States at some point after the mines are laid. It is suggested that procedures ensure rendering safe 'as soon as military exigencies permit' after mines cease being under surveillance. In fact, it might be assessed as innovation pointing toward new age of reducing collateral damages and bending war at sea toward humanity.

Article 4 requires neutral navies when constructing protective minefields in proactive manner to inform ship owners and all security stakeholders prior to mine laying operations. It is an example of implementation of comprehensive approach aiming reciprocity. Another indicative example that Convention was ahead of his time is the provision of article 5. It set requirements States to clear the mines present in their own waters. Further, the Article 6 deals with the arrangements to be taken by newly contracting States whose existing mines could not comply with the provisions of the Convention. Probably, this article contributes to controversy. Article 7 has had a remarkable negative effect. Since 1907 when the Convention was agreed in The Hague, there has been no armed conflict during which it has actually applied.

The rules contained in the 1907 Convention having customary status in relation to contact mines alone, they have also, when combined with other elements of customary law, led to the devising of the rules contained in the San Remo Manual [7]. They apply to all forms of sea mines, defined as 'explosive devices laid off in the water, on the seabed with the intention of damaging or sinking ships or deterring ships from entering an area'.

It could be deduced that it shapes an international regulation which is advantageous both to protect neutral commerce and to uphold the principle of immunity of enemy merchantmen from attack without warning. Impartially said, the 'future' Great Powers made it obvious that they were unable to reach complete agreement on the matter. After much deliberation, they arrived at a compromise. Only Article 1 and article 5 contain clear and unequivocal regulations. Article 2 forbids the laying of contact mines off the coast and the ports of the enemy, with the sole object of intercepting commercial shipping but it is of limited value. It is important to belligerents that mines are laid for a purpose other than merely intercepting commercial navigation or not [8].

Despite its shortcomings as a treaty, the 1907 Convention has, over a period of more than a century, had an influence on the development of customary law on sea mining. It is possible to identify a series of rules that can be applied to sea mines of all levels of sophistication.

In some extent, the convention has contemporary dimensions and could serve as fundament of contemporary international legislation provisions. What are the most important outcomes? It is very important a kind of notification to be promulgated via navigational warnings, messages using Diplomatic channels or accessible information systems. Next, it is important to decision makers to keep into account if mine is moored or floating. The minefield should be designed so that mines to become harmless in time, having limited duration in order to reduce the harm to 'innocent' traffic. Another remarkable requirement is channeling of neutrals. Reading the Convention there is inspiration that mines capable of discrimination should be programmed against legitimate targets. It is unnecessary for them to become inactive if they break free from their moorings. Maritime law traditions prohibit the indiscriminate use of mines and therefore mine-laying operations that are not or cannot be directed against a military objective are unlawful.

4.3. Some Geneva 1949 Conventions provisions applicable to the use of naval mines

Another source of international law is derived by Geneva Conventions. Some of fundamental legal principles for LOAC come from the Geneva Conventions of 1949. The Geneva Conventions consist of four separate but correlated international treaties aiming to protect human beings involved in armed conflict from unnecessary suffering. They also provide legal protection civilians and private property. Geneva Conventions distinguish individuals as lawful combatants, non-combatants, and unlawful combatants.

For a first time, a lawful combatant is determined as an individual authorized by governmental authority to engage in hostilities. A lawful combatant is usually a member of a regular armed force but may be part of a paramilitary force. In spite of that, it is necessary the commander to be responsible for his subordinates, to oblige these individuals to wear uniforms and fixed distinctive emblems recognizable at a distance, to carry arms uncovered, and to conduct combat operations according to specified norms. Accordingly, the crew of mine-laying vessel would be recognized as lawful combatants. In spite of that, it contradicts the paradigm of camouflage, suddenness, etc. Non-combatants are individuals not authorized by governmental authority to engage in hostilities. Civilians accompanying the Armed Forces and certain military personnel who are not authorized to engage in combatant activities, such as medical personnel should be assumed as non-combatants. Normally, non-combatants are not targeted as object of direct attack. Nevertheless, there is a high probability for non-combatants to suffer injury or death incident during direct attack. Unlawful combatants are individuals who directly participate in hostilities without authorization or under international regulations. For example, armed rebels or mercenaries deploying improvised mines violate law and become lawful military targets. Undetermined status is applicable when there is uncertainty as to whether an individual is a lawful combatant, non-combatant, or an unlawful combatant. Then the Geneva Prisoner of War Convention provides a protection until status is determined. The capturing nation should convene a competent tribunal to determine the detained person's status [9].

4.4. The NATO perspective and the nature of LOAC

According to the NATO concept countermine responsibility is divided into allied and national in accordance with the nature of control upon specific area. Within any NATO operational area, the command and control of mine warfare assets may be divided between both NATO and competent coastal state authorities. In general, as a customary rule coastal state authority are responsible for mine warfare operations carried out in their own coastal waters. An example could be given mentioning sea trade lanes in TTW, cabotage routes and critical maritime infrastructure. Sensitive areas are harbours, anchorages and anchorage approaches, coastal shipping routes, and channels. In addition to this, NATO authorities are responsible for countering the mine threat in open ocean or sea areas. Vital to neutralizing the level of threat is the responsibility for the overall co-ordination of mine warfare effort in mine danger areas. They may delegate operational control to a coastal state authorities or local area subordinate command. The paradigm of building security in coalition format gives another viewpoint. Normally, the commander`s mine-laying plan and embodied ROE are pursuant to National and International legislations.

ROE compromise agreement is founded on controversial claims. On the one hand, it expresses an obligation to follow humanitarian and commercial interests. On the other hand, ROE must support military priorities in order to achieve desired military goal. The San Remo Manual is an important document related to the law aspects of mine warfare. The San Remo Manual was the result of a process of consultation and discussion between 1987 and 1994 at two round tables and six meetings of a group of naval experts and lawyers. It is an issue relevant to comprehensively examined law by the Institute of International Humanitarian Law (IIHL) based in San Remo, Italy. The result of the process, published in 1995, was authoritative without being either an attempt to develop conventional law or account for the then current state of customary law. Indeed, it was innovative in several respects in attempting to place the law of naval operations during armed conflict into a late-twentieth century context. As a source of the law it is, therefore, somewhat ambiguous.

ROE are issued by competent commanders after legal review and approval. ROE are relevant to the circumstances and limitations under which forces will engage the opponent. Normally, ROE are promulgated by operations plans and operations orders. ROE provide appropriate use of force according to national policy, mission objectives, and international legal regulations. ROE implement existing LOAC principles. ROE set forth parameters of rights for self-defense. During given military operation, LOAC and specifically designed ROE provide detailed guidance on the use of force. Rules give commanders direction on the use of force in self-defense against a hostile act or hostile intent [10].

LOAC is applicable to international armed conflicts and in the conduct of military operations and correlated military activities. It is an implementation of gathered military experience and international law regulations. LOAC has a specified nature aiming to prevent unnecessary suffering and destruction. LOAC regulates the conduct of hostilities and protects human life. There are three basic LOAC principles - military necessity, distinction, and proportionality.

Military necessity requires engaging in only those acts necessary to accomplish a legitimate military objective. Using naval mines in armed conflict, minefields should be limited strictly against the involved naval ships. It generally means the mine may target those floating physical objects, so if destroyed, would lead directly toward partially or completely achieving of mission objectives. It is vital not to use naval mines or weapons systems that violate international law. At times, some restrictions are required on use in order to increase compliance with the LOAC.

Distinction is the capability of a mine to distinguish between lawful combatant targets and noncombatant targets such as merchant vessels, fishing boats, cruise vessels, offshore supply vessels, etc. The mean concept of distinction is to engage magnetic, acoustic or hydro-dynamic field of valid military targets. Distinction is possible when it is possible to separate military targets from civilian ones as precisely as it is possible. For that reason, it would be a bad idea to transport civilians on board of naval ship during armed conflict at sea.

Proportionality prohibits the use specific weapons or degree of force that would exceed needed effects to reach desired military objectives. Proportionality is an indicator that is an expression of gained military advantage in relation with the harm caused during operation [11].

Once the security threat terminates, armed mines must be removed or rendered harmless while floating mines must be removed or rendered harmless one hour after they are deployed in any circumstance. In peacetime, a state is prohibited from laying mines of any type in the territorial seas or internal waters of another state. To do so would violate the territorial sovereignty of the latter and constitute a use of force unless the mine-laying is done with the consent of that state.

Most states agree that, as a general rule, international law prohibits the laying of armed mines in international waters, international straits and sea lanes during peacetime. The prohibition on laying armed mines in international straits derives from the principle that all ships enjoy the right of unimpeded transit passage through international straits. …It gives 'appropriate publicity to any danger to navigation … within … the strait of which they have knowledge'. Moreover, unlike innocent passage, transit passage cannot be suspended. These obligations also apply to archipelagic sea lanes. Armed mines may be laid in international waters prior to the existence of an armed conflict on the basis of individual or collective self-defence. The law is unsettled as to whether controlled mines may be laid in international waters in peacetime since a state's obligation is to have 'due regard' to the rights of other states. Not all states share this view as some do not maintain that, in peacetime, armed nor controlled mines may be laid in international waters.

5. Conclusion

In conclusion, it is perceivable that the Hague VIII Convention is an example that there are some needs for regulating the use of naval mines at sea. Actually, the Convention has not expired but it falls behind its time. It is true to be stated that it is not among the most popular legislative sources of international law. UNCLOS-82 fails to treat the use of naval mines in details, so there is a vacuum, in this sense. Accordingly, the effects of the Hague VIII Convention should be matter of concern not only for Legal advisors and Lawyers but for all of those people that are involved in maritime security and maritime business. The protection of civilians and innocent passage is one of the main goals of the maritime security system.

Nevertheless, the Hague VIII Convention gives unsatisfactory aged definitions. It has got 'gaps' concerning modern influence mines. So, these blank fields become larger and larger. Contemporary legal aspects of naval mines use are defined successfully by NATO. The San Remo Manual is a good example to produce comprehensive manual for commanders at sea. There is necessity of newly-designed applicable international legal regulations applicable to the use of Naval Mines. Probably it is appropriate for the issue to be under UN jurisdiction. Raised issues should be adequate to contemporary trends in maritime security such as asymmetry, hybrid threats and challenges in delimitation of the sea.

References

[1] Sekimizu K 2012 *A concept of a sustainable Maritime Transportation System* (Rio de Janeiro: IMO) pp 6–17

[2] *Energy, transport and environment indicators. 2017 edition.* (Luxembourg: Publications Office of the European Union) pp 92-139

[3] Escámez S 2010 *Free movement of goods. Guide to the application of Treaty provisions governing the free movement of goods* (Luxembourg: Publications Office of the European Union)

[4] *Goulielmos A Anastasakos A 2005 Worldwide security measures for shipping, seafarers and ports: An impact assessment of ISPS code.* Disaster Prevention and Management: An International Journal 14(4) pp 462-478

[5] Peniston B 2013 *No Higher Honor: Saving the USS Samuel B. Roberts in the Persian Gulf* (Annapolis: Naval Institute Press) pp 115-136

[6] Tileubergenova E 2016 *Defining the Regulations of War in the Hague Convention of 1907* (4th Annual International Conference on Cognitive - Social, and Behavioural Sciences) pp 246-252

[7] Cole Alan 2009 *Sanremo Handbook on Rules of Engagement* (Sanremo: International Institute of Humanitarian Law)

[8] Schindler D Toman J 1988 *The Laws of Armed Conflicts* (Leiden: Martinus Nihjoff Publisher) pp 804–807

[9] Best G 1997 *War and Law since 1945* (Oxford: Clarendon Press) pp 288-293

[10] Fleck D 2001 *The handbook of The Law of Visiting Forces* (Oxford: OUP) pp 512-522

[11] Arribas G 2010 *International law aspects of the EU's security and defense policy, with a particular focus on the law of armed conflict and human rights.* Netherlands International Law Review *57*(3) pp 507-510

Quantifying security risks in the cloud environment in the quest for minimizing energy consumption

S Eftimie[1], M Rogobete[2] and C Răcuciu[3]

[1] Military Technical Academy, Bucharest, Romania
[2] GE Power Romania
[3] Titu Maiorescu University, Bucharest, Romania

Abstract. In this paper we investigate the practices and methods used for quantifying the security risks involved with the usage of cloud services. The usage of external cloud services greatly reduces the consumption of energy related to hardware and the industry needs a way to evaluate those risks and address them accordingly. This work proposes a unified framework for quantifying security risks that enables companies to reduce operational risk related to cloud while minimizing energy consumption and reducing the environmental impact.
Keywords: Cloud, Framework, Risk quantification, Energy consumption

1. Introduction

Cloud computing has become a highly innovative space with an active community and a growing client base. Current developments in technology and the increasing focus towards mobility have changed the way we look at cloud and its inherent issues. Businesses and individuals have begun to use external cloud servers managed by other companies to access data or computing services from any physical location. This move came with a major trust concern related to cloud providers.

Security remains one of the major obstacles to large-scale cloud adoption. SaaS (Software-as-a-Service) providers are working to secure their products to protect their business. A breach may have a devastating effect to a company since image damages can translate to huge financial losses. SaaS providers invest in security products and hire good talent to protect their business and work on preventing attacks to their infrastructure, such as denial of service attacks. Since SaaS providers do not prioritize the security related to the client's access control policies, the companies that use their services should implement their own security measures.

The biggest risks that the consumers of cloud services are facing are related to the data disclosure or data loss. The benefits of cloud computing are significant: low cost, high reliability and immediate availability of additional computing resources when needed. Despite these advantages, both cloud service providers and consumers must be aware of their own set of unique risks of cloud computing, which is usually associated with storing and processing data.

In the recent years, there have been a series of incidents in which customer data hosted in the cloud was released online (for purposes of hacktivism and vandalism) or stolen for criminal purposes. Cloud computing is made possible using technologies such as Internet access, virtualization and third-party data centres. In the case of online access to a cloud service provider, access controls take the form of usernames and passwords. In the case of virtualization, such access controls can be implemented through the logical separation of data. In the case of third-party datacentres, such access controls may take the form of physical access controls or software-based access to prevent unauthorized access to customer data of people working in the organization.

In principle, the access controls mentioned above are solid. However, in practice, such controls have been circumvented. If any of those access controls are compromised, the risk of data leakage is high. However, in the event of a data breach where the associated data is acquired in encrypted form, it is essentially useless for an attacker (unless the encryption algorithm used is weak and / or the attacker knows the associated decryption key).

Otherwise, if a security breach occurs and the associated data in plain text is stolen by the attackers, the effects can be disastrous for a company, from negative publicity and damaged reputations to fines in accordance with the data protection legislation.

A framework that would allow a transparent communication of the security risks that would essentially put actual value on possible security events in the cloud environment would allow stakeholders to make informed decisions and better understand the impact of a potential cyber-attack. For this reason, we have started to investigate ways in which we can quantify risks in ways that can be easily understood by non-technical people.

2. Risk quantification

Risk quantification represents the process of assessing the dangers that have been recognized for a project and building up the information that will be required for addressing them. Risk management is performed during the whole duration of a project. During a project, there are situations where both qualitative and quantitative analysis are required. The target of the risk quantification is to set up a method for arranging the risks in the order of their impact on the business. The next step is to address them using this classification and it is expected that in most projects there won't be sufficient time or money to address every risk that is recognized. One practical measure for evaluating risks represents the severity of the risk. Severity can be though as a pair combination between the risk impact and the risk probability. A quantitative investigation should be made with care since utilizing bad or wrong data can lead to bad decision making regarding the risks. There should also be an assessment regarding to the collection of data because the total cost of the risk can be lower than the collection of data used to quantify it.

In a previous work [1] we have identified the threats and risks involved with cloud computing. Other works such as [2] have categorized those risks into general ones, risks that emerge from the service models and risks that emerge from deployment models as can be seen in table 1.

Table 1. Cloud computing risks

<table>
<tr><th></th><th></th><th>Specific risks</th><th>General risks</th></tr>
<tr><td>Service model</td><td>IaaS</td><td>Malicious insiders</td><td rowspan="8">Data breaches
Data loss
Account hijacking
Insecure APIs
Denial of Service
Cloud abuse
Malicious insiders
Shared technology vulnerabilities
Insufficient due diligence</td></tr>
<tr><td>Service model</td><td>PaaS</td><td>Loss of control over hosted applications</td></tr>
<tr><td>Service model</td><td>SaaS</td><td>Loss of data ownership</td></tr>
<tr><td>Deployment model</td><td>Public Cloud</td><td>Tenant breach</td></tr>
<tr><td>Deployment model</td><td>Private Cloud</td><td>Insufficient resources</td></tr>
<tr><td>Deployment model</td><td>Hybrid Cloud</td><td>Interdependence risk</td></tr>
<tr><td>Hosting type</td><td>External hosting</td><td>Resource relocation
Environments are not isolated</td></tr>
<tr><td>Hosting type</td><td>Internal hosting</td><td>N/A</td></tr>
</table>

IOP Conf. Series: Earth and Environmental Science **172** (2018) 012014 doi:10.1088/1755-1315/172/1/012014

3.Qualitative risk assessment

Qualitative risk assessment is suitable early in the implementation of a solution and its scope should be the categorization of the risks and the appropriate actions that need to be made to address them. This type of analysis does not give an exact value of the risk but it's useful in the case of a short assessment period. When using this type of analysis, quantitative values can be associated with the identified risks to calculate their severity.

Cloud risk management is difficult because the technology is not controlled by the company that uses it. Conventional risk frameworks such as [3] ISACA [4], are not entirely suitable in the cloud environment because they were conceived for other types of risk.

Companies are finding hard to assess the cloud risk for a few reasons. The rapid developments in technology make it hard to identify security controls and risk parameters. Cloud providers also update constantly their service and infrastructures.

This leads to the conclusion that new risk frameworks must be developed to meet the demands of the new advancements surrounding cloud. Risk quantification is particularly useful when interacting with stakeholders. Decisions regarding the usage of different cloud services should be made in respect to the risks involved and a good measure can be represented by the balance between investment and risk.

The European Network and Information Security Agency (ENISA) released a risk assessment framework [5] for the cloud environment that has become a standard in the industry. Two documents are part of the ENISA release, a cloud information assurance framework and a guide for risk assessment. The agency completed an investigation of the risks and advantages of using cloud services and its essential goal is to decrease the identified risks to a satisfactory level. ENISA has distinguished technical, legal and organizational risks that need to be assessed before the cloud adoption.

The risk management is the next logical step after the risk assessment. In traditional IT systems, risks were identified based on critical areas, quantified, evaluated and addressed accordingly. Cloud has added a layer of complexity that has yet to be solved. There are still unanswered questions regarding the segregation of duties between the cloud provider and clients.

In [6] a framework is proposed that uses an incremental approach that can enable, plan, screen and manage information security in cloud services. The framework uses a model that is based on multiple criteria that incorporates static and dynamic measurements. The main objective of this framework is to make the security administration a continuous process.

Another proposal for a risk assessment framework can be found in [7] where the authors include the cloud users in the overall risk assessment by applying a so-called analytic hierarchy process that quantifies the risk.

The starting point for an organization can be any type of these frameworks and an effort should be made to investigate and utilize methods from more than one and create and adapt according to the organization's unique needs.

Moreover, existing frameworks are less prescriptive about internal controls suitable for hybrid cloud setups and center around controls that the cloud suppliers usually set up inside their infrastructures. During the overall risk management, it's important that the information security controls such as encryption and key administration, authentication, data breach warnings and legal controls are performed, and they should also be specified in the service agreements with the cloud providers.

The risk appetite of a company represents a concept that is opposite to risk management. It's basically a measure of the amount of risk that a company is willing to take to pursue its business initiatives. Cloud computing is relatively new in terms of adoption and presents a unique set of risks. Due to the rapid evolution of the technologies in this area, the awareness over a company's risk tolerance can have a strategic importance.

The assessment of a cloud risk appetite needs to take several factors into consideration. The risk tolerances of similar businesses need to be evaluated to have a good balance between risk and innovation on the market. For businesses that are moving to cloud, there is a possibility that risk assessment has already been done in the past and that there are items that can be reused and adapted to the new cloud paradigm. Technology and the future strategy of the company also need to be taken to consideration when assessing the risk tolerance.

The final step is the formal documentation of the risk appetite. This document should be made available to the organization's staff so that they act accordingly.

4. Framework proposal

Cloud access security brokers are cloud-based or on-premises security policy enforcement points, placed between cloud service providers and cloud consumers. CASBs consolidate multiple types of security policy enforcement [8]. One of the features of CASBs is the comparison between different SaaS vendors highlighting different aspects regarding security. There are several CASBs vendors on the market and each one of them has its own set of features. For an organization that wants to adopt different SaaS solutions for its business, it's difficult to assess the risk for different SaaS vendors and it's difficult to choose a CASB vendor to address those risks. This leads to the conclusion that there is a need for a standardized way of assessing risk in cloud which has been proved to be difficult as we have seen in the previous sections of this paper. As a solution, we propose a unified framework for quantifying risk to resolve this issue. These are its core principles:

- Usage of the CSA approved list of cloud risks
- Transparency to stakeholders
- Ease in assessing CASBs
- Cloud community participation to the framework
- Anonymous submission of reported losses
- Cost efficiency

The accurate communication regarding the cyber risk involved with the cloud adoption represents the main challenge that chief information security officers face when dealing with upper management boards. One obvious way to quantify a risk represents the loss in money. This is a familiar concept to stakeholders, it's easy to understand and allows key people to make informed decisions about the SaaS that they want to adopt, evaluate the costs involved and balance the risk and opportunity [9].

Table 2. Quantification example of cloud risks associated with a SaaS solution

Cloud Risk	CASB Costs	Risk Mitigation	Total Investment	Risk Value	Risk Tolerance
Data Loss	€3000	€5000	€9000	€300000	0%
Data Breaches		€1000		€10000	0%
Account hijacking	€1000	€2000	€3000	€10000	10%
Insecure APIs	€5000	€2000	€17000	€20000	30%
Denial of Service		€3000		€3000	30%
Cloud abuse		€5000		€12000	50%
Malicious insiders		€2000		€20000	10%
Shared technology	€0	€1000	€1000	€6000	10%
Insufficient due diligence	€0	€1000	€1000	€6000	10%
Total	**€8000**	**€22000**	**€30000**	**€117000**	**N/A**

In table 2 we can observe an example where risks are assessed for the adoption of a SaaS solution. The CASB Solution can offer discount packages as part of the offer, for example data loss and data breach mitigation security products are bundled together. Although CASB products are advanced, additional security measures need to be implemented on-premise and their costs should be taken into consideration.

IOP Conf. Series: Earth and Environmental Science **172** (2018) 012014 doi:10.1088/1755-1315/172/1/012014

We propose that the value for a specific cloud risk be calculated using the following formula:

$$Rc = \frac{\left(\frac{Li}{Vi} + .. + \frac{Ln}{Vn}\right)}{n} \times Vc \times (1 - Tc)$$

The risk is expressed in an actual amount of money, $\boldsymbol{Rc}$. First, we calculate an average of the monetary losses caused by attacks relative to the respective values of the companies. There are multiple ways to assess the value of a company: Market Capitalization, Times Revenue, Earnings Multiplier, Discounted Cash Flow, Book Value, Liquidation Value and for accurate results we note that the same method should be used for one calculation. $\boldsymbol{Li}$ represents the loss associated with a cloud risk experienced by a company that has the value $\boldsymbol{Vi}$**.** The average result expressed as a percentage is then multiplied by the value of the company $\boldsymbol{Vc}$ that addresses a cloud risk. $\boldsymbol{Tc}$ represents the risk tolerance of the company to that specific cloud risk, expressed as a percentage. For example, if the company's tolerance towards data loss is 10% the value obtained after taking in consideration the company's value will be multiplied by 90% (1-10%).

The value obtained should be a starting point for the quantification of the risk [10] since it expresses real losses that can occur in the case of a security event and can be altered if the CISO for example decides that the respective case is not applicable to the business that he is trying to protect.

Getting the correct loss values and valuating the companies are the most difficult steps. We believe that a platform that would ease the reporting of such events and make the cost transparent is necessary. This would need the participation of the whole cloud community and companies should be transparent about the attacks they suffer which is not an easy thing to do since even news about the attacks would cause financial losses to their business. Multi-party computation (MPC) is used to describe a form of computation that can be performed over a series of private inputs. In an MPC scheme several participants want to compute the value of a public function on their private data while keeping their inputs secret. The goal of MPC is to build an algorithm where the participants can obtain the result of the function without having to rely on a third-party entity. We hope that this will be a subject for future work since it's a plausible way to deal with the anonymous submission of losses.

Once we have an assessment of all the risks applicable to the business and the associated monetary losses attached to them, a comparison can be made between the different security brokers considering the points where the CASBs will address those risks. In some cases, additional security measures will have to be put in place by the CISO department and they should also be quantified.

Once they are presented an overall view of the costs involved, board members can make a cost decision regarding the adoption of a particular SaaS solution and regarding the risks involved and the cost of their mitigation. The assessment part and the valuation of the companies are not easy to perform but we envision a collaboration of the cloud community for this common scope.

5. Conclusions

Although cloud computing has brought many advantages to companies, it has opened the doors for new security issues. The use of traditional models for risk assessment is not suitable for the new cloud computing landscape due to the new types of threats and new attack surfaces. In this paper we have analyzed several risk assessment methods and we have proposed a framework that helps the risk assessment of SaaS solutions in a manner that is easy to communicate to non-technical people. The SaaS market has experienced significant growth and the security breaches continue to fuel cybersecurity market growth as well. CASB's are just one of the technologies that have emerged in this new cloud security landscape. In this context, a method for assessing security risks is essential for a company that wishes to implement several SaaS solutions and wants to take advantage of a CASB. The proposed framework is designed to close the communication gap between the CISO department and the board members of a company. A clear understanding of the cloud risks impact over time can help companies to tolerate a higher level of risk in an efficient way.

6. References

[1] Răcuciu C and Eftimie S 2015 Security threats and risks in cloud computing *Mircea cel Bătrân Naval Academy Scientific Bulletin*, vol **18**, pp. 105-108

[2] Belbergui C, Elkamoun N and Hilal R 2017 Cloud Computing: Overview and Risk Identification Based on Classification by Type *2017 3rd International Conference of Cloud Computing Technologies and Applications (CloudTech), Rabat, 2017,* pp. 1-8

[3] Maneerattanasak U and Wongpinunwatana N 2017 A Proposed Framework: An Appropriation for Principle and Practice in Information Technology Risk Management *2017 International Conference on Research and Innovation in Information Systems (ICRIIS), Langkawi, 2017,* pp. 1-6.

[4] ISACA 2009 The risk IT Framework *Information Systems Audit and Control Association*

[5] Madria K 2009 Cloud Computing Risk assessment *Missouri University of Science and Technology*

[6] Djemame K, Armstrong D, Guitart J and Macias M 2016 A risk assessment framework for cloud computing *IEEE Transactions on Cloud Computing*, pp. 265-278

[7] Cayirci E, Garaga A, Oliveria A and Roudier Y 2016 A risk assessment model for selecting cloud service providers Journal of Cloud Computing, p.14

[8] Eftimie S, Dumitru L and Opriş V 2016 Cloud Access Security Brokers *Education and creativity for a knowledge-based society International conference 2016*

[9] Deloitte 2016 The value of cyber risk quantification, pp. 1-7

[10] Filkins B 2016 Quantifying risk: closing the chasm between cybersecurity and cyber insurance *SANS Institute, 2016,* p.3

IOP Conf. Series: Earth and Environmental Science **172** (2018) 012015 doi:10.1088/1755-1315/172/1/012015

Modeling and simulation of hydro-acoustic signal propagation in the Romanian Black Sea coast

E G Curcă[1], R G Damian[1], O Radu[1] and V Roşca[1]

[1]Research Center for Navy, Ştefăniţă Vodă Street, No. 4, Constanţa, Romania

E-mail: staff@ccsfn.ro

Abstract. Sound energy is the primary method of detecting and locating submarines and mines. The study of the propagation of sound waves in sea water is the most important issue in assessing sonar systems performances. The variation of the sound speed in the marine environment radically influences the propagation paths of the acoustic signals. This paper presents a modeling and simulation model of hydro-acoustic signal propagation for a specific sound speed profile from the Romanian Black Sea shore.

1. Introduction

An important issue for the Navy since it employs sound energy as the primary method of detecting and locating submarines and mines, is the study of the sound waves propagation in sea water. Active sonar systems create a pulse of sound in the water and then listen for echoes bouncing off the target submarine or mine. In passive sonar systems, the searcher listens for the weak acoustic signals generated by the target submarine itself. In either case, the structure of the sound speed environment radically affects the propagation paths of the sound energy. Knowledge of the sound speed structure is required to correctly set the sonar systems and to choose the appropriate tactics to locate the target. In turn, the sound speed depends on temperature, salinity and pressure variations of the marine environment.

Considering marine environment as an acoustic waveguide limited above by the sea surface and below by the seafloor, acoustic wave propagation can be modeled by a set of mathematical equations in relation to a number of parameters specific to the marine environment.

We consider an important issue to simulate the propagation of the sound waves in a marine environment specific to the Romanian Black Sea shore. Software modules developed in the **Matlab R2017b** programming language correlated with an interface built in **GUI** toolbox **(Graphical User Interface tool)** of **Matlab** perform a modeling of the paths of the acoustic signals in a certain area of the marine environment, being designed to accomplish a hydro-acoustic forecast.

2. Theoretical framework of the hydro-acoustic ray tracing problem

For beginning, in the sound waves propagation simulation should be considered an aquatic range delimited by two horizontal positions and two vertical positions. Often the vertical positions are considered the sea surface and the seafloor. The position of the acoustic signal source is established in this range.

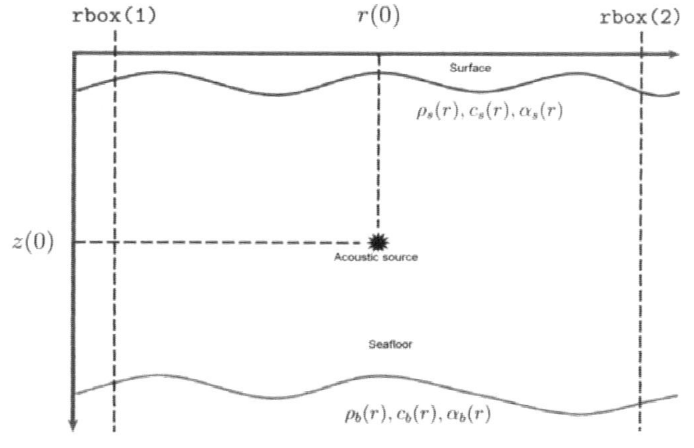

Figure 1. Ray tracing aquatic range [4]

The modeling problem of the sound waves propagation is approached from the perspective of the ray theory. The general strategy of ray tracing is the following:

- establish the launching angles at the source position and then trace the ray for every launching angle, until it is absorbed or exits the range;
- at every step of ray coordinates determination check if the ray is above de surface and below the bottom. For a positive check determine the intersection of the ray with de boundary, reflect the ray, and continue with the ray coordinates determination.

To apply this strategy should be known the sound speed profile in the considered range and then determine the rays coordinates.

The sound speed profile could be determined by direct measurement or by using a calculation method based on CTD sensor – conductivity/salinity, temperature, depth. Analyzing the results of a series of calculation methods (UNESCO-Chen & Millero equation, Medwin equation, Coppens equation), Leroy equation approximates the best sound speed values obtained by direct measurement for different periods and locations in the Black Sea shore. Intervals used for temperature, salinity and depth, in case of Leroy equation, are valid in the entire basin of the Black Sea. Therefore we considered in the simulations the Leroy equation to approximate the sound speed values in a Black Sea shore range:

$$c = 1402.5 + 5T - 5.44\times10^{-2}T^2 + 2.1\times10^{-4}T^3 + 1.33S - 1.23\times10^{-2}ST + 8.7\times10^{-5}ST^2 + 1.56\times10^{-2}D + 2.55\times10^{-7}D^2 - 7.3\times10^{-12}D^3 + 1.2\times10^{-6}D(\Phi - 45) - 9.5\times10^{-13}TD^3 + 3\times10^{-7}T^2D + 1.43\times10^{-5}SD \quad (1)$$

where

- T = Temperature [^{0}C]
- S = Salinity [‰]
- D = Depth [km]
- Φ = latitude [degree].

The expression is valid under the following conditions:

- T [0, 40] ^{0}C
- S [0, 42] %
- D [0, 6000] m.[3]

IOP Conf. Series: Earth and Environmental Science **172** (2018) 012015 doi:10.1088/1755-1315/172/1/012015

To trace the hydro-acoustic rays in a particular range according to the rays' theory is required to solve the Eikonal equations system, which governs the acoustic waves' propagation.

The acoustic rays propagation problems frequently involves punctual sources that have cylindrical symmetry, or at least are considered mediums that vary slowly in azimuth, and azimuth dependence can be ignored when calculating transmission in a particular direction. In this case $\vec{r} = (r, z)$ where r is the horizontal coordinate and z is the vertical coordinate (depth).

In cylindrical coordinates the equations that governs the acoustic rays propagation can be written as

$$\frac{dr}{ds} = c\sigma_r(s), \quad \frac{d\sigma_r}{ds} = -\frac{1}{c^2}\frac{\partial c}{\partial r},$$

$$\frac{dz}{ds} = c\sigma_z(s), \quad \frac{d\sigma_z}{ds} = -\frac{1}{c^2}\frac{\partial c}{\partial z}, \tag{2}$$

where $r(s)$ and $z(s)$ represents the ray coordinates in cylindrical coordinates and s is the arclenght along the ray. The pair (σ_r, σ_z) represents the components of the slowness $\sigma = \frac{1}{c}$, where c represents the sound speed.

The initial conditions for $r(s)$, $z(s)$, $\sigma_r(s)$ and $\sigma_z(s)$ are

$$r(0) = r_s, \; z(0) = z_s, \; \sigma_r(0) = \frac{\cos\theta(0)}{c(0)}, \; \sigma_z(0) = \frac{\sin\theta(0)}{c(0)}, \tag{3}$$

where $\theta(0)$ represents the launching angle, $(r(0), z(0))$ is the source position and $c(0)$ is the sound speed at the source position. The coordinates are sufficient to obtain the ray travel time:

$$\tau = \int_\Gamma \frac{ds}{c(s)}, \tag{4}$$

along the curve $[r(s), z(s)]$. When sound speed depends on depth only the horizontal slowness $\sigma_r(s) = \frac{\cos\theta(s)}{c(s)}$ is a constant and is preserved:

$$\frac{d\sigma_r}{ds} = 0. \tag{5}$$

To determine the numerical solution of the system of Eikonal equations we start by rewriting the system as a vector equation

$$\frac{dy}{ds} = f \tag{6}$$

where

$$y = \begin{bmatrix} r \\ z \\ \sigma_r \\ \sigma_z \end{bmatrix}, \; f = \begin{bmatrix} c\sigma_r \\ c\sigma_z \\ -\frac{1}{c^2}\frac{\partial c}{\partial r} \\ -\frac{1}{c^2}\frac{\partial c}{\partial z} \end{bmatrix}.$$

The vector equation (6) is solved using the Runge-Kutta-Fehlberg numerical method.[4]

3. The initial data setting

To trace the hydro-acoustic rays in a certain marine range are needed to be set, a series of parameters. The parameters were structured in three blocks:

- parameters that characterize the marine range:

IOP Conf. Series: Earth and Environmental Science **172** (2018) 012015 doi:10.1088/1755-1315/172/1/012015

- Horizontal range[m];
- Vertical range [m];
- Sea surface type - cases:
  - Absorbent flat surface;
  - Perfectly reflective flat surface;
- Seafloor type – cases:
  - Absorbent flat surface;
  - Perfectly reflective flat surface;

• parameters that characterize the acoustic signal source (We considered the case of a punctual source):

- Source horizontal coordinate (rs) [m];
- Source vertical coordinate (zs) [m];
- Source frequency [Hz];
- Number of traced rays;
- First launching angle [degrees];
- Last launching angle [degrees];

• parameters that characterize the sound speed profile.

4. Simulations and analyses

The Black Sea basin presents some specific characteristics like the interaction between the upper and lower layers causing a non-uniform structure of the marine environment. The temperature and salinity of the water layers in the area of the continental platform, corresponding to the Romanian shore, does not remain constant during the entire year. It is worth noting the average low salinity specifies the entire basin of the Black Sea, but with limit values in the vicinity of the mouth of the rivers (the phenomenon of dilution).

The existence of significant differences between sound speed profiles in the winter compared to that in the summer, in the same geographic location, determined as, to analyze the propagation of the hydro-acoustic waves for two specific sound speed profiles summer/winter in the same area. The two sound speed profiles were calculated using the Leroy equation. The temperature, salinity and depth values were measured with a CTD sensor.

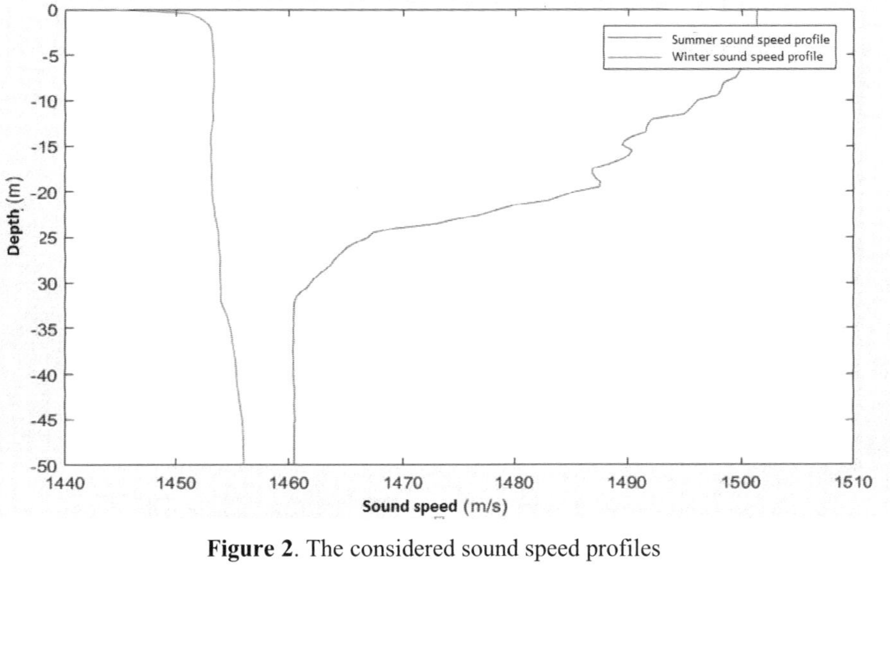

Figure 2. The considered sound speed profiles

IOP Conf. Series: Earth and Environmental Science **172** (2018) 012015 doi:10.1088/1755-1315/172/1/012015

In the following figures are presented the results of the hydro-acoustic signal propagation simulations with the software modules developed for ray tracing. We considered four positions of the source for each sound speed profile.

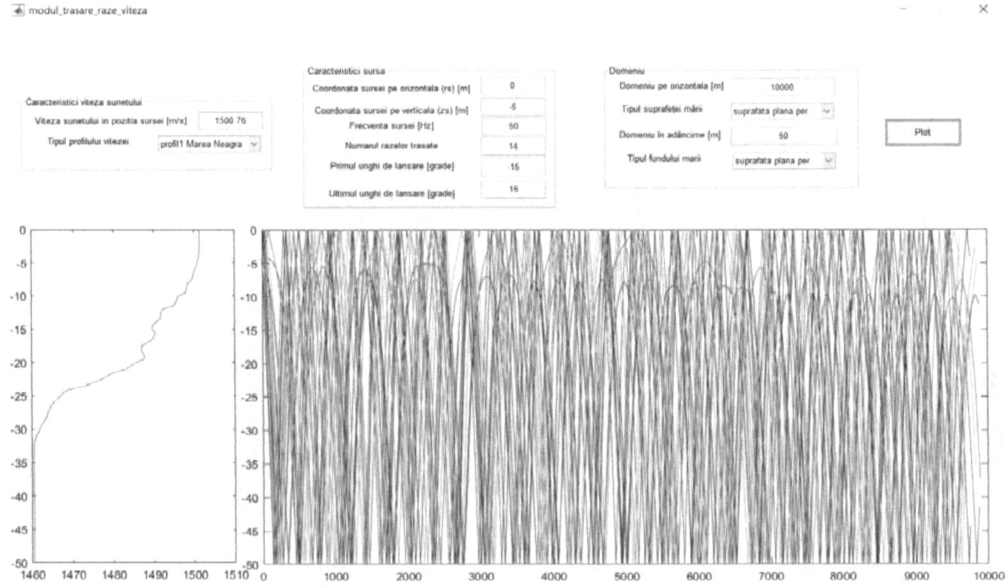

Figure 3. Sound waves propagation simulation – summer profile, source depth 5m

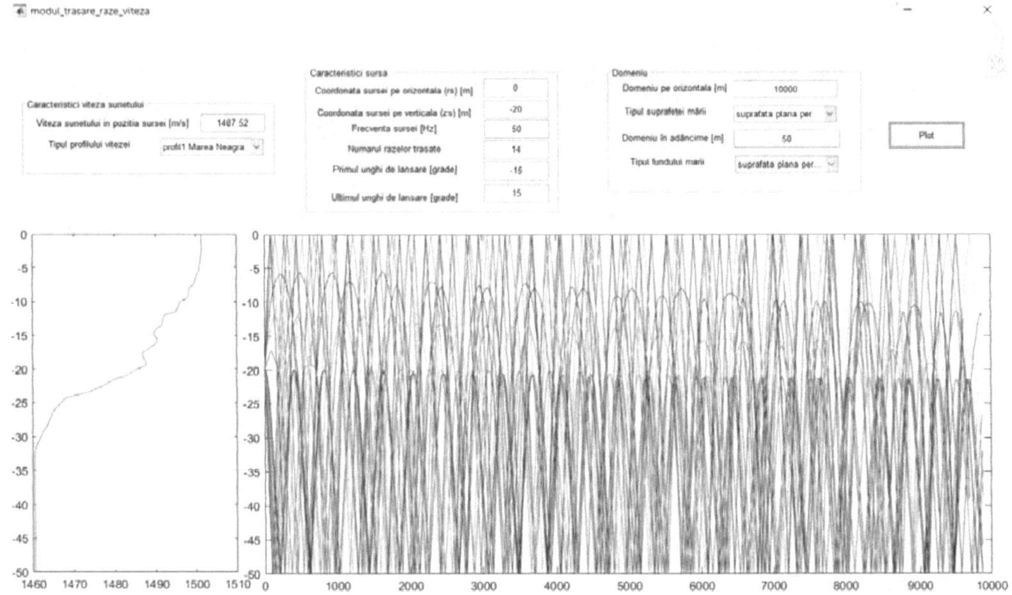

Figure 4. Sound waves propagation simulation – summer profile, source depth 20m

IOP Conf. Series: Earth and Environmental Science **172** (2018) 012015 doi:10.1088/1755-1315/172/1/012015

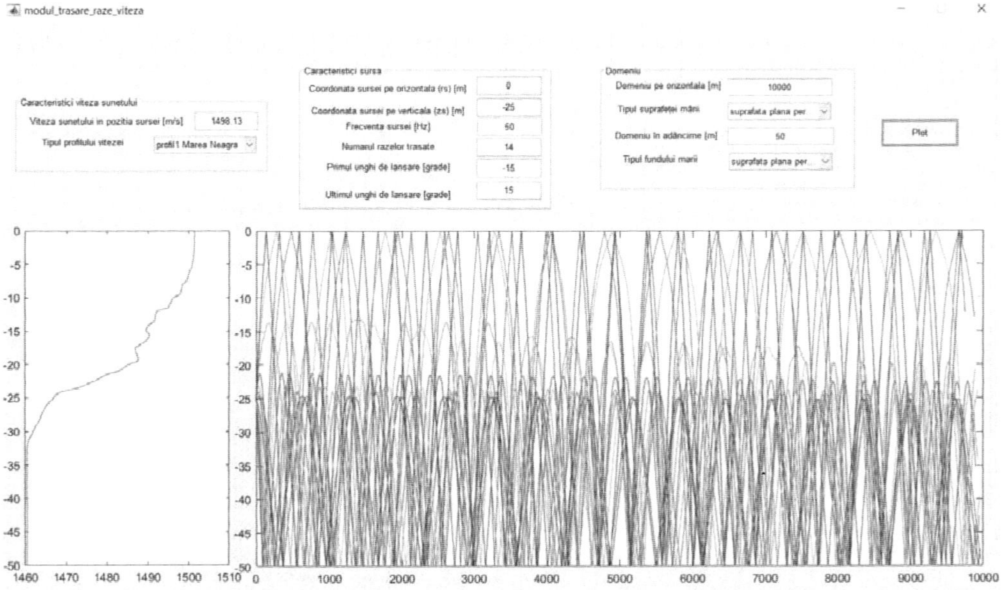

Figure 5. Sound waves propagation simulation – summer profile, source depth 25m

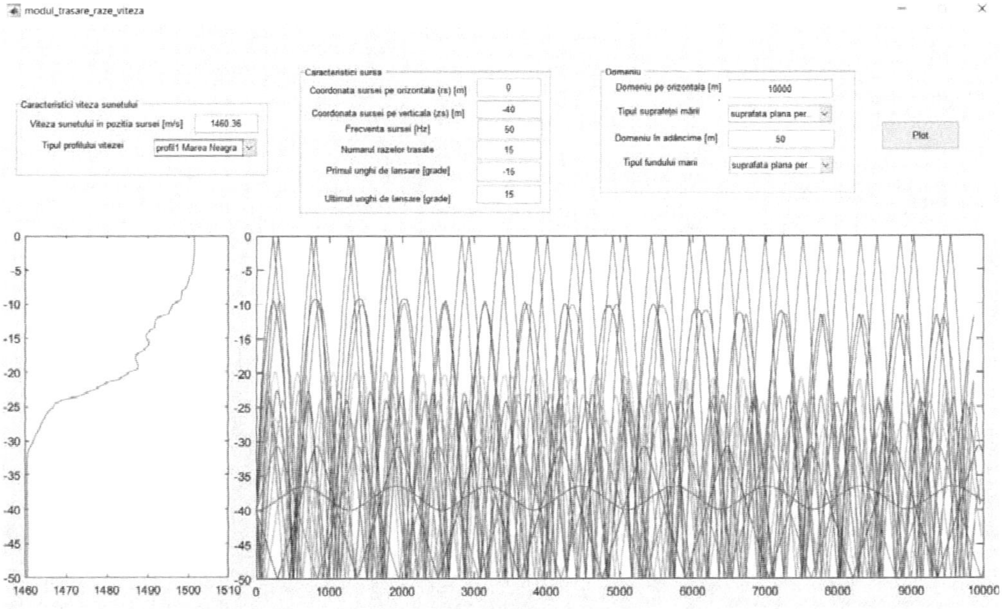

Figure 6. Sound waves propagation simulation – summer profile, source depth 40m

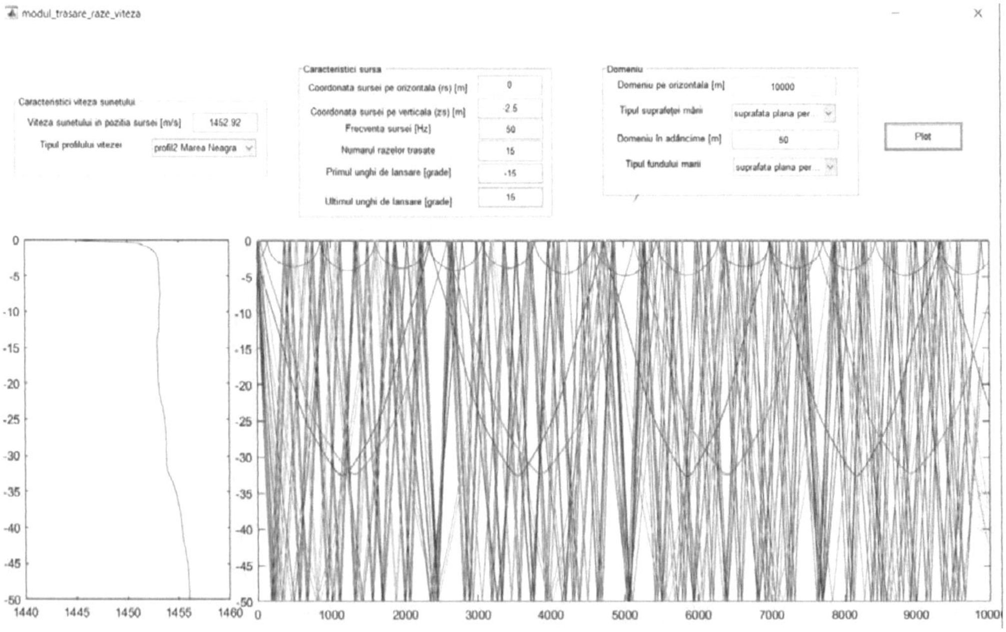

Figure 7. Sound waves propagation simulation – winter profile, source depth 2.5m

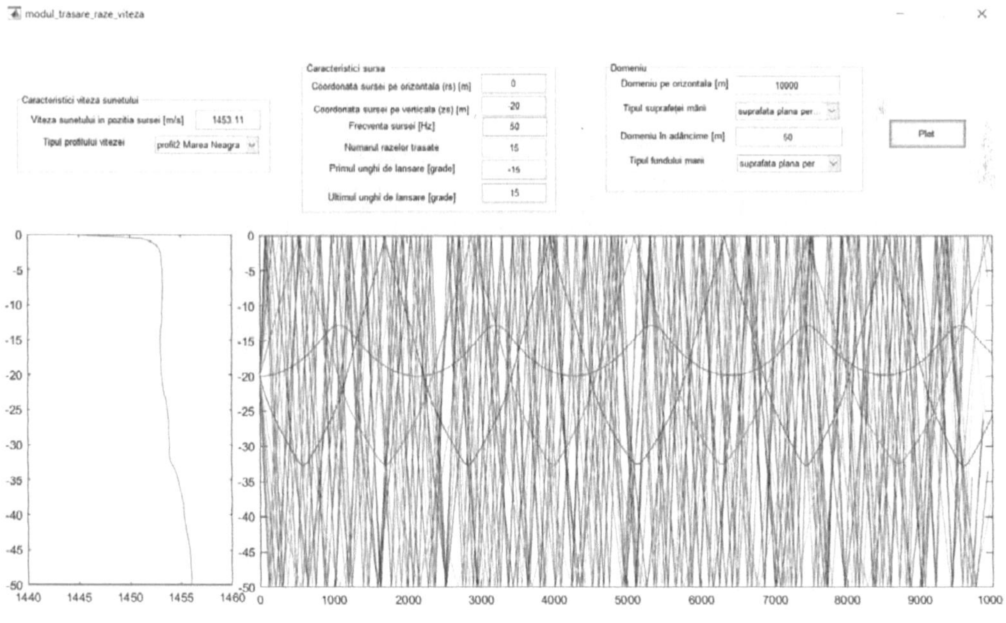

Figure 8. Sound waves propagation simulation – winter profile, source depth 20m

IOP Conf. Series: Earth and Environmental Science **172** (2018) 012015 doi:10.1088/1755-1315/172/1/012015

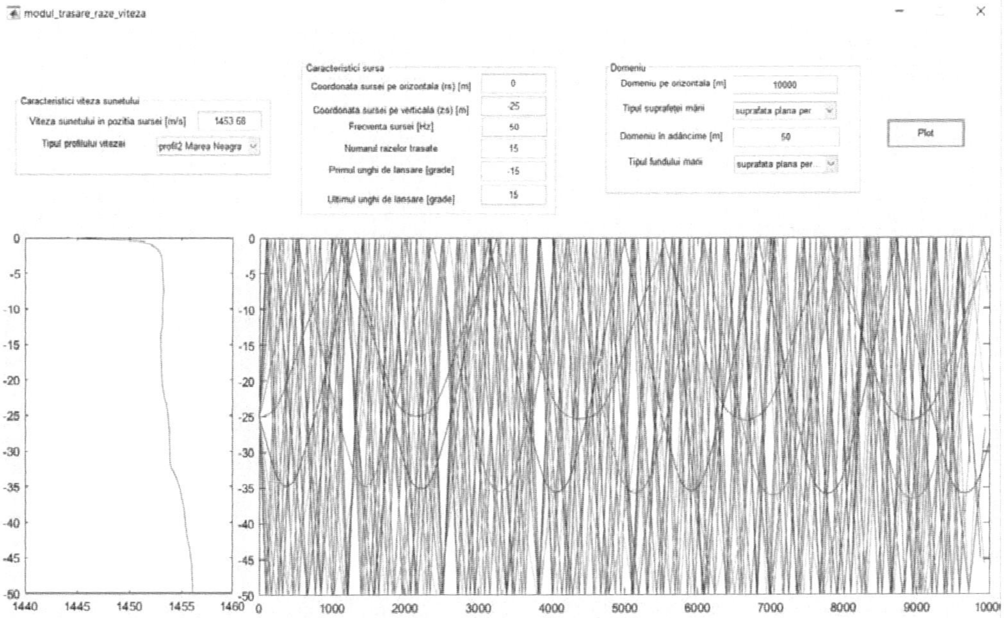

Figure 9. Sound waves propagation simulation – winter profile, source depth 25m

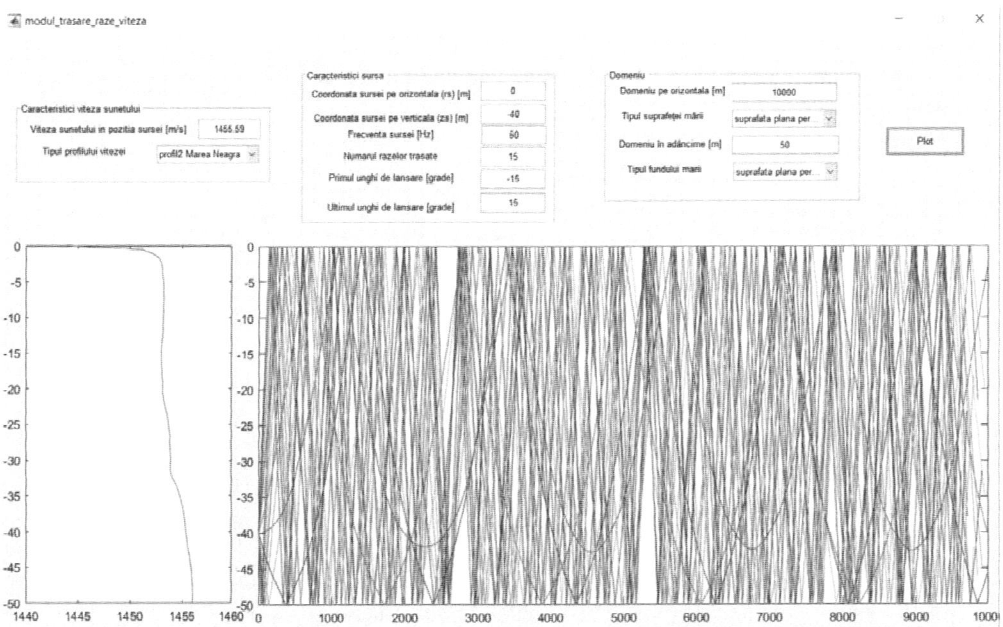

Figure 10. Sound waves propagation simulation – winter profile, source depth 40m

In both simulated situations (summer conditions – figures 3-6 , respectively winter conditions – figures 7-10) it is noted that the acoustic rays tend to concentrate in the areas where the sound speed decreases, fact that confirm the acoustic waves propagation theory. In the case of summer conditions the acoustic rays focus on deeper areas with the immersion of the acoustic source.

Can be observed that the acoustic source immersion corroborated with the sound speed profile offers a quite complete coverage of the explored area with acoustic rays and hence a minimization of areas where a possible threat can be hidden. This fact becomes more visible in the deep depths of the explored area (hundreds of meters depths).

In the developed simulation model we considered the boundaries – the sea surface and the seafloor – perfectly reflective flat surfaces and we didn't included the propagation losses determined by the signal frequency and by the traveled distance of the ray.

If these limitations will be eliminated in the next stages of the software modules development and if will be considered sound speed horizontal range dependent profiles the simulations will offer a better approximation of the sound wave propagation.

5. References

[1] Jensen F B et al 2011 *Computational Ocean Acoustics 2nd ed.*, Springer

[2] Fox D N et al 2002 *The modular Ocean Data Assimilation System* Oceanography **Vol 15 No 1**

[3] Leroy C C et al 2008 *A new equation for the accurate calculation of sound speed in all oceans* J. Acoust. Soc. Am., **124 (5)**

[4] Rodriquez O C 2011 *The Traceo ray tracing program* www.siplab.fct.ualg.pt

IOP Conf. Series: Earth and Environmental Science **172** (2018) 012016 doi:10.1088/1755-1315/172/1/012016

Considerations regarding open-source systems and equipment for UAV's

V Prisacariu[1], I Cîrciu[1] and A Luchian[2]

[1] Aviation Department, Henri Coandă Air Force Academy, 160 Mihai Viteazul street, Braşov, Romania
[2] Engineering and Industrial Management Department, Transilvania University, 5 Mihai Viteazul street, Braşov, Romania

Email: aerosavelli73@yahoo.com

Abstract: Unmanned aerial vectors have become a mature technology that requires multidisciplinary and trans-disciplinary approaches based on airline typology. The technological level implemented has proven to be independent of the size of the air vector, which takes into account the missions, costs and limitations of the system. The article wishes to summarize the information about the automatic open-source APM 2.6 pilot for the construction and connection, programming and operation of the sensors used.

Acronyms:

APM	Ardupilot Mega	PPM	Pulse Position Modulation
PCB	Printed Circuit Board	GCS	Ground Control Station
IMU	Inertial Measurement Unit	EPROM	Erasable Programmable Read-Only Memory
ADC	Analogic Digital Convertor	RC	Radio Command
ESC	Electric Speed Controler	VFR	Visual Flight Rules
VTOL	Vertical Take-off Landing	USV	Unmanned Surface Vehicles
UCAV	Unmanned Combat Aerial Vehicles	MAVLink	Micro Air Vehicle Link
GUI	graphical user interface		

1. Introduction

The equipment level of a UAS is reflected in the cost of design, manufacture, acquisition and operation. Like any technical system as a finished product, the operational UAS encompasses concepts, techniques and design and manufacturing methods and a set of requirements determined by manufacturing, operational and cost constraints, which may even be the objectives conception and design of the air system.

1.1. Brief history, evolution and classification

The first preoccupations in the field of unmanned aircraft are recorded starting with 1483 Leonardo da Vinci and 1754 by Mihail Lomonosov [1, 2], and in 1843 George Cayley conceives and tests the convertoplan carriage, see Figures 1 and 2. [3].

The first pioneering attempt to use in a military conflict is recorded in 1880 when the Austrian bombs floated above Venice and in 1916 the first use of Archibald Montgomery Low's propelled air vehicle [4, 5].

The first attempt to use the autopilot concept was due to Elmer Sperry in 1914 at the Sperry Gyroscope, the functional model was put into practice after two years of research based on proportional control being used on a plane piloted for steering and altitude, the autopilot being connected to a gyroscopic position indicator and an altitude indicator, [12, 13].

Figure 1 Convertoplan carriage, George Cayley, [3]

Figure 2 Ruston Proctor Aerial Target, [5]

1.2. Development Trends

Although the UAS technology consumer is the military domain (data acquisition and retrieval and UCAV) but in the last decade we have serious benchmarks in the civilian market regarding the use of these types of air systems in different areas of economy, tourism, environment and not the last row of education and research.

The main development trends of the UAS domain concern VTOL capabilities to increase flight autonomy and expand operating environments (hybrid systems: UAV-UGV, USV-UAV, innovative batteries), increase autonomy level and intrinsic decision of the system (use of algorithms and techniques complex programming: fuzzy, genetic, probability), see figure 3, [7, 8, 9, 10].

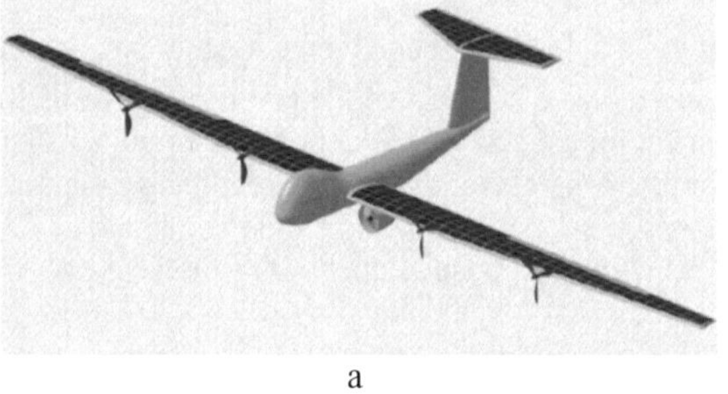

a

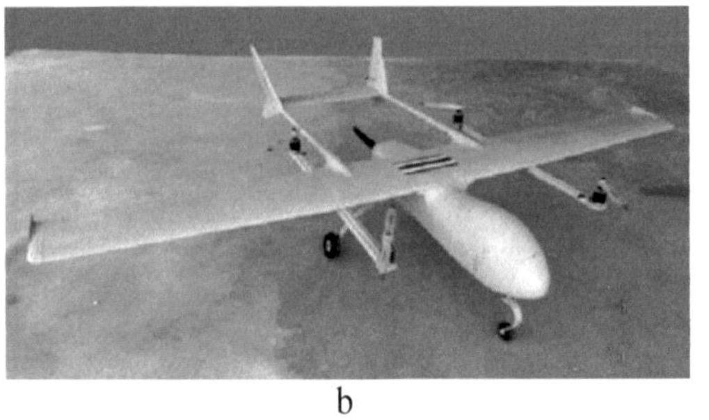

b

Figure 3 Hybrids UAS, a.turbofan/solar hybrid UAV, [7]; b. Falcon-V VTOL, [8].

An unmanned airplane system with a high level of decision-making involves high-performance radio-electronic equipment, equipment that has cost-of-function and reliability in operation, a focus on open-source equipment that offers upgrades and almost limitless programming, a convincing example is provided by ATMega microprocessor development boards that led to the introduction of the autopilot products from the APM and Pixhawk family on the UAV market.

APM 2.6 is open source electronic equipment that allows the transformation of a UAV into a fully autonomous unmanned aerial vector, the system is based on the hardware and software components as follows:

2.APM 2.6 hardware

APM 2.6 is composed of two PCBs on which the pilot components are mounted. The plaque containing the automatic pilot is called ArduPilot Mega, and the board on which the sensors are positioned is called the IMU Shield. Together, these two plates are the brains of the system, the IMU plate compiles and transmits the sensor information, and the ArduPilot Mega plate analyzes this information and acts in accordance with the way the automatic pilot was programmed. The two boards have analog and digital inputs for different sensors and modules that can be connected.

IOP Conf. Series: Earth and Environmental Science **172** (2018) 012016 doi:10.1088/1755-1315/172/1/012016

2.1.ArduPilot Mega

This electronic board deals with routing the UAV, it receives information from the sensors that are available, analyzes the data and acts as planned. From it, all commands that trigger the 3D space of the drones leave. The UAV's spatial motion is accomplished by varying the engine speed (multicopter) or the position of the control surfaces (airplanes). The main components and their layout on this PCB can be seen in Figure 4: 1-pin input, 2-pin PPM, 3-button RESET, 4/6/9-pin IMU, 5-port ATMega 1280, 7-way automatic, 8-pin GPS connection, 10-multiplexer, 11-pin power supply, 12-pin I/O data.

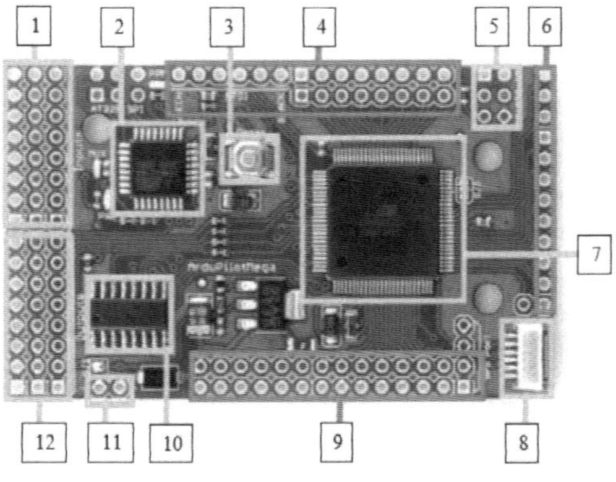

Figure 4 ArduPilot Mega, [11].

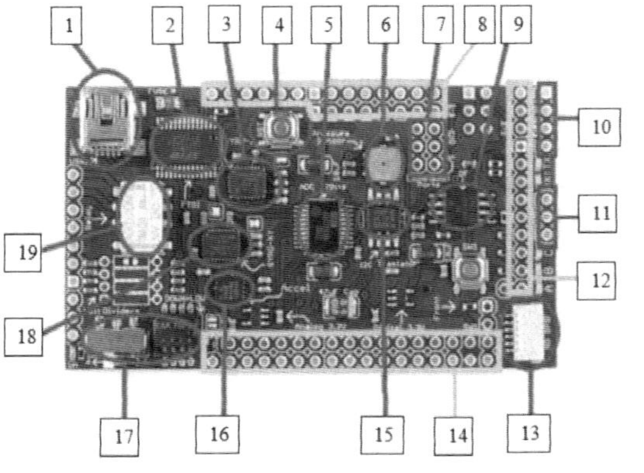

Figure 5 IMU Shield

2.2.IMU Shield

The basic component of the IMU Shield is the PCB board (see the layout of the components in figure 5), the following components are mounted on the board: 1-port USB with 2-chip USB support, Z-axis 3-gyroscope, 4-button RESET, analog-digital converter, 6-pressure sensor, 7-port analog sensors, 8/12/14-pin ArduPilot Mega I/O data, 9-megapixel flash memory, 10-pin Pitot tube connection - 13-port connection external magnetometer, 15-magnetometer, 16-accelerator with three axes, 17-switches, 18-gyroscopes X, Y, 19-trigger axes.

The integrated commercial version of APM 2.6 has a unitary architecture and a protective case as shown in Figure 6.

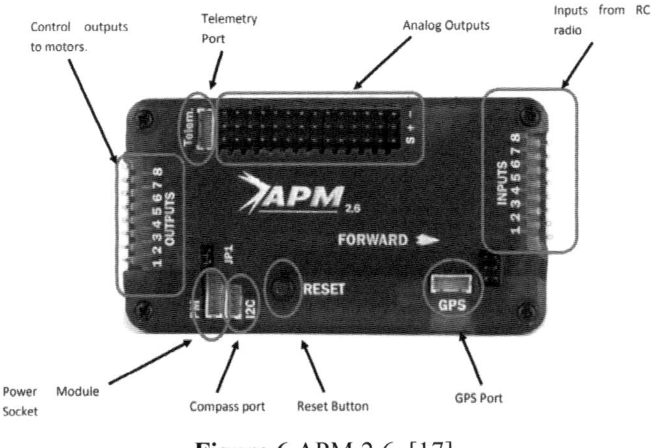

Figure 6 APM 2.6, [17]

3. APM 2.6 software

The software resources used to operate the unmanned air vehicle are divided into programming resources and firmware resources.

IOP Conf. Series: Earth and Environmental Science **172** (2018) 012016 doi:10.1088/1755-1315/172/1/012016

3.1. Programming resources

They consisted of various software components that control hardware components via firmware. The program used in APM 2.6 is the Arduino IDE, [14] needed to update the firmware with new features or to create a new firmware, Arduino uses a C ++ like programming language.

Example of integration of resources in APM architecture. The GPS module.

Putting into operations of the GPS module on the APM 2.6 automatic pilot assembly involves both connecting the GPS receiver connector to the port on the ArduPilot Mega board (Figure 7) and writing the programming code.

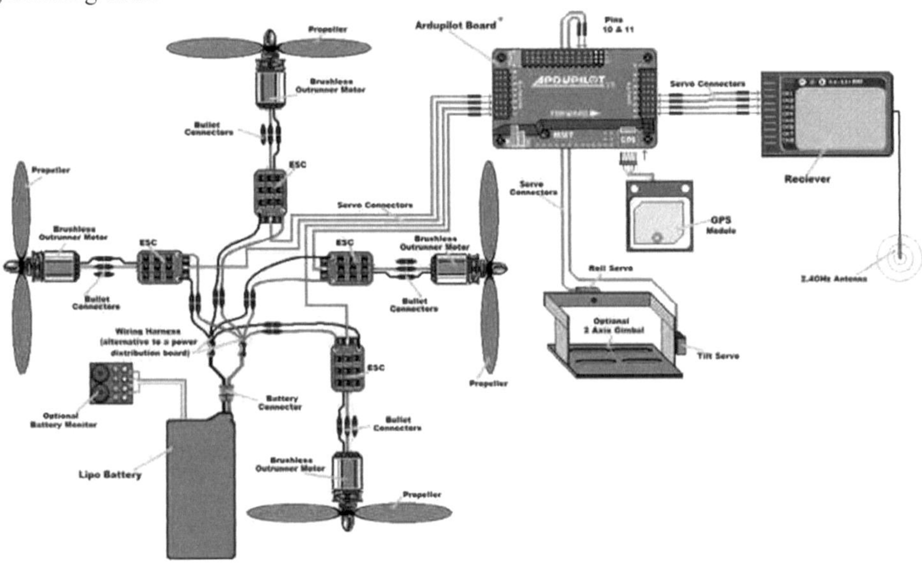

Figure 7 GPS connection in the multicopter architecture, [15]

The "#include" operator inserts the firmware code so that the automatic pilot recognizes the information received from the GPS (Figure 8a). "RxGPS 3" and "TxGPS 5" designate pins 3 and 5 for receiving and transmitting data from and to the GPS receiver. The data transmission rate is 4800 bps. "Void loop" provides the read cycle for receiver location, and "Serial.println (s);" allows the display of coordinates in the "Serial Monitor" window (Figure 8b).

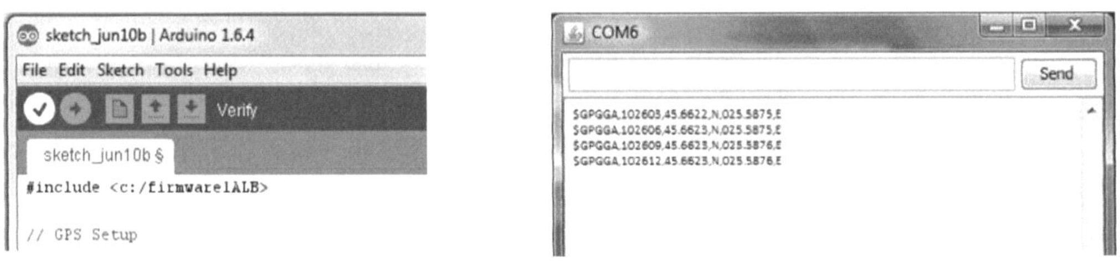

a b

Figure 8 Integration of GPS module into APM 2.6, [14]

The "Serial Monitor" window opens from the "Tools" menu of the Arduino program. Before the GPS firmware is loaded, press the "Verify" button to check for errors in the code, and then click the "Upload" button.

To check if the GPS interacts with the autopilot, it should be noticed if the green light is on and the blue light is blinking. The green LED alerts the user that the GPS receiver is connected to the

ArduPilot 2.6 board and receives electricity, the flashing blue LED cautions the user that the GPS receiver has located its position. The source code is highlighted below:

```
#define rxGPS 3
#define txGPS 5
GTPA010 serialGPS = GTPA010 (rxGPS, txGPS);
String stringGPS = "";
void setup() {
pinMode(rxGPS, INPUT);
pinMode(txGPS, OUTPUT);
Serial.begin(9600);
Serial.println("Started");
serialGPS.begin(4800);
digitalWrite(txGPS,HIGH);
while(serialGPS.available())
if (serialGPS.read() == '\r')
break; }
void loop() {   String s = checkGPS();
if(s && s.substring(0, 6) == "$GPGGA")
{    Serial.println(s);   }}
```

3.2. Firmware resources

The firmware is a set of basic instructions written on the non-volatile memory chip. These instructions enable a small number of input/output (I/O) hardware components, but more importantly, it makes possible communication between hardware and software that control the hardware. Depending on the type of memory used, PROM or EPROM, the firmware can be deleted and rewritten to add new features.

In case of APM 2.6, the firmware resource instructs the pilot automatically how to recognize input data from different sensors and how to use them. For example, input data from gyros are used to stabilize the three axes of the drones. The course of these data is as follows: the electronic gyroscopes on the IMU board measure the attitude difference from the reference axes and continuously send signals to the ADC signal converter, it takes the signals and converts them into binary data by sending them further to the automatic pilot on ArduPilot Mega electronic board.

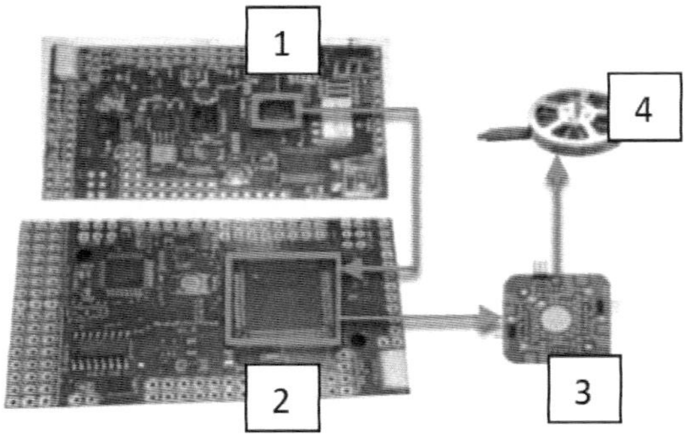

Figure 9 Sensor-microcontroller data connections and execution element:
1-gyroscope, 2- automatic pilot, 3-board distribution current, 4-electric motor.

It analyzes the information and sends pulses to the power distribution board from where the energy is distributed, according to the instructions of the autopilot, to the electric motors. A schematic of this example can be seen in Figure 9.

APM Plane, APM Copter, APM Antenna Tracker, Mission Planner, APM Planner, is the APM planner resources for APM applications, [16].

Mission Planner

Mission Planer as APM Planner is a software tool installed on the ground control station (GCS) for the APM 2.x and Pixhawk series for low cost open-source systems. Mission Planer allows: updating the automatic pilot firmware (Figure 10a); live reading of telemetry data and reading offline sensor data (Figure 10b); setting up and adjusting the autopilot system and planning a flight mission; connecting with a flight simulator; downloading log files and analyzing them, [17, 18].

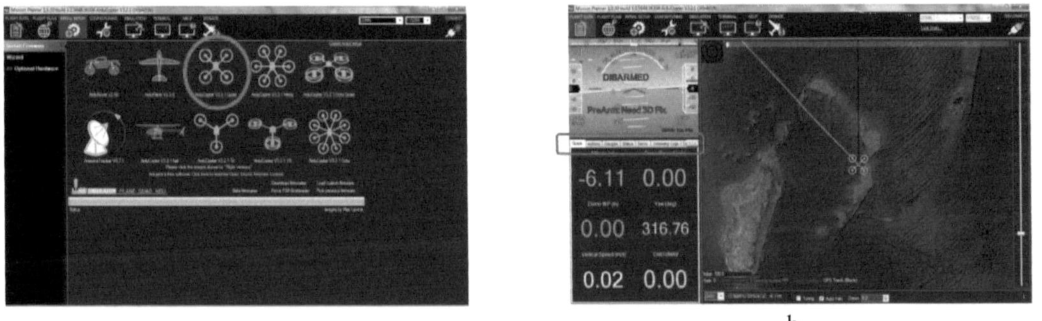

a b

Figure 10 Mission Planner GUI, a. Firmware choice, b. Live telemetry data read

Reading telemetry data allows you to view some operating parameters such as: UAV (pitch, yaw, roll) and speed data on the three axes; data about GPS parameters (position, speed, altitude, compass head), see Figure 11; mission type data; data on navigation and servo output signals (see Figure 12); data about radio signals.

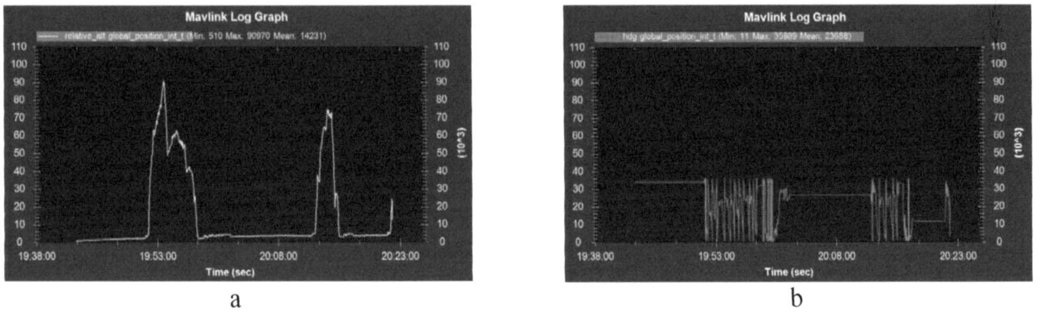

a b

Figure 11 GPS flight parameters, flight altitude, b. compass

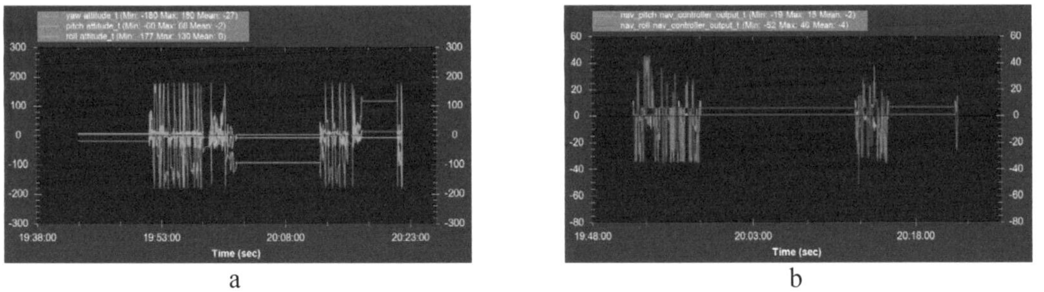

a b

Figure 12 GPS flight parameters, a.pitch-yaw-roll, b. output signals

Flight data can be viewed and analyzed from dataflash logs to diagnose mechanical defects, vibrations, magnetic interference, or engine operation, these flaws occurring in the form of pitch, roll, or girth leaps. Vibrations with significant values can be read by displaying data provided by accelerometers (AccX, AccY, AccZ values) of IMU messages for horizontal position control. Magnetic interferences that may be generated by the power distribution board, motors, accumulators, ESCs can cause malfunction of the compass that causes unwanted UAVs, these interferences can be viewed at CUSTOM / VFR_HUD.

The Flight Data View interface also allows data export in *.txt, *.csv, and Matlab files for post-processing and numerical analysis.

Conclusions

Unmanned aerial vectors have become a mature technology that requires multidisciplinary and transdisciplinary approaches based on airline typology. The technological level implemented has proven to be independent of the size of the air vector, a level that takes into account the missions, costs and limitations of the system.

Opensource systems and equipment such as Ardupilot Mega (APM) opened new opportunities in the UAV market and the nonprofit segment with significant effects on operating and mission concepts such as data acquisition in areas of interest in completely autonomous platforms.

APM has opensource features and a simple process to load and install the firmware, programming knowledge is only needed if new lines of code are created using Arduino. APM is capable of both bidirectional telemetry with the MAVLink protocol and possibilities for setting ground and air parameters for mission planning and re-planning, on-board video display and data storage. APM is designed and designed for easy integration of most types of UAVs, unmanned land-based vehicles (UGVs), unmanned surface / underwater vehicles (USV/UUV) and ground station control.

Acknowledgment

The National Authority for Scientific Research, Romania supported this work – CNCS-UEFISCDI: with PN-III-P2-2.1-PED-20161972, MAPIAM project, contract 65PED/2017.

References

[1] Valavanis K.P., *Advances in Unmanned Aerial Vehicles,* USA, 2007, ISBN 978-1-4020-6113-4, www.springer.com

[2] Dalamagkidis K et al., *On Integrating Unmanned Aircraft Systems into the National Airspace System*, Intelligent Systems, Control and Automation: Science and Engineering 54, DOI 10.1007/978-94-007-2479-2 2, Springer

[3] http://www.aviastar.org/helicopters_eng/cayley.php, accessed at 26.02.2018

[4] Barnhart R.K., Hottman S.B., Marshall D.M., Shappee E., *Introduction to unmanned aircraft systems*, CRC Press, 2012, ISBN 978-1-4398-3520-3, 215p.

[5] https://skyward.io/marilyn-monroe-and-the-origin-of-drone. accessed at 28.02.2018

[6] Prisacariu V., Boşcoianu M., Luchian A., *Innovative solutions and UAS limits*, REVIEW OF THE AIR FORCE ACADEMY, 2(26)/2014, Braşov, Romania, ISSN 1842-9238; e-ISSN 2069-4733, p51-58.

[7] Erdinç Mermer, Serkan Özgen, *Conceptual design of a hybrid (turbofan/solar) powered HALE UAV*, 7th European Conference for Aeronautics and Space Sciences (EUCASS), doi: 10.13009/EUCASS2017-200, p19.

[8] *Flacon-V VTOL UAV*, disponibil la http://www.top-enggroup.com/product.htm, accessed at 14.03.2018

[9] Daeil Jo, Yongjin Kwon*, *Analysis of VTOL UAV Propellant Technology*, Journal of Computer and Communications, 5/2017, p. 76-82, ISSN Online: 2327-5227, https://doi.org/10.4236/jcc.2017.57008

[10] Office of the Assistant Secretary of Defense for Research and Engineering, Rapid Reaction Technology Office, *Lighter-than-air vehicles*, 90p, available at http://www.defenseinnovationmarketplace.mil/resources/Final_LTA_report.pdf, 2012

[11] http://newlab.westernwillow.com/component/ardupilot-mega-2560-oilpan-h, accesed at 12.02.2018

[12] S. Bennett, *Nicolas Minorsky and the Automatic Steering of Ships*. 1984, p10-15, available at http://ieeecss.org/CSM/library/1984/nov1984/w10-15.pdf

[13] Ingrid Hagen Johansen, *Autopilot Design for Unmanned Aerial Vehicles,* Norvegian University of Science and Technology, 2012, p.108, available at https://daim.idi.ntnu.no/masteroppgaver/007/7001/masteroppgave.pdf
[14] https://www.arduino.cc/en/Main/Software, accessed at02.03.2018
[15] http://ardupilot.org/copter/docs/connecting-the-apm2.html, accessed at05.03.2018
[16] http://firmware.ardupilot.org/, accessed at 05.03.2018
[17] Oborne M., Samantha Nelson, *Practical Work II Introduction to Mission Planner and APM Autopilot*, 24p., available at www.comp.nus.edu.sg/~ctank/uav/Practical%20Work%20II.pdf, accessed at 08.03.2018
[18] Deploying ardupilot, 29p., available at www.bandung-aeromodeling.com/tutorials/APM-TS40.pdf, accessed at 08.03.2018

IOP Conf. Series: Earth and Environmental Science **172** (2018) 012017 doi:10.1088/1755-1315/172/1/012017

Modelling simulation of the thermal field for the heat transfer in environments with discontinuities

B Dumitru, I Severin and G M Dumitru

Department of Materials Technology and Welding, Faculty of Engineering and Management of Technological Systems, University "Politehnica" of Bucharest, Splaiul Independentei, no. 313, RO-060042, Bucharest, Romania

E-mail: bogdan_dmt@yahoo.com

Abstract. The paper proposes a simulation model made using the method of finite element analysis for the case of thermic transfer by infrared thermographic control with impulses for specific sample parts with induced defects. Results showed a correlation between the finite element analysis and the images obtained by infrared thermography with a maximal error of 5-6%. This correlation proves the viability of the model and could represent the basis for the developing of a working instrument for non-destructive testing of parts that are conditioned or reconditioned by thermal spraying technology.

1. Introduction

The conditioning and reconditioning process by thermal spraying is complex and involve a multitude of parameters [1, 2]. Any deviation of these parameters can cause the failure of the process, that is non-conform properties for the layers deposed by thermal spraying. The test samples can at most attest the validity of the proposed thermal spraying procedure, but they cannot ascertain that the parts conditioned or reconditioned by thermal spraying are on the same level, non-conformity risks being rather significant.

The analogies of type model – reality are important instruments for the study of different phenomena taking place in environments with discontinuities [3, 4]. The present paper proposes the finite element analysis of the thermic transfer in case of the infrared thermographic control with impulses for test samples covered by thermal spraying technologies, samples that have induced shape and dimension defects. The finite element analysis of the test samples or of the real parts, correlated with the images obtained by infrared thermographic control with impulses has as objective the proposal of a working instrument for the non-destructive control of the parts conditioned or reconditioned by thermal spraying.

2. Modelling with the method of finite element analysis

Finite element analysis of the thermic transfer phenomena is a complex process containing a geometric model, a numerical model with finite elements and the software of software packages used to solve the problem. In order to simulate the thermic field for the heat transfer in environments with discontinuities, we have chosen ANSYS 14.5, software allowing the usage of complex mathematical models with multiple analysis possibilities.

The model proposed for the finite element analysis of the thermic field in case of heat transfer in environments with discontinuities uses the following calculus hypothesis [5]:

- Material hypothesis;

- The dependence of the thermo-physical properties of the temperature;
- The existence of thermic exchange with the environment by radiation;
- The existence of thermic exchange with the environment by convection;
- Perfect contact between the base material and the deposed layer.

The geometric model used by the finite elements analysis of the heat transfer in environments with discontinuities is presented in Figure 1 and corresponds to test samples P1, P2 and C1, with defects induced artificially for the purpose of the study. The geometric configuration of the used model is presented in Figure 1 with the appropriate mashing volumes in Figure 2.

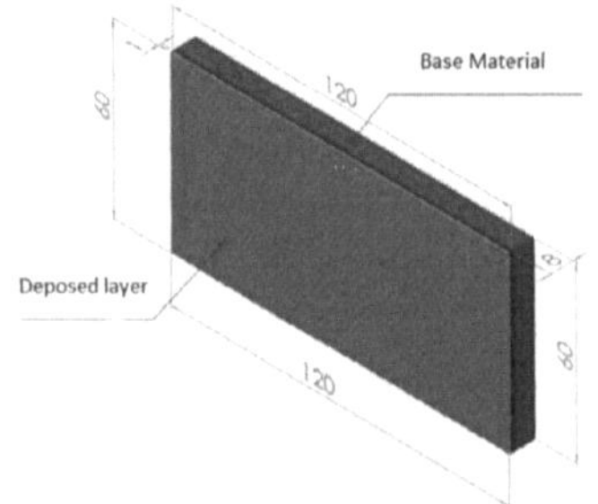

(a) Rectangular samples

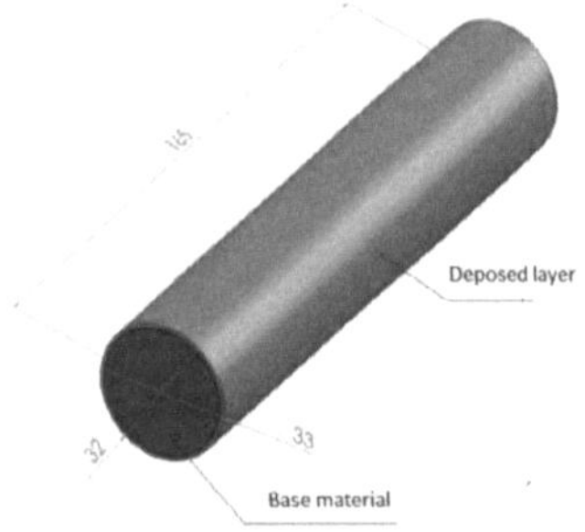

(b) Cylindrical samples

Figure 1. The geometric configuration of the model

The analysis with finite elements of the thermic transfer in environments with discontinuities used the software module *Transient Thermal* of ANSYS software, the main mashing element being the element SOLID227. To simulate the loses by radiation during the heat transfer process was used the overlaying on the surfaces that emit radiation of elements of type SURF152. The superficial emission coefficient is considered 0.5 and the constant value Stefan-Boltzmann is 5,667 E-8.

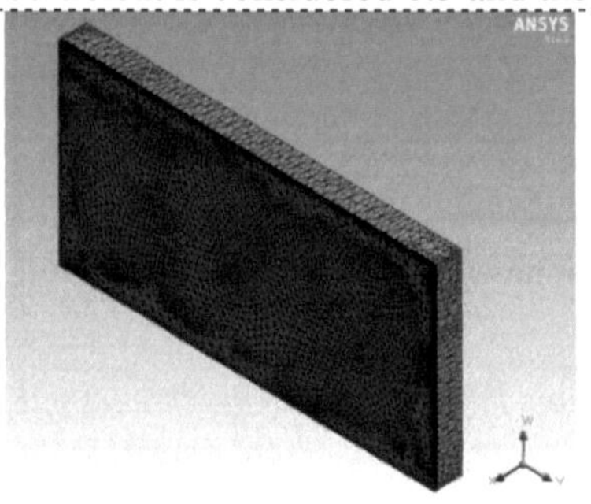

(a) P1 isometric view

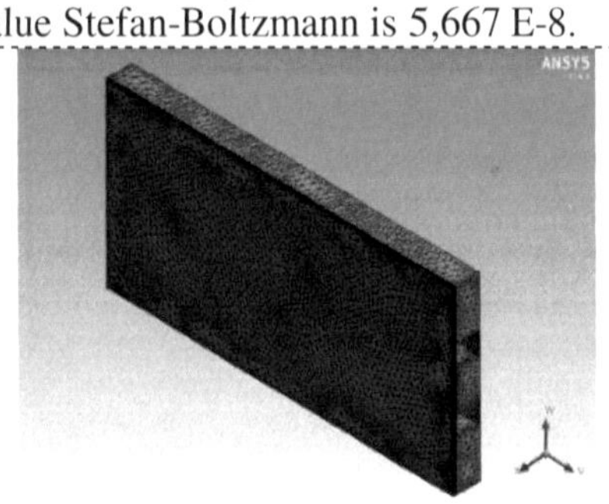

(b) P2 isometric view

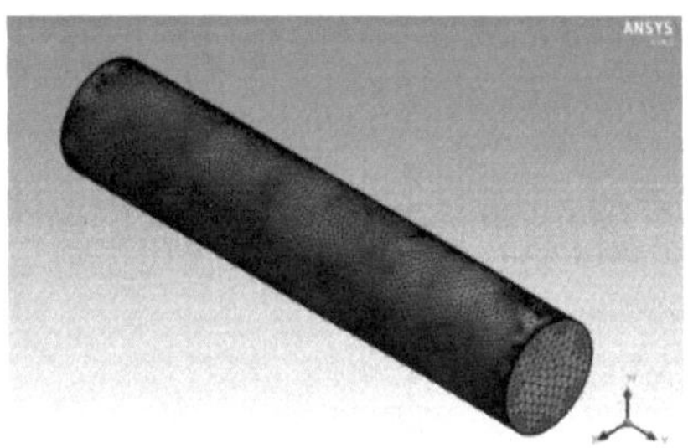

(c) P3 isometric view

Figure 2. Mashing volumes

IOP Conf. Series: Earth and Environmental Science **172** (2018) 012017 doi:10.1088/1755-1315/172/1/012017

The thermal analysis in transient regime is based on the particularity that the values of the temperature and / or thermic flow loadings vary over time, this variation being given by an equation or a loading curve with finite number of steps [6].

Table 1 presents the material and loading parameters used in the analysis process.

Table 1. Material and loading parameters used in the analysis

Material properties	Thermic conductivity for deposed layer	45 W/m· ^{0}C
	Thermic conductivity for base material (test samples P1, P2)	60.5 W/m· ^{0}C
	Thermic conductivity for base material (test sample C1)	55.5 W/m· ^{0}C
Loads	Emissivity within environment	0.50
	Thermic flow	300 W/cm^2
	Temperature of the environment	22 ^{0}C
Stefan Boltzmann constant		5.667 E-8 Wm^{-2}K^{-4}

Figure 3 lists the results obtained by the analysis of sample P1, having as induced defects holes of different diameters filled with epoxy resin. The thermic field was recorded at 1 second intervals. The maximum temperature observed during the thermographic process was 92.55^{0}C. As time passed was observed that the extremities start to participate at the heat exchange with the environment by radiation.

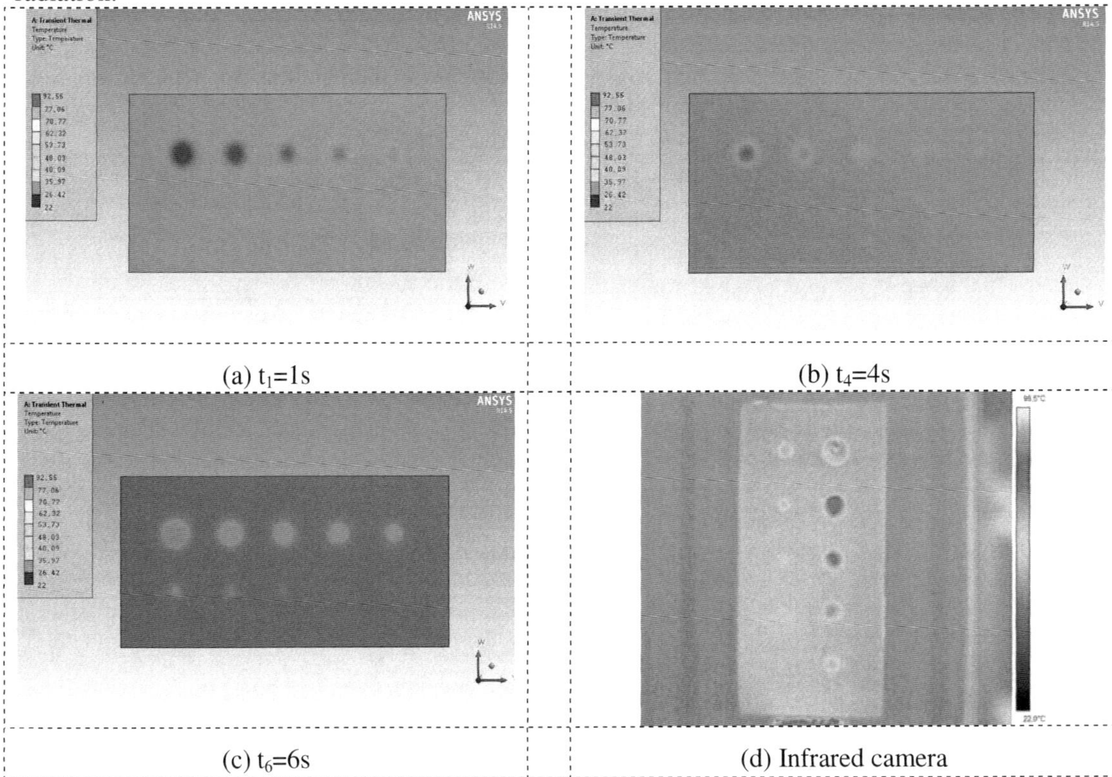

(a) t_1=1s

(b) t_4=4s

(c) t_6=6s

(d) Infrared camera

Figure 3. Temperature distribution for sample P1, during the time frame 1÷6s

The model of the finite element analysis presented in figure 3.a) is almost identical with the thermografic image presented in figure 3.d). The median values of the temperatures are 42.44^{0}C for the infrared thermographic examination with impulses and 42.01^{0}C for the finite element analysis, which leads to a model error of 1.01%. The temperature evolutions for sample P1 presented in Figure

IOP Conf. Series: Earth and Environmental Science **172** (2018) 012017 doi:10.1088/1755-1315/172/1/012017

4 show that the model chosen follows closely the real sample behavior, thus the model being considered adequate for the situation.

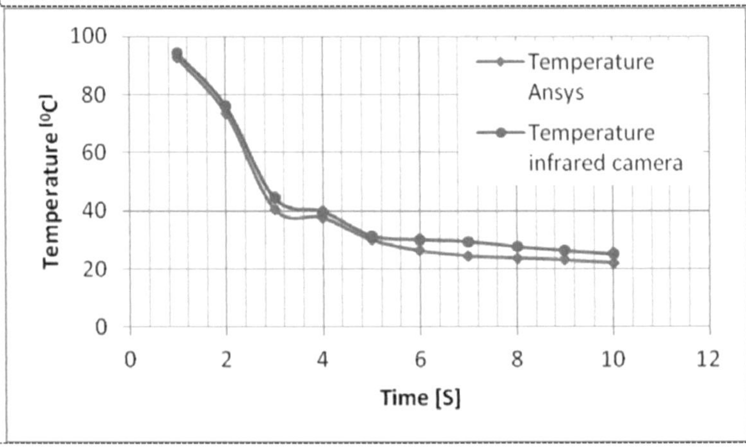

Figure 4. Temperature evolution on the surface of the sample P1

Figure 5 presents the results for sample P2, in form of isotherms of the transient thermic field, in frontal view. The sample P2 has induced as defects two channels that are filled with epoxy resin, one with constant depth, and one with variable depth. The thermic field was recorded at 1 second intervals. The maximum temperature reached during the thermography was 95.19^{0}C. The heat exchange with the environment by radiation is made on the posterior side and through the lateral sides. This can be observed graphically as time is elapsing, by gradual heating of the base materials and the cooling of the deposed layers [6].

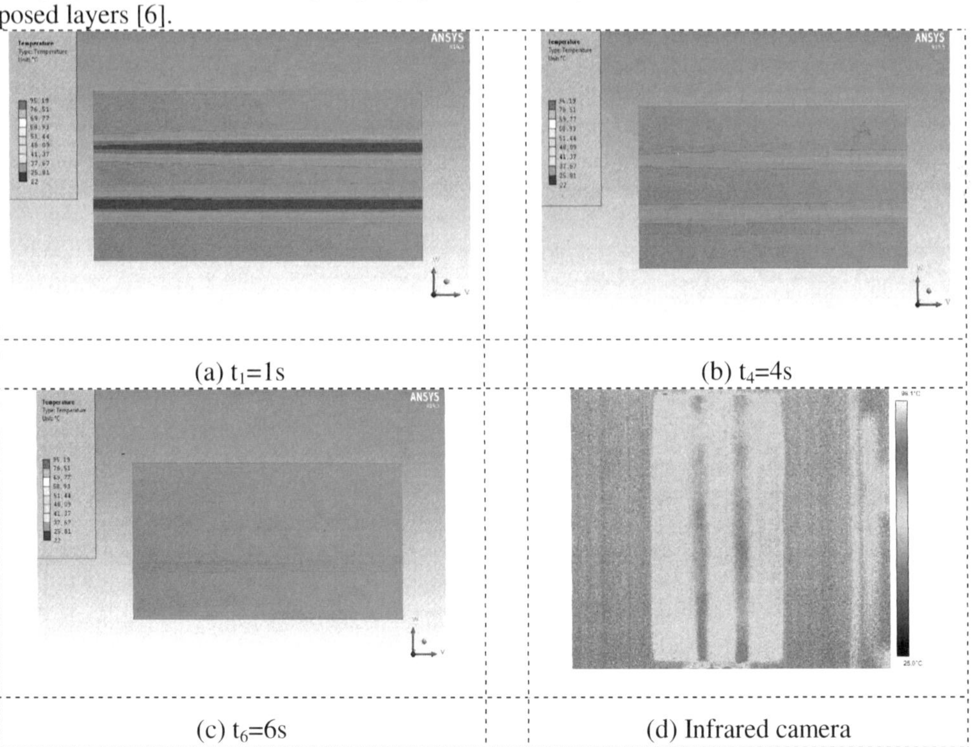

(a) t_1=1s (b) t_4=4s

(c) t_6=6s (d) Infrared camera

Figure 5. Temperature distribution for sample P2, during the time frame 1÷6s

A minimum difference between the results obtained by thermography and the finite element analysis was observed. The average temperature value was 53.06^{0}C for the examination with infrared camera and 49.81^{0}C for the finite element analysis. Thus, the error of the finite element, for the medium temperature, is 6.12%.

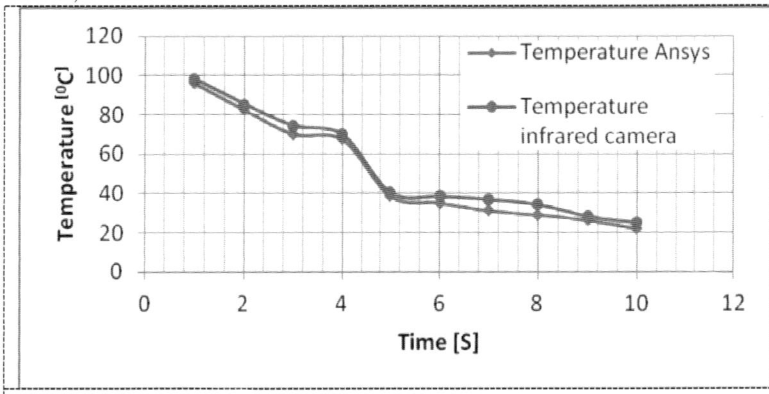

Figure 6. Temperature evolution on the surface of the sample P2

Figure 7 lists results for sample C1, as isotherms of the transient thermic field, in isometric view. The thermic field was recorded at 1 second intervals. The maximum temperature observed during thermographic control process was 60.21 ^{0}C.

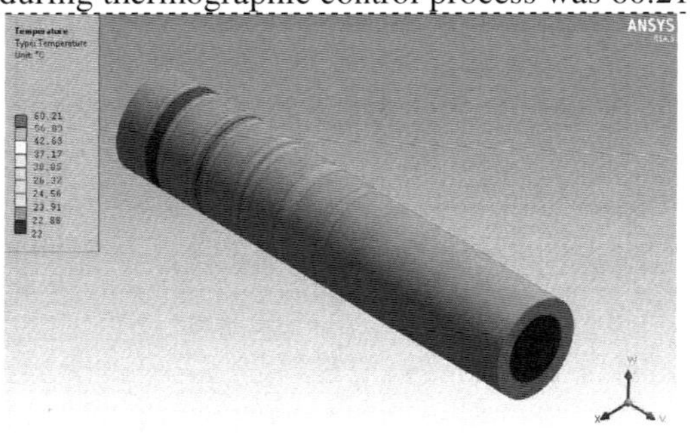

(a) t_1=1s

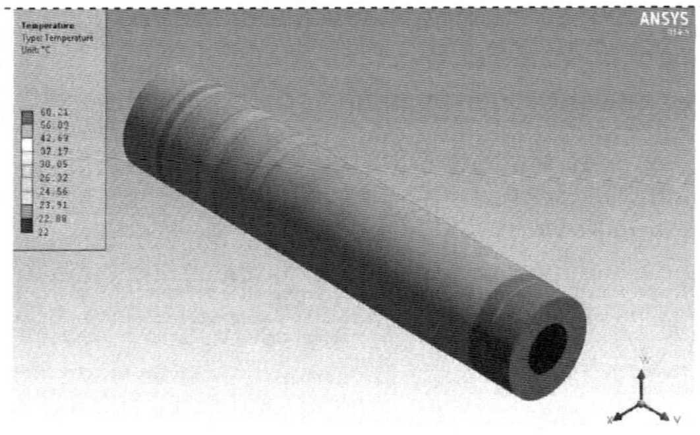

(b) t_4=4s

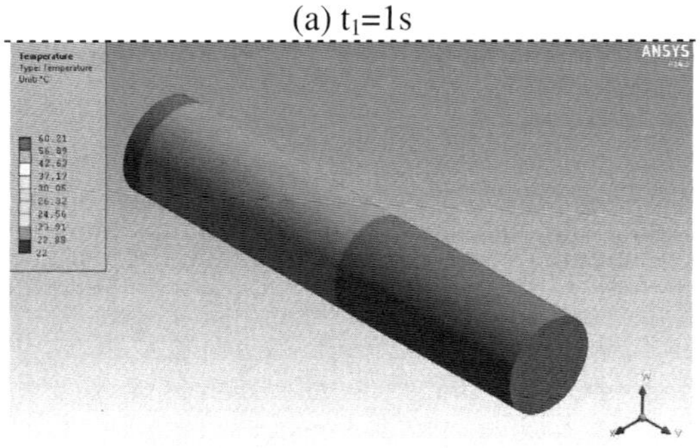

(c) t_6=6s

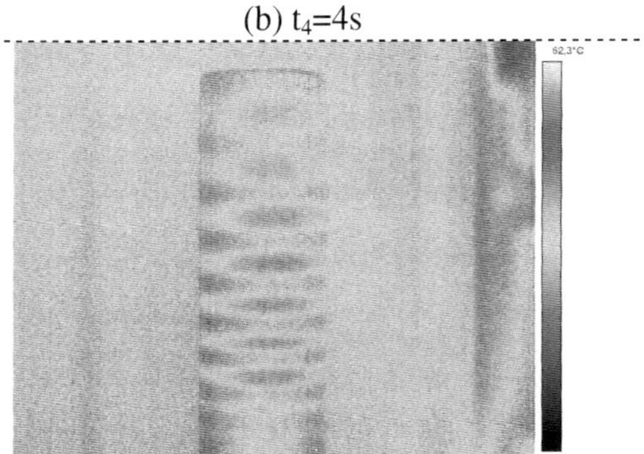

(d) Infrared camera

Figure 7. Temperature distribution for sample C1, during the time frame 1÷6s

It was observed a minimum difference between the results obtained by thermography and the finite element analysis. The medium value of the temperature is 36.38 ^{0}C for the examination with infrared camera and 34.20 ^{0}C for the finite element analysis. Thus, the error of the finite element, for the medium temperature, is 5.99%.

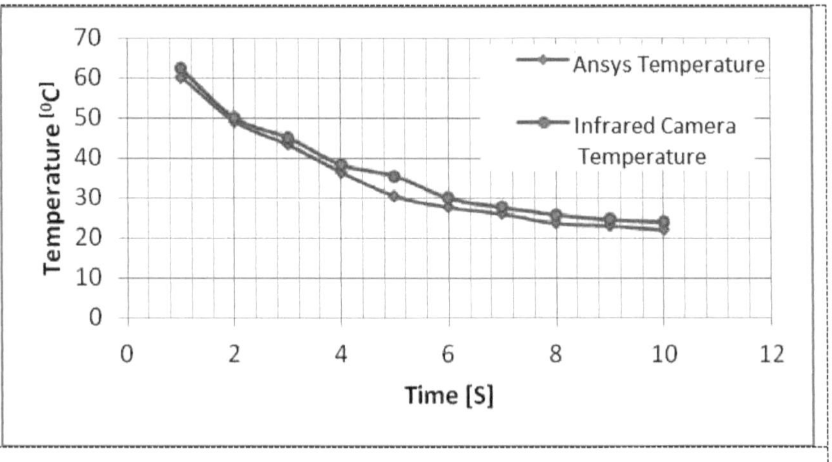

Figure 8. Temperature evolution on the surface of the sample C1

3. Conclusions

The finite elements analysis of the thermic transfer process for the infrared thermographic control process with impulses emphases for the studied cases the following conclusions:

- The righteousness of the model with reference to the circular defects of different shapes and dimensions on the surface of the base material or on the interface deposed layer by thermal spraying – base material;
- The finite element model fidelity is acceptable (maximum 5-6% error);
- In case of defects with dimensions comparable to the part itself, the heat transmission is tridimensional, which makes the images for defects that are close to be slightly blurred;
- In order to obtain a better fidelity with cylindrical samples, the usage of a uniform circumferential heating system is recommended.

4. References

[1] Pawlowski L, 2008 *The science and engineering of thermal spray coatings, Ed.2* (New York J. Wiley & Sons)

[2] Dumitru G M, Constantin R, Dumitru B and Mihăilescu A, 2013 *Reconditioning and repairing of products* (Bucharest - Printech Editure)

[3] Mihai A, 2009 *Nondestructive Examinations* (Bucharest - Printech Editure)

[4] Ghelmez M, Lazar B, Toma C, Dumitru B and Slavnicu E, 2006 Bulk and Surface Ultrasonic Waves Generated in Metallic Solids by Pulse Shaped Sources, *Rev.Chim.* **57** P722

[5] Nitoi D, Dumitru G M, Dumitru B, Mihailescu A and Cismas S C, Reconditioning by Metal Deposition Technology - Temperature Finite Element Modeling, 2014 *Advanced Materials Research* **09 Vol. 1029**, P130-133

[6] Botez S C, Dina V C, Dumitru G M, Dumitru B, Iacobescu G, 2014 Characterization of Metallic Layers Deposited through Thermal Spraying, *UPB Sci. Bulletin*, **Series D, Vol. 76**

IOP Conf. Series: Earth and Environmental Science **172** (2018) 012018 doi:10.1088/1755-1315/172/1/012018

Study on cutting temperature and surface roughness during the milling process of aluminium alloys

E N Patru, N Craciunoiu, D Panduru, M Bica

Faculty of Mechanics, University of Craiova, Calea Bucuresti, 107, Craiova, Romania

dumitru_panduru23@yahoo.com

Abstract. In this paper some experimental determinations on the temperature and surface roughness during the milling process of aluminium alloys (5083, 6082 and 7075) was conducted, using different cutting parameters (depth of cut, feed and rotational speed). The results are presented as graphical dependencies of temperature and surface roughness as function of cutting parameters also as a screen capture of the values obtained using an adequate technique for temperature of the milling process of aluminium alloys.

1. Introduction

Aluminium alloys are some of most used materials in technique due to low weight and appropriate physico-mechanical properties.

The study of the cutting temperature during the milling process is very important due to the high thermal conductivity of aluminium alloys, and, as a consequence, a large amount of the heat is absorbed by the work piece. On this way, the workpiece temperature increases, and can affect the nominal dimensional, tolerance, and sometimes the surface damage occurs.

In [1] the authors predict the temperature during milling process of aluminiun alloy 7050-T745 using an infrared thermal image system and finite element method, and show that the highest cutting temperature increase with an increase in the flank wear.

Using the infrared thermography, in [2] the influence of cutting parameters (spindle speed, axial and radial depth, and feed per tooth) on the average cutting temperature is evaluated. The authors remark the difficulty of measuring the temperature of the materials due to the emissivity and their strong reflexion power.

The temperature analyze using different cutting conditions (flood cooling, near dry machining and dry cutting) is performed in [3], and to perform these analyze a non contact infrared thermometers was used. The temperatures were simultaneous measured for object (cutting area) temperature, internal temperature, and the external one. The maximum temperature was obtained for dry cooling conditions.

Taking in account for milling the thin-walled components of the airplanes, in [4], the influence on stability and heat generation is analyzed, using experimental conditions and also simulation. The authors perform the experiments with a chamfered cutting edges and different process parameters. They conclude that the increasing of feed per tooth, the maximum temperature in worpiece can be reduced.

In [5], studying the dry end milling operation on 6083, aluminium alloys, show that the rise in the temperature of the work piece, which may lead to dimensional inaccuracies, surface damage and deformation of the workpiece and try to optimize the milling parameters, and they identify the depth of cut as the most relevant factor affecting the workpiece temperature.

In order to measure the temperature during milling process, different methods are used. So, the authors, studying the temperature during milling magnesium alloys, show that the optical pyrometry is suitable only for monitoring surface temperature of the workpiece, thermocouples is not suitable for temperature measurement in the cutting area, but infrared camera proves to be most suitable in terms of recorder temperature values, [6].

From point of view of the surface roughness during milling of aluminium alloys, using minimum quantity lubricant and different cutting parameters, in [7], the authors conclude that Surfaces roughness is better using minimum quantity lubricant techniques rather than using coolant by wet techniques.

In [8], the authors analyze the influence of the feed on surface roughness for 6061 aluminium alloy and give some practical recommendations to improve the surface quality.

This paper is focused on on the measurement and analysis of work piece temperature and surface roughness for three types of aluminium alloys.

2. Experimental setup

2.1 Materials and cutting tool

Materials used in this study are three types of aluminium alloys, 5083, 6082 and 7075, with composition and mechanical properties presented in table 1 and table 2.

The probes was prepared for the experiments are prepared as in figure 1(b), and the shape and dimensions of the probes are shown in figure 1(b).

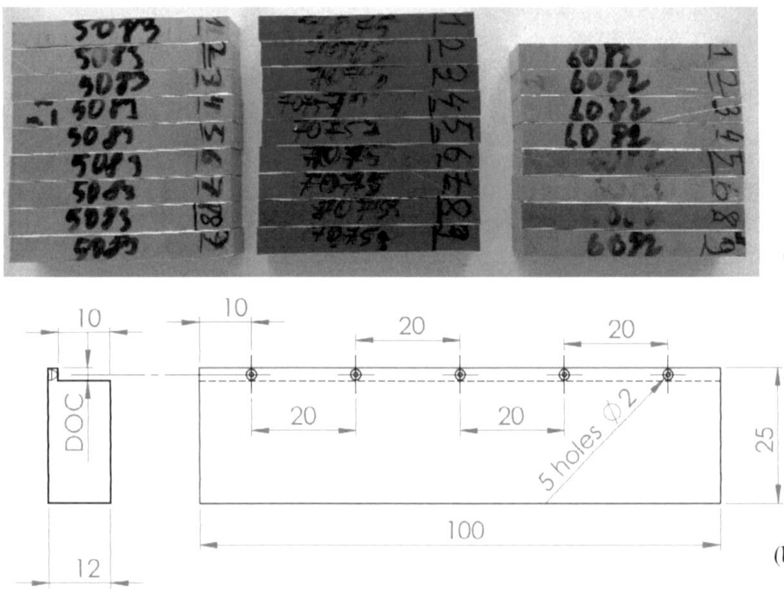

(a)

(b)

Figure 1. Materials used for temperature study in milling process and shape and dimensions of the probe

Table 1. Mechanical properties and composition of 7075 and 5083 aluminium alloys

Material	Mechanical properties			Chemical composition									
	Rp0,2, MPa	Rm, MPa	A5 %	Si	Fe	Cu	Mn	Mg	Cr	Zn	Ti	Others each	Others total
EN AW-7075 EN 485	532	594	10	0,1080	0,1630	1,6460	0,1090	2,6960	0,2020	5,8810	0,0282	0,0112	0,0430

EN AW-5083 EN 485-	302	176	20	0,26	0,39	0,07	0,60	4,80	0,10	0,08	0,03	0,02	0,05

Table 2. Mechanical properties and composition of 6082 aluminium alloy

Material	Mechanical properties				Chemical composition, %											
	Rm, MPa	Rp0,2 MPa	Elong, % 50 mm	HB	Si	Fe	Cu	Mn	Mg	Cr	Ni	Zn	Ti	Ga	V	Al
EN AW-6082 EN 485-	295	240	8	89	1,1	0,42	0,1	0,7	0,85	0,17	0,013	0,095	0,016	0,009	0,2	REM

The cutting tool, carbide end mill, with ϕ10 mm diameter and 4 flutes, is presented in figure 2.

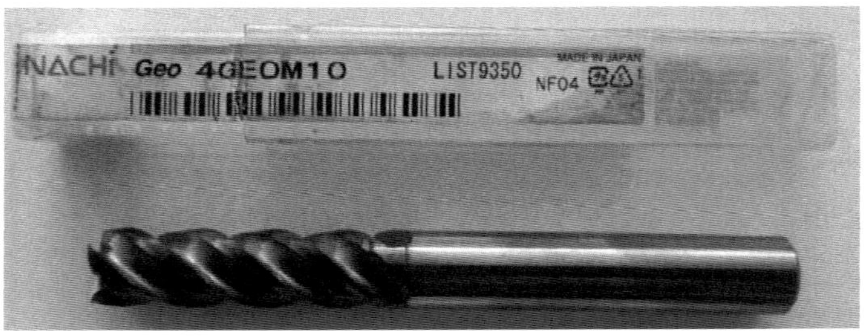

Figure 2. Carbide end mill, with ϕ10 mm diameter and 4 flutes

2.2 Experimental assembly for temperature measuring
In order to measuring the cutting temperature, in figure 3, an adequate experimental assembly was realized.

The experimental assembly consists in: Portable infrared thermometer, Compact Connect type [9], with the measuring temperature range (-32^{0}C – 760^{0}C), emissivity adjustable between (0.1-1.00), precision ±1% and ratio distance/spot size 20/1, coupled with K-type thermocouple.

2.3 Cutting conditions
Rotational speed, n = 360 rpm, v_f = 100 mm/min) and depth of cut: a_p = variable (a_p = 2 mm, a_p = 1.5 mm a_p = 1 mm) for all experiments on aluminium alloys.

2.4. Measuring of the surface roughness
In order to measuring the surface roughness a Portable Surface Roughness Tester - TR200, [10], as it is shown in figure 4.

3. Results
3.1 Temperature results
The data were collected using the cutting conditions, portable infrared thermometer above described. The measured data was obtained using a Compact Connect software, and figure 5 presents a capture of the Compact Connect display.

For all data collected, using the ".dat" files, the graphical dependencies of the temperature as function of time, for n = 360 rpm, v_f = 100 mm/min, and a_p - variable are represented, as it shown in figure 6, figure 7 and figure 8.

In the all three graphical dependencies the data was represented using the values of the temperature registered for each material and determination (5083, 6022 and 7075, and also a_p = 2 mm, a_p = 1.5 mm and a_p = 1 mm, respectively).

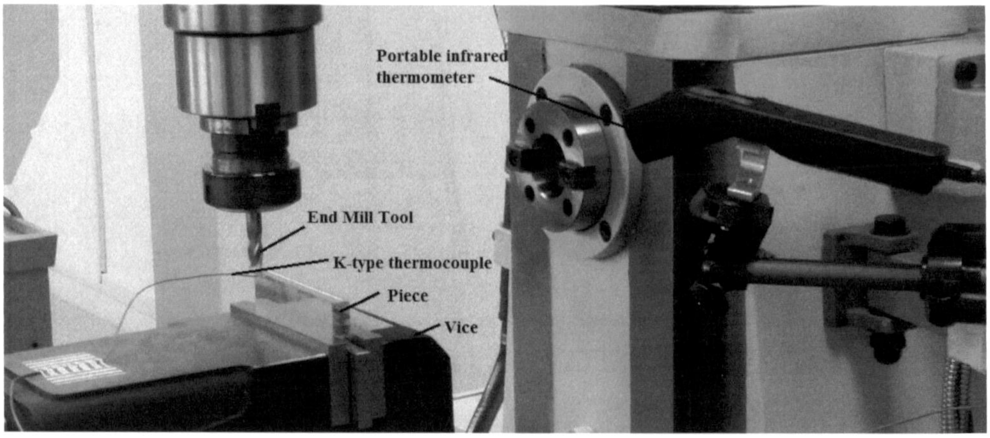

Figure 3. Experimental assembly

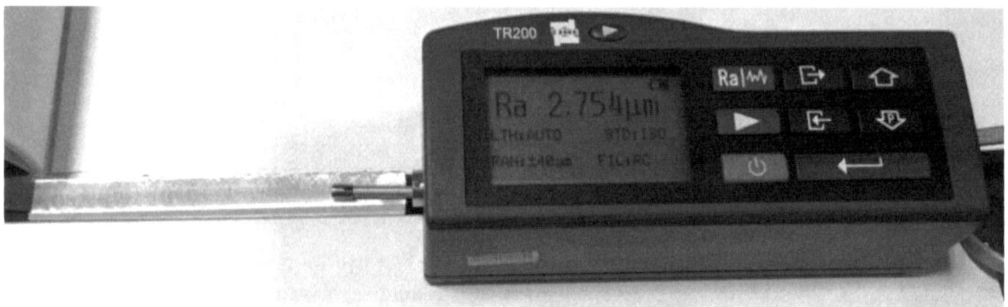

Figure 4. Portable surface roughness and workpiece

3.2 The surface roughness results

For each piece (see figure 4), using the roughness tester TR 200, the roughness parameters was measuring, and the results are presented in figure 9, figure 10, and figure 11.

Using the data collected for milling process, in figure 12 is graphically represented the values of the surface roughness as function of the machined material (aluminium alloys 5083, 6082 and 7075) using the same conditions: n = 360 rpm, v_f = 100 mm/min; a_p = variable).

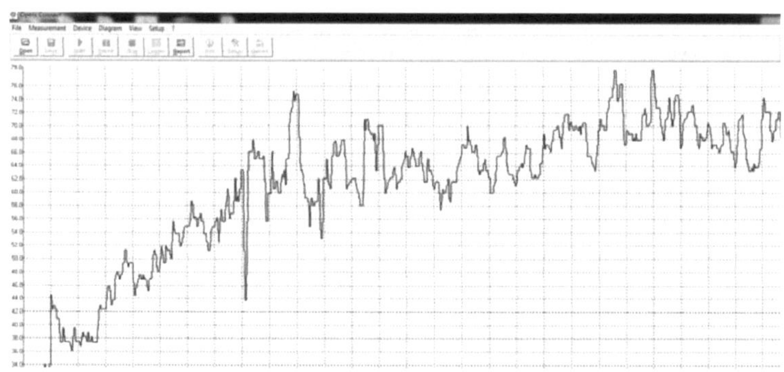

Figure 5. Temperature registered during milling process of aluminium alloys with Portable infrared thermometer

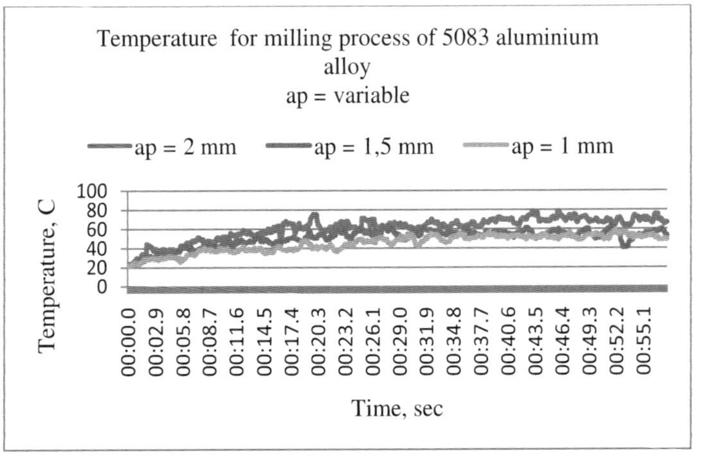

Figure 6. Temperature for milling process of aluminium alloy 5083, n = 360 rpm, v_f = 100 mm/min; ap = variable

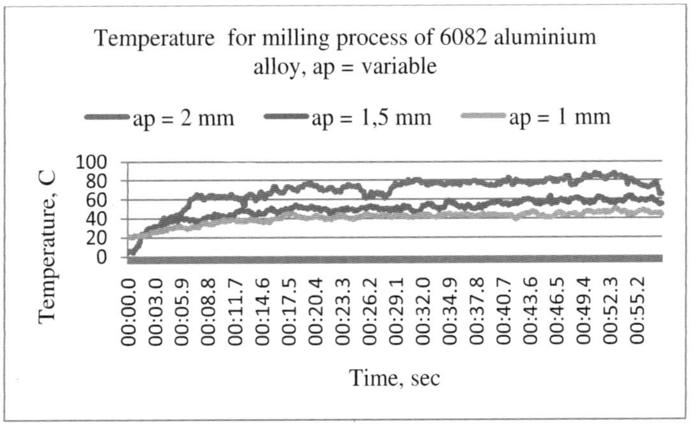

Figure 7. Temperature for milling process of aluminium alloy, 6082, n = 360 rpm, v_f = 100 mm/min, a_p = variable

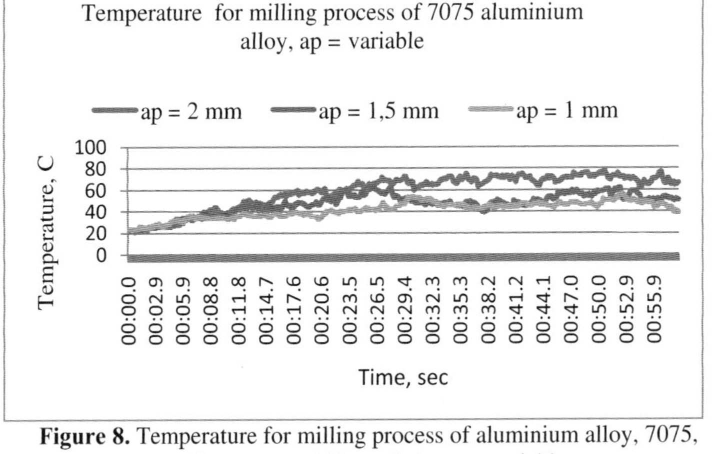

Figure 8. Temperature for milling process of aluminium alloy, 7075, n = 360 rpm, v_f = 100 mm/min, a_p = variable

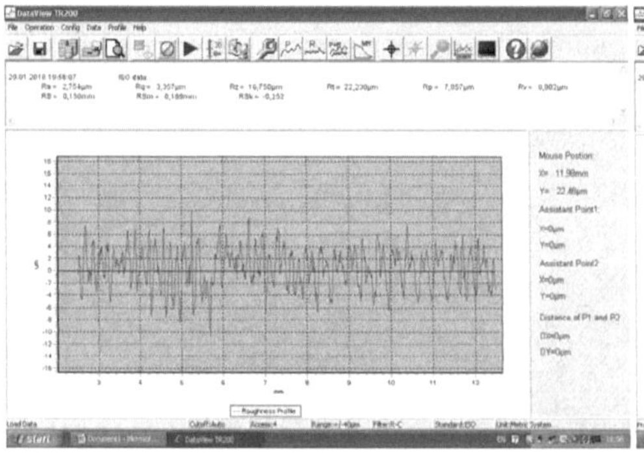

Figure 9. The Surface roughness for 5083, aluminium alloys
(n= 360 rpm, v_f = 100 mm/min, a_p = 1 mm)

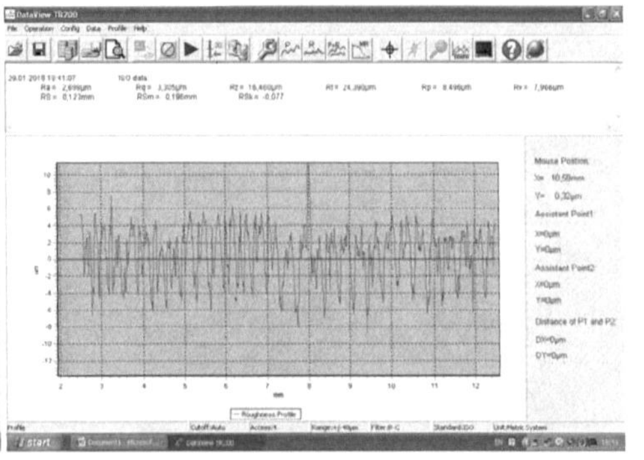

Figure 10. The Surface roughness for 6082 aluminium alloy
(n = 360 rpm, v_f = 100 mm/min, a_p = 1 mm)

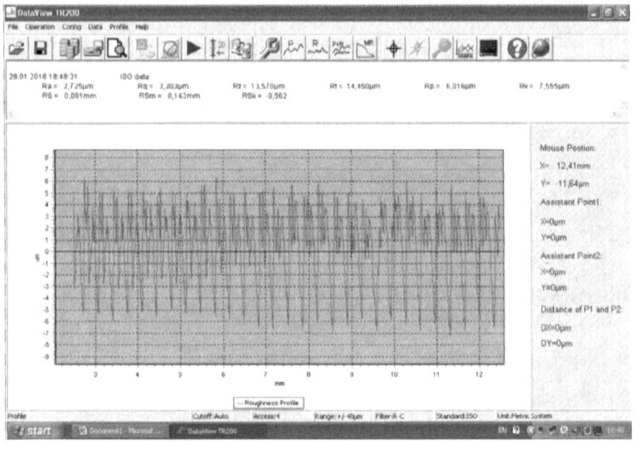

Figure 11. The Surface roughness for 7075 aluminium alloy
(n= 360 rpm,v_f = 100 mm/min, ap = 1 mm)

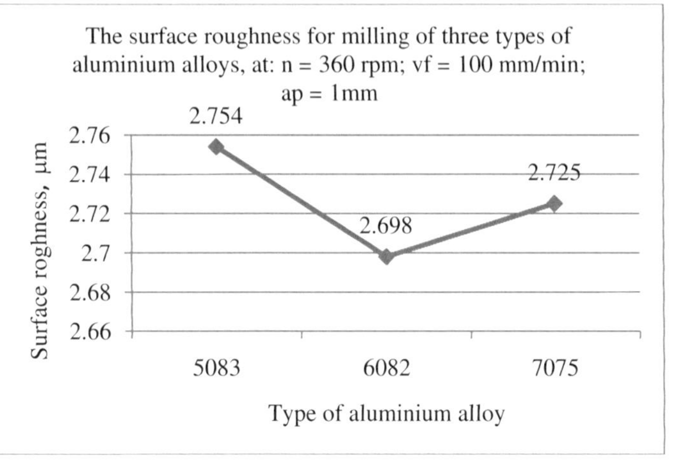

Figure 12. Surface roughness for milling of aluminium alloys 5083, 6082 and 7075, for n = 360 rpm; vf = 100 mm/min; ap = 1mm

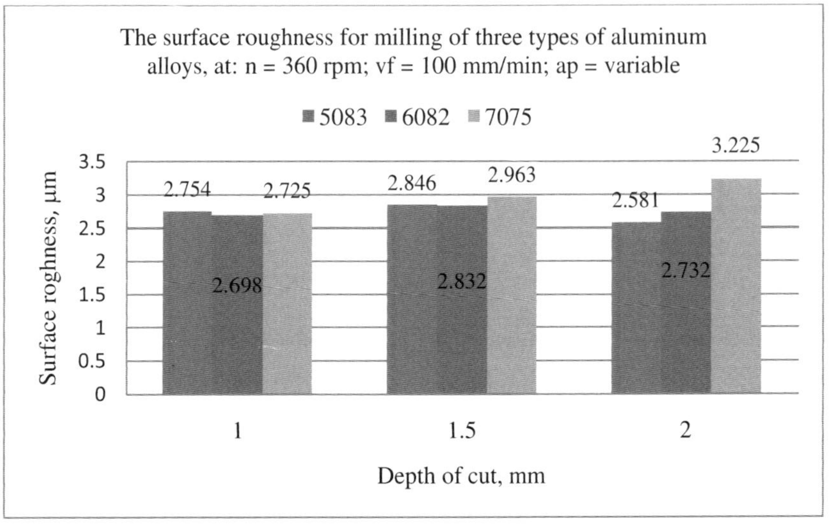

Figure 13. Surface roughness for milling of aluminium alloys 5083, 6082 and 7075 for different depth of cut

4. Conclusions

From point of view of temperature as function of depth of cut, the maximum value (figures 6, 7 and 8), for all aluminium alloys, was measured at depth of cut, a_p = 2 mm. If is consider the type of aluminium alloy, the maximum value of temperature was 84.7^{0}C, for 6082, greater than 5083 (77.2 ^{0}C) and 7075 (75.6 ^{0}C).

For the surface roughness maximum value measured, for a_p = 2 mm and 7075 aluminium alloy was R_a = 3.225 µm Also for a_p = 1,5 mm and 7075 aluminium alloy, was R_a = 2.963 µm, but no for a_p = 1mm, where the maximum value was R_a = 2.754 µm, for 5083, aluminium alloy (figure 13), while the minimum value for the surface roughness was R_a = 2.581µm, for milling of 7075 aluminum alloy.

But, due to the difficulties of cutting process of aluminium alloy, especially built-up-edge phenomena, both the temperature and the surface roughness are negatively influenced.

5. References

[1] Z.T. Tang, Z.Q. Liu, Y.Z. Pan, Y. Wanb, X. Ai, The influence of tool flank wear on residual stresses induced by milling aluminum alloy, *Journal of Materials Processing Technology*, **209** (2009) 4502–4508.

[2] Medina, N., et al. Evaluating Temperature in Faced Milling Operations by Infrared Thermography. *THERMAL SCIENCE*, Year 2017, Vol. **21**, No. 6B, pp. 3051-3061.

[3] Fratila D.Fl., Assessment of Cutting AreaTemperature to the Face Milling using Several Cooling Methods, *Acta Mechanica Slovaca*, Vol. **15,** No. 1, 2011, pp. 50-54.

[4] B. Denkena, J. Brüning, D. Niederwestberg, R. Grabowski, Influence of Machining Parameters on Heat Generation during Milling of Aluminum Alloys, *Procedia CIRP* **46** (2016), pp. 39 – 42.

[5] Józef Kuczmaszewski, Ireneusz Zagórski, Methodological Problems of Temperature Measurement in the Cutting Area during Milling Magnesium Alloys, *Management and Production Engineering Review*, Vol. **4**, No. 3, Sept. 2013, pp. 26–33

[6] N.L. Bhirud, R.R. Gawande, Optimization Of Process Parameters During End Milling And Prediction Of Work Piece Temperature Rise, *Archive of Mechanical Engineering*, Vol. **LXIV** 2017 Number 3

[7] A. Imthiyas, D. Jayasuriya, M. Jothi Mani and A. Manova Abharam, Experimental Investigation of Milling Operation on Aluminum Alloy 6063 by MQL Technique, *Int. J. Chem. Sci.:* **14**(4), 2016, pp. 3113-3118.

[8] Mathew A. Kuttolamadom, Sina Hamzehlouia, M. Laine Mears, Effect of Machining Feed on Surface Roughness in Cutting 6061 Aluminum, https://cecas.clemson.edu/manufacturing-lab/documents/publications/kuttolamadom%202010b.pdf

[9] ***Optris infrared sensing, Basic principles of non-contact temperature measurement

[10] *** Portable Surface Roughness Tester - TR200. Innovatest, Manual TR200-Surface roughness tester

IOP Conf. Series: Earth and Environmental Science **172** (2018) 012019 doi:10.1088/1755-1315/172/1/012019

Reinforcement learning: The application to autonomous biomimetic underwater vehicles control

J Magalhães[1], B Damas[1] and V Lobo[1]
[1] CINAV, Portuguese Navy Research Center, Almada, Portugal

E-mail: candeias.magalhaes@marinha.pt

Abstract. The Autonomous Biomimetic Vehicles have been increasing in popularity in the past few years. Controlling such type of vehicles is not trivial: due to its complex dynamics and kinematics, it is complex to analytically derive controllers that can efficiently perform a given task, such as reaching a given position target in a minimum time. In this paper we will evaluate the results of the implementation of a reinforcement algorithm in autonomous biomimetic underwater vehicles, providing a new way to control this type of vehicles in which the algorithm is in constant learning.

1. Introduction and motivation

Biomimetics is a field where various principles are applied to mimic biological processes, such as the humanoids robotics field where human-like robots are developed [1]. In the context of underwater vehicles the kinematics and dynamics of the vehicles try to emulate the motion of different sea creatures.

The study of autonomous Biomimetic Underwater Vehicles (BUV) have been increasing in the past few years in a military context due to its furtive locomotion capabilities: they have the ability to camouflage within the environment and a silent undulating propulsion that generates a very different acoustic signature when compared to conventional Unmanned Underwater Vehicles (UUV) [2].

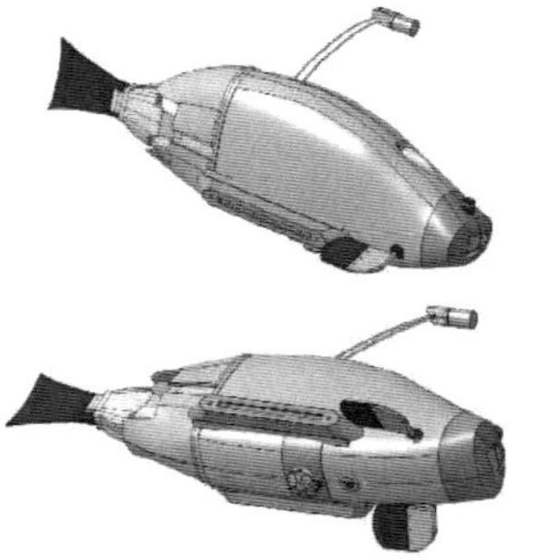

Figure 1. 3D model of fish-like vehicle

This work is part of the international project "Swarm of Biomimetic Underwater Vehicles for Underwater Intelligence, Surveillance and Reconnaissance (ISR)" (SABUVIS), whose main objective is to use BUVs in missions for stealth data collection and surveillance, where Portugal, Poland and Germany and the countries part of the consortium. Where the Portugal collaborators are Escola Naval, LSTS and OceanScan.

Under this project two different vehicles are being built, one mimicking a fish (Figure 1) and the other replicating a seal. Both vehicles have the same basics characteristics, similar form, two side fins and a tail, the major differences being their size and tail. The seal-like tail is composed by two smaller tails that mirror the movement of each other, trying to mimic a breaststroke or "frog" stroke style of swimming, while the fish-like tail is a simple tail sectioned in two parts to give more fluidity.

Figure 2. Seal-like vehicle during tests

Under this project the generated acoustic noise developed by the BUVs, as well as their autonomy capabilities, will be measured and compared to conventional UUVs.

These different mechanical structures that are inspired in biological systems present a much more complex kinematic structure that makes the task of controlling the motion of the vehicle a non trivial one. For a fish-like BUV [3], for example, it is not straightforward to develop controllers that actuate the fins and the tail of the vehicle in order to make it follow a desired trajectory in an efficient way [4][5].

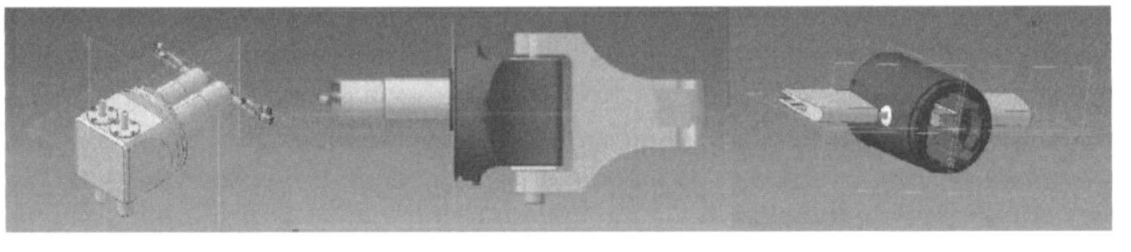

Figure 3. Model of the seal-like tail, fish-like tail and fins models, respectively

The purpose of this work is to develop adaptive controllers for these kind of vehicles based on Reinforcement Learning (RL) techniques. Reinforcement learning is an Artificial Intelligence technique, also biologically inspired, that revolves around "trial and error" theory. In RL an agent is supposed to choose its actions in a way to maximize an external reward that it gets from its interaction with the environment: a positive reward is given when it fulfills certain conditions defined by the programmer, and a lower reward is obtained when that condition is not met. This reward is tightly linked to the desired behaviour for the agent and should be chosen carefully as the agent, or robot, will solely learn based on the rewards it gets. For instance, when learning to navigate from a position to other the agent should get higher value rewards as it gets closer to the desired position.

To learn to perform a given task the agent has an internal policy that maps states to actions. Given the history of visited states, associated performed actions and received rewards the agent continuously adapts its policy in order to maximize its expected accumulated reward in the long run, this way progressively starting to exploit the actions that get the higher positive rewards, until it reaches the best way to perform the task.

There is no need to worry about programming every action to perform in every possible state when using a RL scheme, as these actions are learned from trial and error; it is even possible to the algorithm to adapt to unexpected situations, because the agent will teach itself in a continuous process of adaptation. This is a major advantage of using RL schemes when compared to traditional controllers, where a good knowledge of the process to control is typically needed.

2. Reinforcement learning in underwater biomimetic vehicles control

To implement a RL controller in the BUVs the work will be divided in two different stages. In the first stage a simulator will be developed and the RL algorithms will be tested and applied in a simulated environment. As the BUV will learn how to control its movements from scratch, the learning phase will start in such a simulated environment to prevent actuator wear, collisions and other damage to the physical robot that would inevitable result from the initial exploratory random movements it would perform in this phase. Afterwards, when the simulated robot can already control its movement in a satisfactory way, the learned model will be transferred to the real vehicle, where it will undergo a second learning phase. Such adaptation is required to take into account the differences between the simulated and the real environment.

2.1. BUV Q-learning simulation

Q-Learning is a Reinforcement Learning algorithm that estimates, from its interaction with the environment, the utility of performing a given action in a particular state, given by Q(S, A) [6]. This policy can be learned from this interaction, without the need to have a model for this environment, using the update rule.

$$Q(S,A) = Q(S,A) + \alpha[R + \gamma max_a Q(S',a) - Q(S,A)], \qquad (1)$$

where Q denotes the expected accumulated future reward obtained if action A is performed in state S. S' is the observed next state after executing action A and R the corresponding reward. $0 < \alpha \leq 1$ is a step-size parameter and $0 < \gamma \leq 1$ denotes the discount-rate factor: the lower this value the lesser the importance given to distant future rewards, *i.e*, the more myopic the agent is regarding future rewards.

The vehicle in training is a fish-like composed by two lateral fins, one on each side, and one tail. Each motor is actuated by a sinusoidal signal with variable frequency (F) [mHz] and deflection (K), the average value in which the tail or fins cycle around, where F and K denote the tail variables, F1 and K1 the left fin and F2 and K2 for right fin. Together these 6 controlled variables define the action vector [7].

The tabular nature of the vanilla Q-Learning algorithm requires a discretized set of actions and states. With respect to actions we define A as the variables whose columns contain the discretized action values for each component of the action vector

$$A = \begin{bmatrix} 3000 & 500 & 3000 & 500 & 3000 & 500 \\ 2250 & 250 & 2250 & 250 & 2250 & 250 \\ 1500 & 0 & 1500 & 0 & 1500 & 0 \\ 750 & -250 & 750 & -250 & 750 & -250 \\ 0 & -500 & 0 & -500 & 0 & -500 \end{bmatrix},$$

where each column represents F, K, F1, K1, F2 and K2 respectively and where we set F = F1 = F2, in steps of 750, and K = K1 = K2, in steps of 250. This way, we have a set of 5^6 different actions from where to choose at each iteration.

We consider the state to be the position in each axis, in meters, and the vehicle angle in the xy plane (in radians). The state variables S, that contain that discrete values, is then defined as

$$S = \begin{bmatrix} 1 & 1 & 1 & 0.6 \\ 2 & 2 & 2 & 1.2 \\ \vdots & \vdots & \vdots & \vdots \\ 9 & 9 & 9 & 5.4 \\ 10 & 10 & 10 & 6 \end{bmatrix},$$

where each column contains the possible values for x, y, z, in steps of 1, and ψ, in steps of 0.6, respectively. This makes 10^4 distinct states.

The objective of this simulation is to reach a goal point and stay there. To fulfill it, the reward function is defined as

$$R = -\sqrt{\Delta x^2 + \Delta y^2 + \Delta z^2},$$

so that the reward will be higher the closest it gets to the goal.

The reward had to be made negative due to Q, that is initialized with zeros, selecting the action with the maximum value to the present and future state. By making it negative, it is guaranteed that each action is selected at least one time for each state, in the infinite number of episodes.

The simulation is divided by episodes. Each episode starts the vehicle in a random position with a 4 meter distance to goal and with the vehicle pointed at it, each episode it has 200 iterations, each one representing 0.1 seconds, and it end when 20 seconds have passed, starting a new episode.

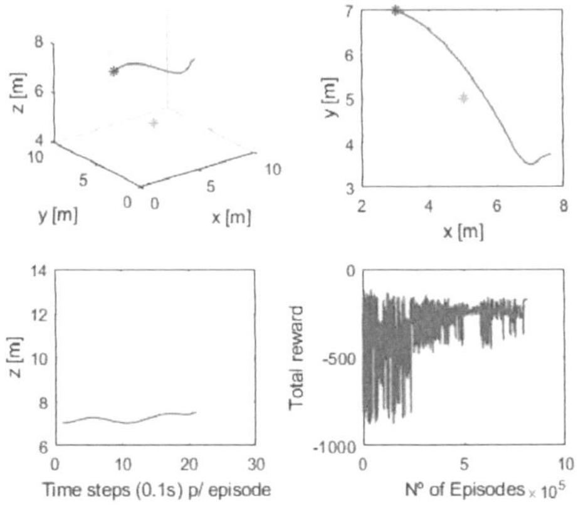

Figure 4. Graphics of early results of the Q-learning algorithm

The graphics above show one of the trajectories of the vehicle, in a 3D dimension and 2D xOy axis, its altitude during the episode and its accumulative reward in the past episodes. The red star being the starting position and the green star the goal.

It is noted that the accumulative reward has already started to converge, which means that is starting to reach a satisfactory solution, but it is taking a long time to do it.

2.2. LSTS

The Laboratório de Sistemas e Tecnologia Subaquática (LSTS) is an interdisciplinary research laboratory part of Faculdade de Engenharia da Universidade do Porto (FEUP) that specializes on the design, construction and operation of unmanned underwater, surface and air vehicles and on the development of tools and technologies for the deployment of networked vehicle systems.

LSTS created the software toolchain Neptus-IMC-DUNE that will be present in the project SABUVIS. DUNE is the system for vehicle on-board software, it has a C++ programming environment and it is responsible for navigation, code for control and access to sensors and actuators; IMC is the communication protocol; Neptus is a command, control, communications and intelligence framework for operations with vehicles, systems and human operators.

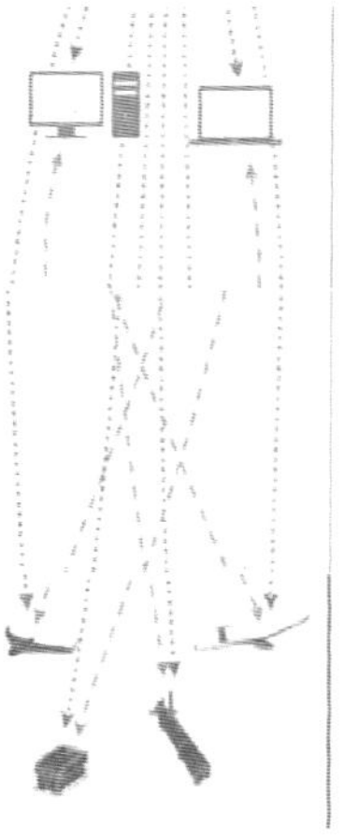

Figure 5. LSTS toolchain, Neptus-IMC-DUNE

In the project SABUVIS, LSTS is responsible for the integration of software and sensors on the vehicle.

3. Conclusions and future work

The convergence of the Q-learning algorithm shows us that we are headed in the right direction with these results, we can already see that the simulation is trying to reach the goal as soon as possible, also it is already starting to learn to turn back and try to return to the goal to obtain a better reward. There are still some problems like needing an adjust in the state variables to decrease the convergence time and the main problem being the discretization of the values. A different approach may resort to function approximation or algorithms based on policy improvement {*e.g.*, PI^2 or POWER), in order to circumvent the discretization of the state and action space. We also expect to implement the algorithm in the physical BUV, using the software toolchain that runs on the vehicle developed by FEUP.

IOP Conf. Series: Earth and Environmental Science **172** (2018) 012019 doi:10.1088/1755-1315/172/1/012019

References

[1] BAR-COHEN, Yoseph, "Biomometics: Biological Inspired Technologies", California, Jet Propulsion Laboratory (JPL), California Institute of Technology, 2006

[2] LISTEWNIK, K., "Sound Silencing Problem of Underwater Vehicles", in Solid State Phenomena, Vol. 196, Trans Tech Publications, 2013, pp. 212-219

[3] LOW, K. H., "Modelling and parametric study of modular undulating fin rays for fish robots", in Mechanism and Machine Theory, Vol. 44, 2009, pp. 615–632

[4] COLGATE, J. E. and LYNCH, K. M., "Mechanics and Control of Swimming: A Review", in IEEE Journal of Oceanic Engineering, Vol. 29, No. 3, 2004, pp. 660-673

[5] MALEC, M. et al., "Analysis of Parameters of Traveling Wave Impact on the Speed of Biomimetic Underwater Vehicle", in Solid State Phenomena, Vol. 210, Trans Tech Publications, 2014, pp. 273-279

[6] BARTO, Andrew G. and SUTTON, Richard S., Reinforcement Learning: An introduction, 1st edition, Cambridge, MIT, 1998

[7] SZYMAK Piotr, "Mathematical model of underwater vehicle with undulating propulsion", Poland, Polish Naval Academy, Institute of Electrical Engineering and Automatics

Aspects regarding the evolution of the transport of containerized goods through the port of Constanta in context of European containerized trade

A R Anghel[1] **and C Ciobanu**[1]
[1]*Navigation and Naval Management Departament, "Mircea cel Batran" Naval Academy, 1st Fulgerului Street, Constanța, 900218, Romania*
Email:arlette.anghel@anmb.ro

Abstract. In the context of a continuous transformation of maritime freight transport, we considered it is important to conduct a study of the evolution of the transport of containerized goods. The aim of this research was to present the evolution of transport of containerized goods through the port Constanta between 2009 and 2017 and also to highlight the comparative analysis for the containerized trade through some of the most important countries which have ports on the Black Sea, the Adriatic Sea, the Baltic Sea, the Mediterranean Sea, the Aegean Sea, the Norwegian Sea and the North Sea. The period analyzed for these was between 2010 and 2016. The methodology used consists in data processing and graphic representation of the comparative analysis of the results.
Keywords: Maritime Transport, Port, Central and Eastern Europe, Black Sea, Containerized Transportation

1. Introduction

It is obviously that the maritime trade is one of the oldest and most economical ways to transport goods at the European level. Maritime transport has developed with the growth of the economy and gradually became a real industry. Today The European containerized trade marine is represented by well-organized fleets, modern port infrastructures, high-performance communications and, last but not least, high-class specialists.

From an economic point of view, the transport of containerized goods has dramatically reduced transport costs, mainly as a result of faster loading and unloading. Considerable intensification of capital investment along the entire transport chain was necessary to ensure the efficient use of containers. At the same time, it can be said that labour intensity has been reduced in a sustainable way, as fewer workers were required to the operations that means that of loading and unloading. Containerization was a radical change in freight transport, this meaning among other that prior to containerization the shipment speed was much lower and the acute delays led to inefficient transhipment resulting an incorrect management of the operational time of goods distribution. In the containerized trade, freight transport has reached a level at which time-based management realizes that the speed of distribution is very important. Any improvement in freight speed may in the future be based on the use of a better port infrastructure.

Talking about containerized maritime trade we used the study from [1] related to the period between 2009 and 2017 with the aim to obtain graphical comparative results and all of this are structured in the five next sections.

So, in the first section we study the evolution of containerized trade transport through Constanta Port between 2009 and 2017.The second section is about comparative aspects of containerized trade transport through the ports of Romania, the Russian Federation, Ukraine and Bulgaria. The third section in the context of closer cooperation between EU member states we present some of comparative aspects of containerized trade transport through the ports of "The Three Seas Initiative" which follows to provide support for better cooperation and interconnection of the economies of the states in the geographical area of the three seas - the Adriatic, the Baltic and the Black, including in the field of transport. In the fourth section we present some comparative aspects of containerized trade transport through some of the important ports from Romania, Greece, Italy and Spain. In the last section we present comparative aspects of containerized trade transport through the ports of Romania, Netherland, Norway and Sweden. It is necessary to point out that in the last three sections, we considered the study period between 2010 and 2016 taking into account the fact that starting with the third quarter of 2010 most of the countries affected by the recession had already been on an upward trend. The comparative parameters which have taken into consideration in this paper are the container ships size, the containerized sea trade volumes and also the ports infrastructure.

Constanta Port [2]

2. The evolution of containerized trade transport through Constanta Port between 2009 and 2017

In Romania, the transport of containerized trade dates back to the early 70's of the 20th century. At that moment work began with the project of the first container terminal in the port of Constanta. After 1989 there was a new reorganization of the port activity and, implicitly, the transport of containerized goods through the port of Constanta had a spectacular expansion. Thus, in 2001, the port of Constanta reached a new record, of 118 645 TEUs, in the context of the transport of containerized goods.

Constanta Port is located at the intersection of the trade routes between Central and Eastern Europe and Central Asia and the Far East, it is a multifunctional port that offers facilities for the operation of large tonnage ships and connections through all modes of transport with customs specific services for the Free Zone legal status. It is an important centre for the transit of containers to the Black Sea ports also it is the main Romanian port on the Black Sea playing a highly important role as the transit node for the land locked countries in the Central and South-East Europe [3]. The Port of Constanta is composed of the Northern Port completely in operation and the Southern Port partly operational. The Northern Port has terminals specialized in the trafficking of petroleum products, ore, coal, cereals, containers, rolled goods and general cargoes. The Southern Port has terminals specialized in chemicals, general, cargo, ferry terminal, Ro-Ro and on the berths of the maritime river basin there are arranged transhipment sectors for ore, coke and coal directly into barges which have access to the Danube-Black Sea Canal [4].

At the end of 2003, Constanta harbour opened the largest specialized container terminal in the Black Sea, operated by Dubai Ports World.

With a total area of over 76 ha (52 ha operational), the terminal offers an annual operating capacity of approximately 1 500 000 TEUs. Due to the depths, which have a minimum of 14.5 meters, the terminal allows the operation of Post-Panamax port-container vessels and the operating facilities ensure the effective operating rate of the mooring vessels.

Another important container terminal which is operated by SOCEP, is located on the north part of the port and it operates on two berths of a combined length of 467 meters and a water depth of 13.5 meters. The nominal capacity of the terminal is 300 000 TEUs. The facilities, depths provided by Constanta Port made it possible in 2007 to achieve a record yearly traffic of containers in the Black Sea region, totalling 1 411 387 TEUs [5].

It is knew that container terminals in the Port of Constanta has an annual capacity over one million TEUs and is the most important container terminal at the Black Sea. The port of Constanta has the capacity to handle significant cargo flows, yet this capacity is currently unused to its full potential. Starting from 2009 however, volumes slumped to around 600 000 TEU. This volume decrease coincides with the repercussions of the financial crisis that impacted the Romanian economy at that moment. The container traffic in Constanta harbour dramatically reduced, reaching 556 694 TEUs in 2010 to climb to 689 066 TEUs in 2015. While the growth has a slowly recovering it could be said that it is far away from the previously results achived between 2004 and 2007.

Based on data provided by [6], we have drawn up the graph represented below in the Figure 1 showing the evolution of containerized trade transport through Constanta Port during 2009-2017. The data presented in the Table 1 are in TEUs.

Table 1

Port/ Year	2009	2010	2011	2012	2013	2014	2015	2016	2017
Constanta	595 303	556 694	662 796	684 059	661 124	668 349	689 066	711 339	696 438

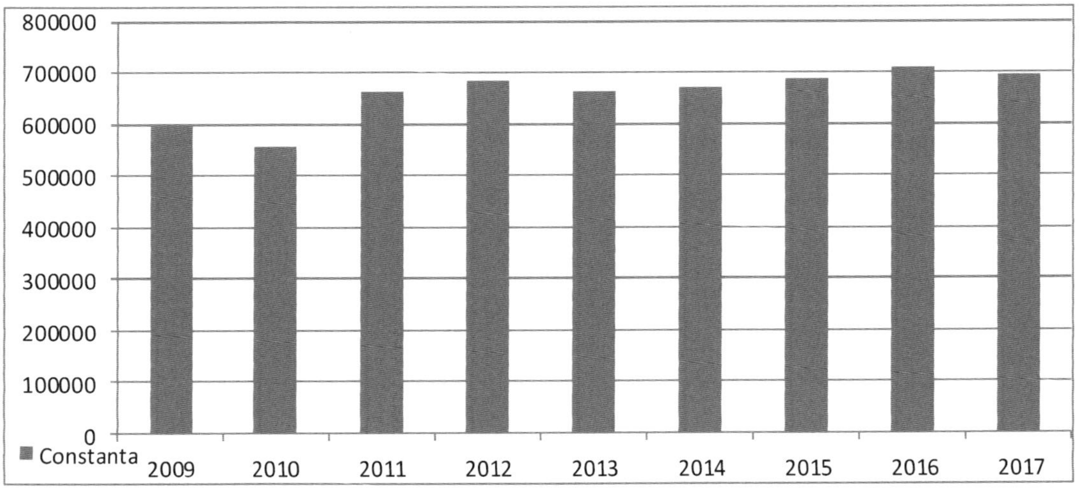

Figure 1

3. Comparative aspects of containerized trade transport through the ports of Romania, the Russian Federation, Ukraine and Bulgaria

The container turnover in Russian ports has been growing since 2005 except in 2009 when it dropped 36% due to the global economic downturn. The trade volumes recovered quickly and in 2012, the container flow through Russian ports reached a record 4 500 000 TEUs. Based on industry reports, export/import container traffic through Russian ports grew by 10% in 2013 but dropped substantially in 2015, like the table below has shown.

The most important commercial ports are those from St. Petersburg and Vladivostok.

The majority of the berths of "The Greater Port of St. Petersburg" can accept vessels up to 9.8 meters of maximum draught, and several berths can serve vessels with 11 meter draught. The Port

IOP Conf. Series: Earth and Environmental Science **172** (2018) 012020 doi:10.1088/1755-1315/172/1/012020

operates year round. During the winter, the Port Administration provides 11 icebreakers for clearing the way for trade vessels. According to the "Greater Port St. Petersburg" Administration, this is adequate to ensure normal port operation during winter months. It has necessary to point out that "The Greater Port of St. Petersburg" belongs to the Russian government and 28 licensed stevedoring companies rent 200 berths at the port and handle cargo loading/unloading works and other operations accompanying the cargo transportation via port [7].

The commercial Port of Vladivostok has 15 berths with a total length of more than 4 kilometres. It is one of the largest multi-purpose and the best-equipped ports in the Russian Far East. Its modern cargo handling equipment enables to provide the customers with the innovative solutions in cargo handling [8].

The closest competitors for Constanta are the Bulgarian ports situated in the Black Sea range Varna and Burgas. Both ports are still under government administration via their respective Port Authorities. The port of Burgas is a small port which handled, like example, an annual throughput of 49 000 TEUs in 2013. It is also the only port that does not have gantry cranes for loading and unloading operations and relies on traditional level lifting cranes.

Both Burgas and Varna Ports are direct competitors for the port of Constanta for hinterland cargo due to the geographic proximity within the range.

Based on the data offered by Ukraine's officials which participated at "Black Sea Container Summit-2016" [9] we can say that in 2016 there was an upward trend. For example, marine terminals in Ukraine during the first 8 months of this year handled 320 000 TEUs, i.e. 18% more than in the same period of 2015. According to its reloading capacities today Ukraine is capable to tranship up to 3 000 000 TEUs each year. At the same time, experts say, these powers are involved only at 16.5%. In the next we have performed a comparative analysis presented through the Table 2 and we drew a graph shown in the Figure 2 for the containerized cargo transport through four chosen Black Sea countries considering a seven-year period starting with 2010 and using statistical data in TEUs from [10].

Table 2

Country/Year	2010	2011	2012	2013	2014	2015	2016
Romania	555 674	662 796	684 059	661 124	668 349	699 000	721 339
Bulgaria	142 611	152 293	169 922	179 840	195 800	201 000	198 400
Russian	3 230 676	3 786 066	4 416 215	4 486 538	5 252 000	3 909 400	3 923 662
Ukraine	605 508	60 640	682 848	732 987	665 000	480 000	588 000

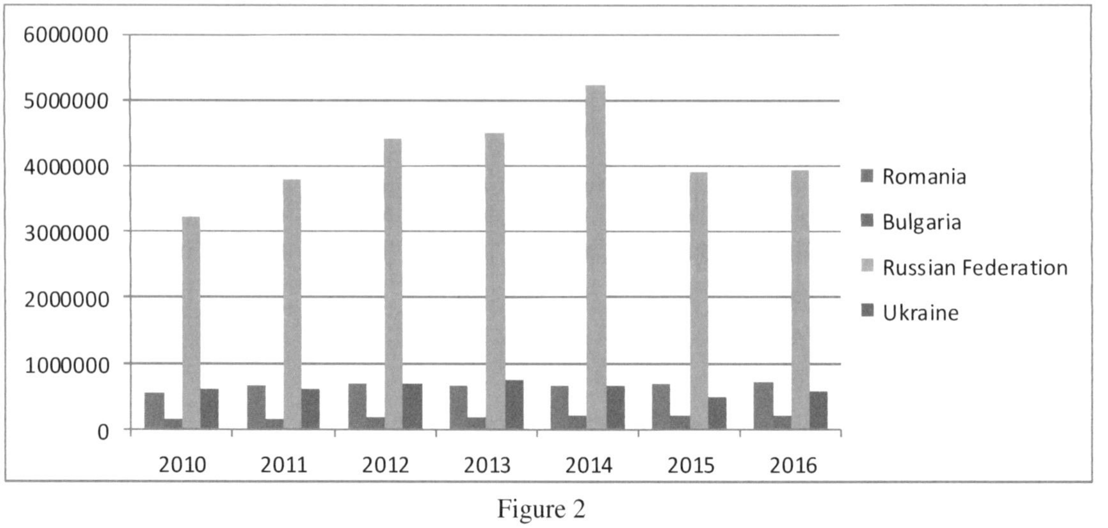

Figure 2

Based on information from [11] it could say that the containers terminals from the Black Sea for Romania, Bulgaria and Russia Federation saw growth of nearly 13% in 2017. The highest growth was achieved by Russia with 23%.

4. Comparative aspects of containerized trade transport through the ports "The Three Seas Initiative"

In the context of closer cooperation between EU member states in 2016 at the summit of "The Three Seas Initiative" - the Adriatic Sea, the Baltic Sea, the Black Sea, one of the relevant aims of the sea countries will be to stimulate economic cooperation and their economic development. The main goal of the summit was to begin plans for the implementation of the initiative that hopes to increase cooperation on trade, energy, and infrastructure between the EU member states of Central and Eastern Europe [12].

The sea trade transport could have a great influence on this, so we aimed to highlight the evolution of containerized cargo transport through the ports of the countries involved in "The Three Seas Initiative": Romania, Bulgaria, Lithuania, Latvia, Estonia, Poland, Croatia, Slovenia, during 2010-2016 based on the data in TEUs provided by [10] and using them in the graph from Figure 3.

Table 3

Country/Year	2010	2011	2012	2013	2014	2015	2016
Romania	555 674	662 796	684 059	661 124	668 349	699 000	721 339
Bulgaria	142 611	152 293	169 922	179 840	195 800	201 000	198 400
Lithuania	295 226	382 185	381 276	402 536	450 400	393 000	443 300
Latvia	254 560	302 973	362 297	381 099	391 200	359 390	388 433
Estonia	226 457	197 717	227 809	253 627	260 293	208 784	208 327
Poland	1 053 634	1 317 182	1 657 433	1 959 537	2 139 616	1 869 800	2 028 577
Croatia	137 048	150 677	152 016	169 943	211 900	181 800	196 850
Slovenia	476 731	589 314	570 744	600 441	647 000	791 000	844 000

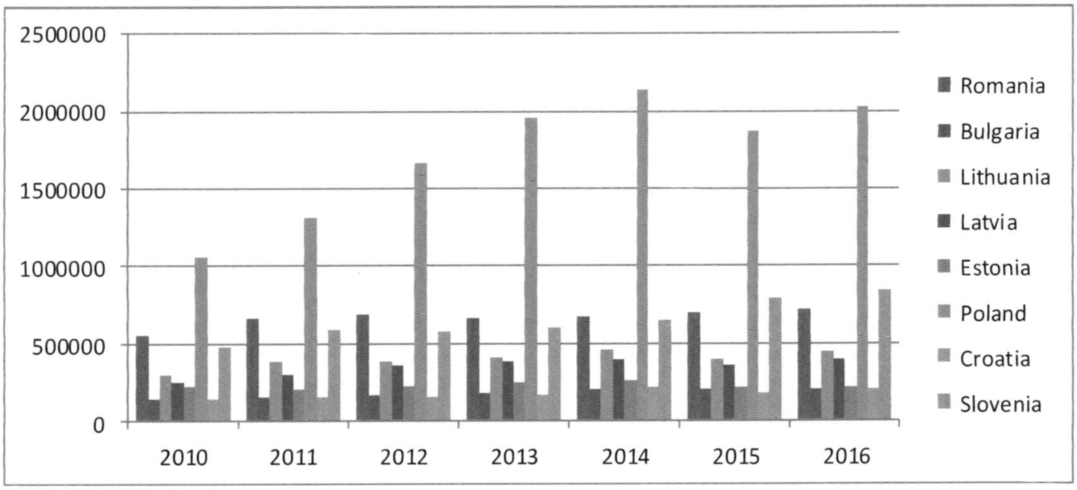

Figure 3

5. Comparative aspects of containerized trade transport through the ports of Romania, Greece, Italy and Spain

About Greece, a study of National Bank of Greece says that container handling covers around 1/5 of port revenue from trade activities (where containers account for 13% of freight volume). The same

IOP Conf. Series: Earth and Environmental Science **172** (2018) 012020 doi:10.1088/1755-1315/172/1/012020

study relived that the container market in Greece is dominated by two ports, Piraeus and Thessaloniki, which mainly depend on container handling (almost 75% of their merchant port operations) [13].

When it comes to referring to Italy it could be said that the specialised port sector (containers) is in full expansion in Italy- nine of the 30 busiest ports in the EU, in terms of container traffic, are located in Italy. The role of maritime transport has been expanded in Italy thanks to the project of 'sea highways' in the framework of the development of the trans-European transport network [14].

When we are talking about the Spanish containerized maritime trade we have to focus on the fact that the most important ports which are in the first 50 world container ports and those are Valencia, Algeciras and Barcelona.

Algeciras port is one of the top five ports of the European Union when analyzing total freight transport. Their positions at the Strait of Gibraltar, as well as the natural conditions of the bay help keep it at this stage of the hierarchy. In 2015, due to the large number of containers handled, "The Journal of Commerce Review" awarded him a prize being considered the third largest container port in Europe.

So, in 2016, Algeciras confirmed its status of the country's largest container hub by generating throughput of 4 759 000 TEUs, which is equivalent to an increase of 5.4% related to 2015.
Valencia achieved the record of 4 722 000 TEUs in 2016, that means an increase of 2.32% after that of previous year of 3.9%. At the same time Barcelona saw spectacular growth of 14.5% in 2016, this being the highest in the last decade, all of this was obtained from the Port Authority's press release [15]. Using data in TEUs from [10] we had drawn the graph from Figure 4.

Table 4

Country/ Year	2010	2011	2012	2013	2014	2015	2016
Romania	555 674	662 796	684 059	661 124	668 349	699 000	721 339
Spain	12 396 923	13 877 789	14 036 891	13 850 009	14 245 488	14 291 755	14 993 941
Greece	1 089 607	1 976 003	3 051 755	3 486 310	3 934 713	3 679 000	4 026 000
Italy	9 736 654	9 464 031	9 562 765	9 989 308	10 260 000	10 217 316	10 574 791

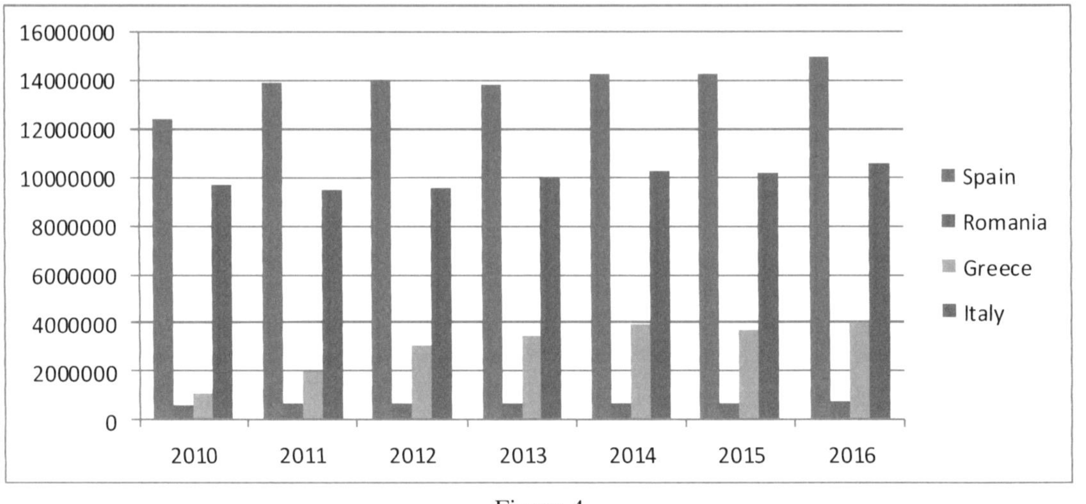

Figure 4

6. Comparative aspects of containerized trade transport through the ports of Romania, Netherland, Norway and Sweden

The Netherlands has a good geographical position on the coast of Western Europe and this had offered them the possibility to play a significant role in the world trade for hundred years, more

IOP Conf. Series: Earth and Environmental Science **172** (2018) 012020 doi:10.1088/1755-1315/172/1/012020

exactly from centuries. It is necessary to remark that the strong infrastructure and the large ports that The Netherlands owns have made them become "The Gateway to Europe". The European Commission published, every year a study by EUROSTAT on commercial activity in EU ports. The port of Rotterdam is the most important transit port of Europe and also it is the largest container port which is still expanding. So, 396 000 000 tonnes of freight was loaded or unloaded at Rotterdam in 2012, a quantity that was approximately equal to the one recorded in 2011, using EUROSTAT data. In addition, Amsterdam Port is ranked 5th in the European ranking and is, for example, the largest cocoa port in the world [16].

The most important ports for Norway are Oslo, Lavrik, and Kristiansand. The Port of Oslo can be described as the gateway to Norway and is Norway's largest port by far. With its modern facilities, central location and short approach, the Port of Larvik represent an important and natural hub between sea and land freight. A total of 61 800 TEUs passed through the port in 2012, making it Norway's second largest container port. The Port of Kristiansand is southern Norway's main port for cargo and ferry traffic and offers the shortest sailing distance to Central Europe and has seen formidable growth, so in 2012 growth exceeded 12%. The terminals handle containers effectively using mobile cranes and are staffed 24 hours a day [17].

When we talk about Sweden it is important to point out that there are 52 cargo ports. Port of Gothenburg is one of the stronger in the container sector where it handles almost 60% of Swedish volumes. Using the data from EUROSTAT the number of containers (TEUs) handled in Gothenburg was 837 000 in 2014, four times more than the second Swedish container port, Helsingborg. Another container port that is the largest container port on the Swedish east coast is Port of Gavle that has announced plans to double its capacity in a project worth approximately US$ 83 200 000 like the authority of Gavle Port had announced [18]. The expansion, which will increase the capacity to 600000 TEUs annually, is expected to be completed in the fourth quarter of 2019. Strategically located just north of Stockholm and right next to the industrial Central Sweden region, Port of Gavle is a natural east coast hub for import and export. Based on data in TEUs from [10] we realized the Table 5 and had drawn the graph from Figure 5.

Table 5

Country/ Year	2010	2011	2012	2013	2014	2015	2016
Romania	555 674	662 796	684 059	661 124	668 349	699 000	721 339
Netherlands	11 420 107	12 266 249	12 282 447	12 046 068	12 671 464	12 609 016	12 800 244
Sweden	1 219 255	1 264 876	1 343 778	1 394 989	1 446 200	1 438 700	1 503 400
Norway	335 291	273 027	273 396	276 414	671 400	735 600	739 663

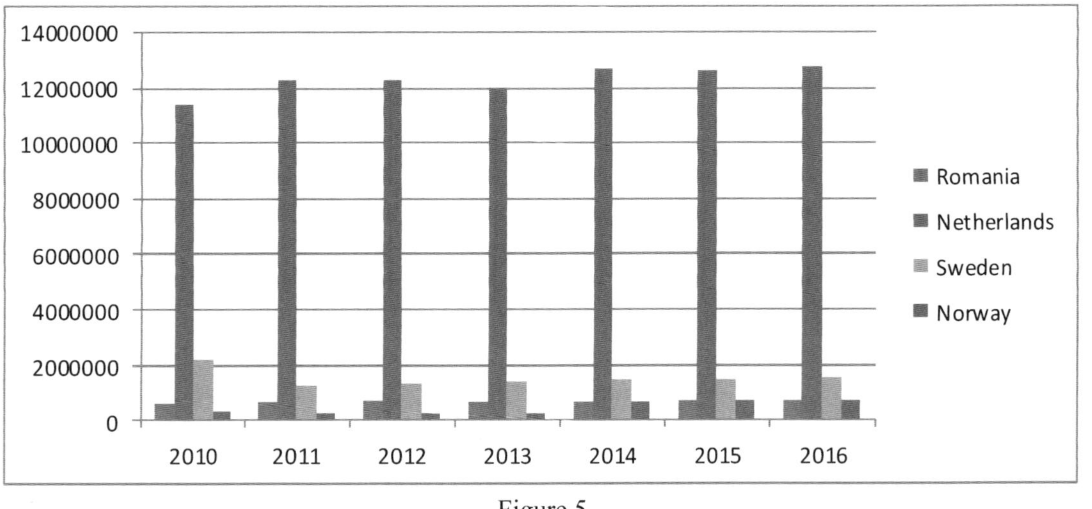

Figure 5

7. Conclusions

Our study, presented in this paper, could be finished with some conclusions which we highlighted in the furthered.

Constanta Port obtained in 2017a growth of 14.53% compared to 2009.

Between 2010 and 2016 Romania obtained an increase of 21.74% while Bulgaria realized a growth of 28.12% and Russian Federation of 17.67%. In the same time Ukraine experienced a downward trend of 2.9%

When we compared the results that give the evolution of containerized transport through the ports of countries involved in "The Three Seas Initiative" , we can say that: Lithuania knew a growth of 33.41%, Latvia one of 34.47%, Estonia has decreased with 8.70% and Poland, Croatia and Slovenia increased with 49.07%, 30.38% and 43.52% respectively.

In the same time Greece increased with 82.94% which represent by far the best grow from our study. On the other hand, Spain and Italy obtained an increase of 17.33% and 7.93% respectively.

When we studied Netherlands, Sweden and Norway we observed that they have obtained an increase of 10.79%, 18.91% and 54.67% respectively.

Finally, we can say that this work can be a starting point for developing a future study in the field both for students from the Degree Master Program and for informing port authorities about taken possible measures related to development, efficiency and growth traffic of sea trade containerized goods.

References

[1] Review maritime Transport 2017 http://unctad.org/en/PublicationsLibrary/rmt2017_en.pdf. Accessed January 22, 2018.

[2] Terminale/Containere Portul Constanta http://www.portofconstantza.com/apmc/portal/static.do?package_id=term_containere&x=load Accessed February 9, 2018.

[3] Raport investigatie servicii de transport maritime (document elaborate de Consiliul Concurentei in colaborare cu Autoritatea de Concurenta din Italia in baza unei initiative a "Componentii Concurenta din cadrul Proiectului de Twinning RO/02/IB/IF/02 "Continuarea dezvoltarii capacitatii autoritatii de concurenta de implementare a acquis-ului comunitar in domeniul concurentei si al akutorului de stat la standardele UE de aplicare" http://www.consiliulconcurentei.ro/uploads/docs/items/bucket6/id6540/raport_investigatie_servicii_de_transport_maritim.pdf . Accessed January 29, 2018.

[4] Guide to Port Entry, Shipping Guides Limited Reigate, England, 2009/2010.

[5] Terminale/Containere Portul Constanta http://www.portofconstantza.com/apmc/portal/static.do?package_id=term_containere&x=load Accessed February 9, 2018.

[6] Statistici generale Portul Constanta 2009-20017 http://www.portofconstantza.com/apmc/portal/static.do?package_id=st_generale&x=load Accessed February 9, 2018.

[7] GAIN Report Number RSATO 1306 https://gain.fas.usda.gov/Recent%20GAIN%20Publications/Exporting%20to%20Russia%20via%20the%20Greater%20Port%20of%20St.%20Petersburg.%20A%20Logist_St.%20Petersburg_Russian%20Federation_4-22-2013.pdf. Accessed February 20, 2018

[8] GAIN Report Number RFATO007 https://gain.fas.usda.gov/Recent%20GAIN%20Publications/Russian%20Port%20of%20Vladivostok%20Snapshot_Vladivostok_Russian%20Federation_3-10-2015.pdf. Accessed February 20, 2018.

[9] Black Sea Container Summit 2016, http://summit.portsukreine.com/en. Accessed January 29, 2018.

[10] Containerized port traffic http://stats.unctad.org/teu. Accessed January 22, 2018.

IOP Conf. Series: Earth and Environmental Science **172** (2018) 012020 doi:10.1088/1755-1315/172/1/012020

[11] Port Strategy Insight For Port Executives http://www.portstrategy.com/news101/world/europe/13-growth-for-black-sea-container-terminals. Accessed February 23, 2018.
[12] Summit "The Three Seas Initiative" https://www.europeaninstitute.org/index.php/ei-blog/305-july-2017/2233-three-seas-initiative-for-eastern-europe-8-2. Accessed February 12, 2018.
[13] National Bank of Grece, Sectoral Report, April 2013 https://www.nbg.gr/greek/the-group/press-office/e-spot/reports/Documents/Container_Ports_2013.pdf. Accessed February 23, 2018.
[14] Italy: Exporting Products, https://en.portal.santandertrade.com/international-shipments/italy/exporting-products. Accessed February 16, 2018
[15] Spanish container volumes up in 2016, https://port.today/spanish-container-volumes-up-in-2016 Accessed February 15, 2018.
[16] EUROSTAT http://ec.europa.eu/eurostat/statistics-explained/index.php/Maritime_ports_freight_and_passenger_statistics#The_Netherlands_is_EU.E2.80.99s_largest_maritime_freight_transport_country. Accessed February 8, 2018.
[17] Statistics Norway, Maritime Transport, https://www.ssb.no/en/transport-og-reiseliv/statistikker//havn/kvartal/2016-06-29. Accessed February 14, 2018.
[18] About Port of Gavle, http://gavlehamn.se/EN/about-port-of-gavle. Accessed February 14, 2018.

IOP Conf. Series: Earth and Environmental Science **172** (2018) 012021 doi:10.1088/1755-1315/172/1/012021

Pulse system for evaluation of parameters of electro-acoustic transducers in a hydroacoustic tank

Iliyan Iliev[1], Nikolai Kolev[2]

[1]Assoc. professor at Electronic Department, Nikola Vaptsarov Naval Academy, 9026 Varna, Vasil Drumev street, 73
[2]Assoc.professor at Electronic Department, Nikola Vaptsarov Naval Academy, 9026 Varna, Vasil Drumev street, 73

E-mail: ilieviliyan@ymail.com
n.kolev@naval-acad.bg

Abstract: In the report an applicationof the pulse method is presented for hydroacoustic test tank measurements. The basic test tank hydroacoustic equipment is from Bruel&Kjaer with National Instruments based control and measurement hardware and software platform. A LabVIEW software is used for synthesis, transmission and measurement of acoustic signals with different pulse lengths and intra-pulse modulations. A number of VI's are designed for digital signal processing and visualization of the signals and results. Capabilities are shown for test tank measurements of sonar transducers pattern and quality factor.

1. Introduction.

The basic parameters to be measured for characterization of an electro-acoustic transducer are radiation and reception sensitivities, electroacoustic efficiency, directional characteristics, quality factor Q, and maximum output power. These parameters can be measured in a free from reflections isotropic and uniform underwater space. Several approaches can be used for underwater electroacoustic measurements according to Bobber [1]. They fall in the general category of free-field far-field measurements recommended by the theory. Facilities to ensure this kind of measurements are of various kinds-piers, bridges, barges, or other semi mobile floating structures, and boats or ships. Even this facilities however are far from ideal in providing the conditions assumed by the theory. A true free field or a uniform boundless medium is only an Idealized concept in this case. Even in relatively big facilities there are surface and bottom boundaries contributing reflections, influenced by meteorology changes in the temperature and water boundaries thus preventing accurate measurements and introducing errors in real transducer directivity pattern. Artificial pools and indoor tanks are used also, but they usually require pulsed sound measurement techniques. The basic requirements are: 1) enough space so that interference from boundary reflections can be eliminated by pulsing techniques, anechoic boundaries, or spreading losses due to long distances; 2) a low ambient noise level; and 3) a water medium that is relatively free of anything that will cause refraction and scattering, like currents, temperature gradients, marine life, bubbles, and pollutants. A more satisfactory method of eliminating the effects of reflected waves in measurements is by the use of pulses. Instead of a steady constant waveform (CW) signal, a pulse corresponding to a sinusoidal CW signal of finite duration is emitted.

This pulse reaches the receiver by the direct path before the reflected pulses arrive. If the response of the transducer can be measured in the interval before their arrival, their effect is entirely eliminated. Usually measurements of this kind are made in test tanks which ensures also controllable environment.

2. Description of impulse measurement system in a hydroacoustic pool developed in Naval Academy.

The sonar laboratory, developed at Bulgarian Naval Academy under National Science Fund infrastructure project [2] is suitable for transducer parameters measurements using the pulse method. Pulsed measurements are fundamental for characterizing transfer functions of systems and objects in free-field and simulated conditions, which are available in a sonar test tank. [1, 3] It is a common technique for eliminating the effects of reflection (reverberation) from boundaries.

In order to have adequate experimental results it is necessary to do the pulse measurements in a free from reverberation and far for the electroacoustic transducer zone (where the sound wave front is assumed flat in the place of the receiving transducer). The measurement hydrophones are situated in the test tank measurement volume. The parameters of the positioning are as follows: depth – h and distance – d between them. The size of the limiting volume boundaries are: - length L and width B – 'Figure 1'.

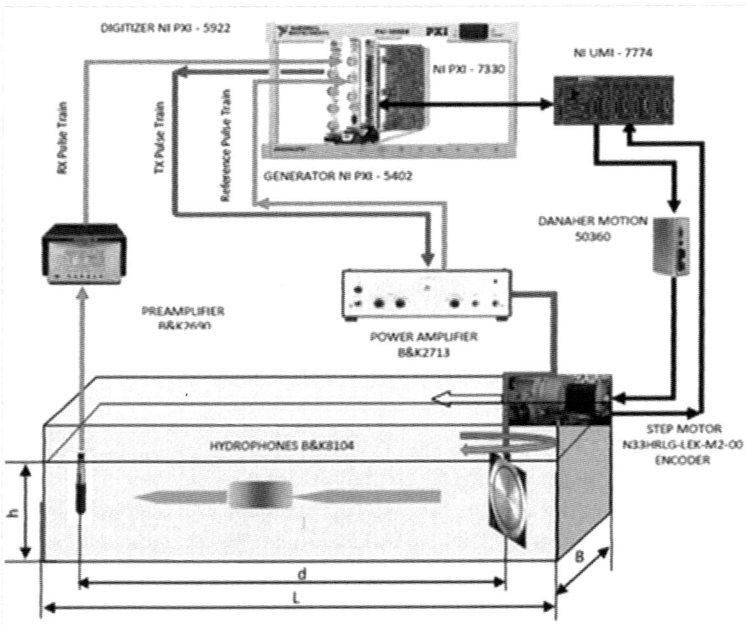

Figure 1. Schematics of the test tank pulse experimental setup.

An electroacoustic transducer (projector) is driven with a pulse or short burst CW signal –'Figure 2'. The transmitter and receiver are controlled so that an acoustic signal will be measured only during the short period (gate) of time when the direct projector-to-hydrophone signal pulse impinges on the hydrophone. Pulses arriving prior to this time (like cable crosstalk) and pulses arriving after this time (like reflections from boundaries) are rejected by the receiving system. This "gating" process is necessary for adequate pulse measurements. For far field condition the distance should be [2]:

$$d \geq \frac{a^2}{\lambda} \tag{1}$$

where, a is the biggest dimension of the hydrophone and λ is the wave length.

The pulse length or duration τ must be long enough for a steady-state condition of oscillation. The practical criterion for this transient period is that it consists of at least Q_m cycles, where Q_m is the

IOP Conf. Series: Earth and Environmental Science **172** (2018) 012021 doi:10.1088/1755-1315/172/1/012021

quality factor of the electromechanical system. Rise time (or build up time) of a projector is the time required for a pulse to reach its steady state amplitude, and can be estimated by its:

$$\tau' = \frac{Q_m}{\pi f_0} \tag{2}$$

where τ' is a time constant, $\tau' = \frac{2L}{R}$, L - inductance (equivalent mass), R - resistance (damping constant), f_0-resonant frequency, Δf is -3dB bandwidth.

In order to eliminate influence of reflections during measurements the following conditions should be met for the pulse width:

$$\tau < \frac{2d}{c} \tag{3}$$

to eliminate reflections between transducers;

$$\tau < \frac{L-d}{c} \tag{4}$$

to eliminate reflections from the walls of the test tank and

$$\tau < \frac{\sqrt{x^2+d^2}-d}{c} \tag{5}$$

to eliminate reflections from the water surface.

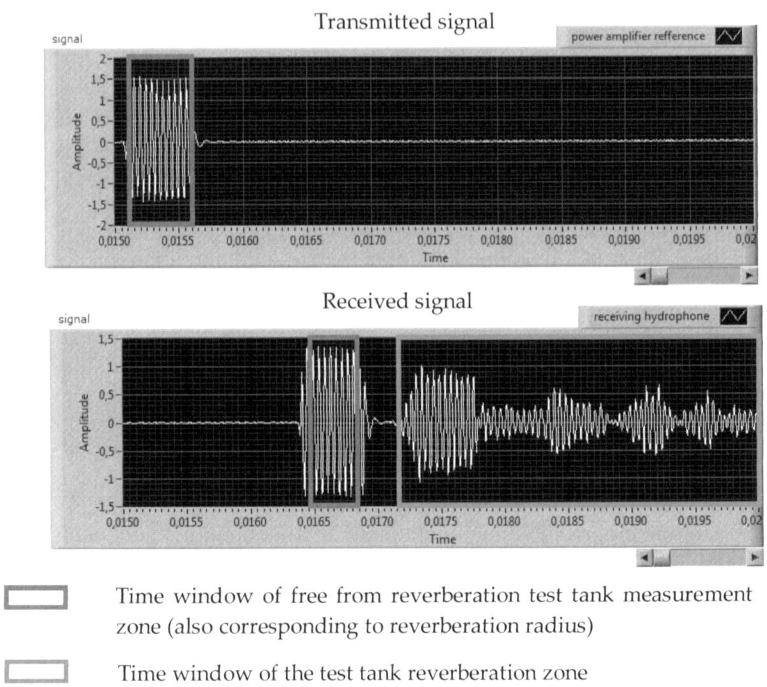

Figure 2. Test tank pulse gating measurement principle with direct pulse in red.

3. Transducer pattern and Q factor measurements.

3.1. Theoretical model of a circular transducer radiation.

The sound pressure of a circular transducer with uniform normal velocity (called a circular piston because of the uniform velocity) in the far field is given with the following expression in cylindrical coordinates [5]:

$$b(\theta) = \frac{2J_1(ka \sin \theta)}{ka \sin \theta} \tag{6}$$

where $J_1(x)$ is Bessel function of first order, k is the wave number and a is the radius of the transducer.

It must be pointed out that this expression is valid for circular piston vibration in an infinite baffle under constant harmonic mechanical excitation. This is also analytical approximation of the far field acoustical pressure field or for very small vibration displacements of the flat transducer surface. It can be expected that the measured radiated pressure field of a real transducer will be approximated by this function.

3.2. Experimental results of a circular transducer radiation.

The pattern of piezoelectric circular piston with diameter of the diaphragm 70 mm is measured. Pulse train signals with frequency 30 *KHz*, width 2 *mS*, pulse repetition period 50 *mS*, transmitted through the water and recorded at the output of the receiving hydrophone are shown on –'Figure3'. [4] The one-tenth of the reference (transmitted) pulse train (red) is sampled at the control output of power amplifier 2710.

Fig. 3 (a)

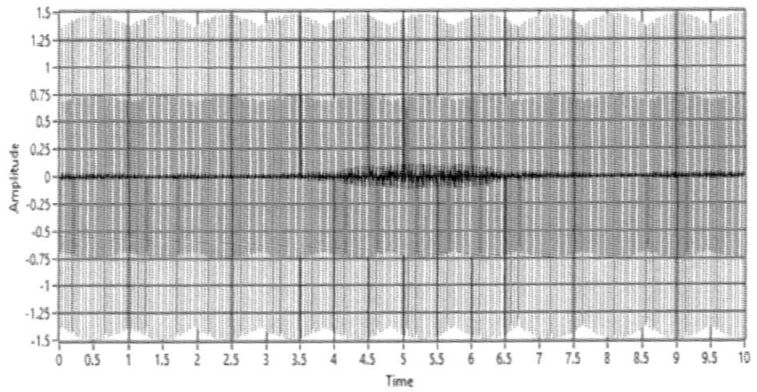

Fig. 3 (b)

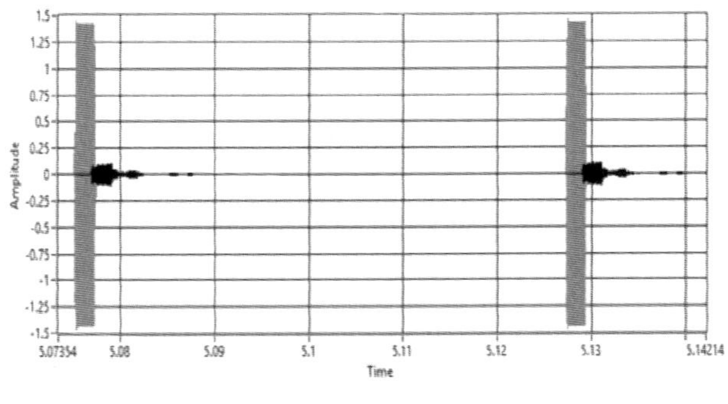

Fig. 3 (c)

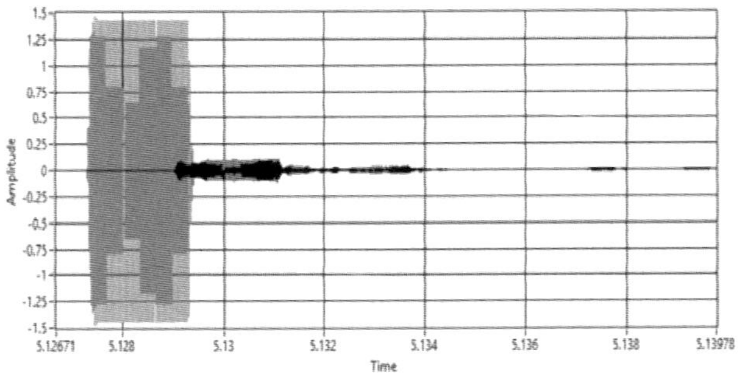

Figure 3. Transmitted and received pulse train at frequency 29.5 KHz in 10 s total time interval of full transducer rotation on 360° in 3 different time scales: Fig. 3(a), Fig. 3(b), Fig. 3(c).

The recorded pulse train was preprocessed and measured in LabVIEW and exported in MS Excel where the final pattern was obtained.

Comparison of the normalized measured pressure pattern and theory pattern of the piston transducer, according to expression (6), are shown on – 'Figure 4'. The received RMS pressure field was measured only for direct signal in time window 0.5 *ms* from the initial direct pulse front to escape from any reverberation.

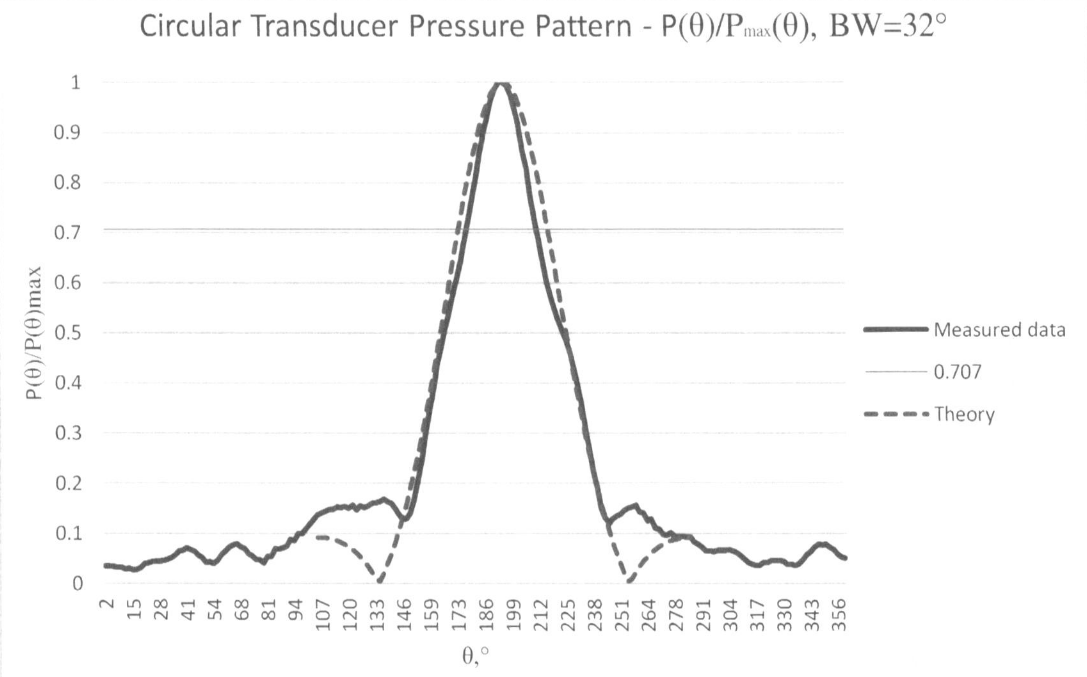

Figure 4. Comparison of the normalized measured pressure directivity pattern and theory pattern – expression (6), of the piston transducer at resonance frequency 29.5 KHz.

3.3. Theoretical model of a transducer Q factor measurement.

The acoustic transducer with the power cable located in an aquatic environment forms a complex oscillating system that has a qualitative factor Q. Measurement of the Q factor can be done by using the "ring-down" method or by using its frequency response.

The process of determining the pressure of the transmitted pulse from the transducer is described by the expression:

$$p(t) = p_{max}\left(1 - e^{-\frac{\pi}{TQ}t}\right) \tag{7}$$

where, T – period of transducer own mechanical oscillations;

$p(t)$ - the pressure created by the transducer in an unsettled mode;

p_{max} – the amplitude or maximum pressure created by the transducer system in an established (steady) mode;

The equipment of the acoustic tank allows observation of the increase of the sound pressure level of the oscillations, and counting how many cycles it takes it to rise to the halve of the amplitude. According to formula (7) if the pressure rise to the half amplitude at time $N.T$ the expression for the pressure is equal to:

$$0{,}5p_{max} = p_{max}\left[1 - e^{-\frac{\pi}{TQ}NT}\right] \tag{8}$$

where, N is the number of cycles (periods T) it takes the pressure to rise to the half of the amplitude.

From the last expression after conversion the quality factor of the electro-acoustic converter can be found:

$$Q = \frac{\pi}{\ln(2)}N = 4{,}53N \tag{9}$$

This expression is the same as the expression of the "ring-down" method.

A second method to find Q factor is to use the frequency response of the transducer and mathematical approximation such as:

$$Q = \frac{f_o}{f_2 - f_1} \tag{10}$$

where f_o is the resonance frequency, f_2 - f_1 is the frequency range between the signal's power is decreased to half the size of the maximum power at resonance frequency.

3.4. Experimental results of a transducer Q factor measuring.

The 'Figure 5' shows a transmitted pulse from a disc piezo ceramic transducer with 7 cm diameter: the horizontal yellow dense line is the reference level of the amplitude, the dashed yellow line is the half of amplitude rising.

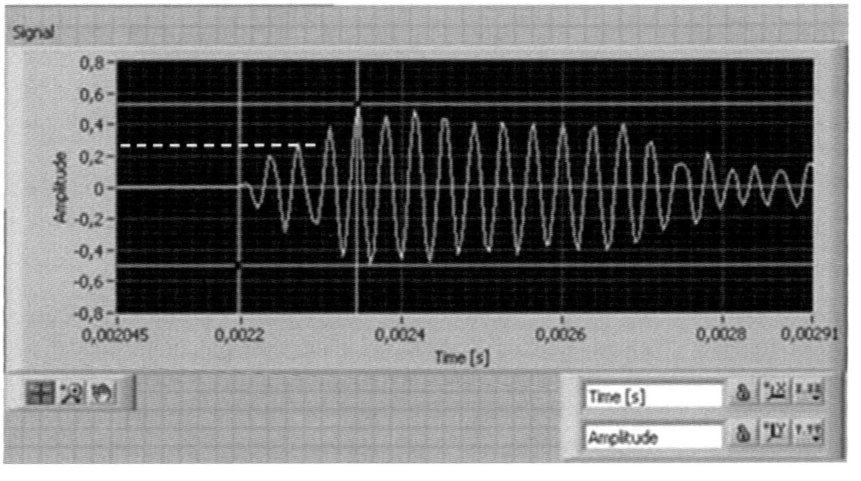

Figure 5. Transmitted pulse from disk piezo ceramic transducer with a diameter of 7 cm, pulse length $\tau = 0{,}00046$ s and a frequency $f = 29{,}5$ KHz.

Here, the transmitted pulse take it 100% reference at $t = 0{,}0023$ *(s)* and we see that it takes about 2 cycles for the amplitude to rise by 50%. Simply by multiplying *2.4,53* we find quality factor $Q \cong 9{,}06$.

The 'Figure 6' shows the normalized frequency power response of the transducer measured in test tank with pulse method. The data has exported and processed in Excel.

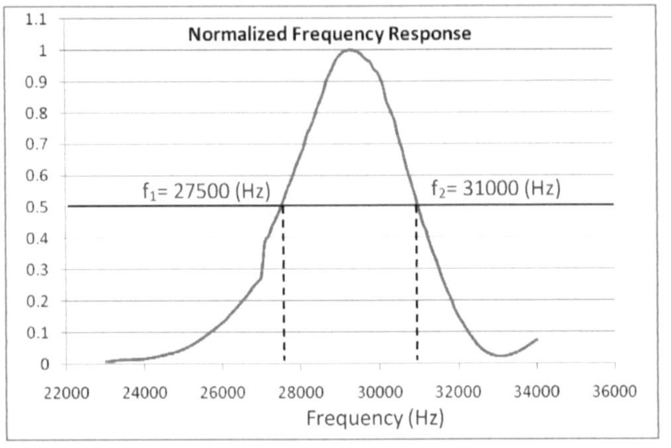

Figure 6. The normalized frequency response of the disk piezo ceramic transducer with a diameter of 7 cm.

From the frequency response we find that the resonance frequency of the transducer is 29,5 KHz and -3 dB bandwidth is between 27,5 KHz and 31 KHz. After applying (12), the Q factor is obtained:

$$Q = \frac{29500}{31000-27500} = 8{,}4 \tag{11}$$

4. Conclusion.

Based on the B & K and National Instrument equipment and NI's LabVIEW software environment, a system and methodology has been developed to enable the implementation of the impulse measurement method in a hydroacoustic test tank. Virtual instruments have been developed based on this method for measuring the direction diagrams and quality factor of electroacoustic transducers.

References

[1] Bobber R.J., *Underwater electroacoustic measurement*",Naval research laboratory, Washington D. C.,1970.

[2] Kolev N., Ivanov P., Iliev I., *An investigation of the pulse method for hydro-acoustic measurements and Naval Academy test tank performance estimation*, National Conference Acoustics 2010, Conference Proceedings, Vol.13, pp. 10-18, ISSN 1312-4897.

[3] Levin P.A., M.Sc., *Calibration of Hydrophones, Brüel&Kjær Technical Review*, №1, 1973, pp. 3-17.

[4] Kolev N., Ivanov P., *Circular piston transducer test tank directivity pattern measurement*,National Conference Acoustics 2015, Conference Proceedings, Vol.17, pp. 100-103, ISSN 1312-4897

[5] Robert J. Urick, *Principles of underwater sound*, McGraw-Hill, Book Company, 1975

REFRESHER TRAINING IN MARITIME QUALIFICATION

B Belev[1], M Dimitrova[1] and A Meczkowska-Christiansen[2]

[1] Naval Academy Nicola Vaptsarov, Varna, Bulgaria
[2] Polish Naval Academy, Gdynia, Poland

E-mail: bl.belev@naval-acad.bg

Abstract: Maritime education and training have followed strict regulations since 1978 when the International Convention on Standard of Training, Certification and Watchkeeping was established. Since then, many training courses were developed. The International Maritime Organization divided them in two groups, which are both contents from STCW Part A and Part B. Since the competences in Part A are mandatory for all seafarers and the competences in Part B are recommended, maritime education and training are perceived from the students and trainees in different ways. The refresher training is part of the above problem. The main reason behind the distrustful attitude towards refresher courses as recommended training is the question: "Who has to pay for them".
The next problem, which usually arises, concerns the time that the seafarers need to spend for additional training. Both questions make shipping companies cautious in their requirements towards the crewmembers regarding refresher training.
The third problem concerning refresher training is the level of competency of the trainers and instructors. Many seafarers neglect this kind of training just because they have more experience and skills than the instructors do.
The article summarizes the authors experience in the educational and training processes in the field of postgraduate education and training. It analyzes research from questionnaires carried out in different shipping companies with regard to seafarers operating on board of different types of vessels.

1. Introduction

The STCW-78 and its amendments made after the IMO`s Assembly in Manila in 2010 highlight the issue concerning refresher training of ship crews to the world's maritime community. This subject is not new but until the 27th Assembly, this type of training was considered optional. An example is the onboard training, which some ship-owners have been providing to their crews so far. Ship crews are familiar with the Seagull and Videotel training films as:

- Ship security – basic training;
- Transport of dangerous, harmful and hazardous goods;
- Operational use of ECDIS, etc.

Is this form of training effective? Is the acquired knowledge sufficient? Do Seagull, Videotel and other films help improving the crew's qualification enough?

The answers to the questions which onboard training raises are not constant. The reason for this is that there are the multiple factors, which affect the training process of engineers and officers, and there are the merchant interests of ship-owners and charterers.

Hiring qualified officers and engineers is the priority of every ship-owner. Qualification is the set of knowledge and skills used on board in order to ensure the safe navigation of ships carrying out their merchant tasks. The factor called "Qualification" includes:

- Knowledge, acquired during the basic maritime education.
- Skills, acquired during the initial sea practice together with the mandatory post-graduation sea practice.

- Knowledge from qualification courses.
- Knowledge and skills acquired by self-training.

Knowledge and skills alone are not effective, until put into practice. An expression of the qualification of an officer or an engineer is his ability to make decisions adequate to the working environment, the surroundings of the ship and the performed tasks.

The notion of "qualification" has another meaning. From the STCW-78 Convention, it consists of the compulsory knowledge and skills that each seafarer applying for a certificate of competence must acquire and demonstrate. The Convention does not provide guidance on how to acquire knowledge and skills. This issue is solved at a state level.

2. Provisions for refresher training

The immense importance of the human factor in modern shipping is an indisputable fact. Analysis of a number of navigational accidents has lead the investigators to the impact of decisions made by a single person – the Captain, the OOW or the Engineer, to the development of the events.

Highly qualified personnel in the maritime industry are the priority of education and qualification in the maritime training institutes. In his address to the participants in the international assembly of the maritime universities (IAMU) in October 2011, the Secretary-General of the IMO, Efthimios Mitropoulos, stated that the education and the qualification of maritime personnel are of immense importance to the safe operation of ships and preserving maritime environment. The process, according to the secretary-general Mitropoulos, is based on the scientific and academic discipline of creating a bridge between the practical and managing skills of the trainees, and a strict focus on the quality of education.

This statement expresses the official opinion of the IMO, in relation to the place of training and qualification of maritime personnel. The organization supports the critical part of the maritime educational institutes in the development of the maritime industry. Its success is only possible via the close cooperation between all involved – the government, the education, the maritime industry, the seafarers and their families [1].

After the 27th assembly of the IMO, the organization added the STCW convention regulations with requirements regarding refreshment training for some types of training. For the officers and engineers of the management and operational level the refresher courses for basic training according to regulations VI/2, VI/3, VI/4 of the convention are mandatory when the period of validity of the competency certificates extends after 2017. Apart from these trainings, the STCW also controls the regulations and the refresher training periods for additional training for officers, engineers and crew of bulkers and passenger ships [2, 3, 4].

Charterers, inspectors and others, who are interested in guaranteeing safe navigation, often require additional and refresher training for seafarers. In the qualification center of the "Navigation" faculty in the Nikola Vaptsarov Naval Academy, several types of refresher training can be acquired:

- Bridge team and resource management
- Ship handling and maneuvering
- Transport of dangerous, harmful and hazardous materials.

The meetings with the participants in the mentioned training courses and their opinions on the offered services gave a reason for the conducting of a questionnaire regarding the necessity and the benefit of refresher training.

3. Contents and analysis of the questionnaire

A questionnaire of 15 questions was filled in, which gathered opinions from 168 participants occupied as officers or engineers. The questions were composed and arranged in such a manner as to create a gradation of the impression from the maritime education – from the general case of elementary maritime education to the personal case of qualification education and refresher training. In addition, respondents were asked about their position on the ship, the type of ship on which they were sailing and the total years of seagoing service. The survey included the idea of life-long learning, which has begun to gather followers and supporters in many maritime organizations and institutions. More and more companies in the industry are taking steps to apply this type of education to the crews through onboard training, fleet officer meetings and refresher courses.

While seafarers accept onboard training without any particular objections, the seminars and refresher trainings are likely to have certain negative responses, mainly because of the engagement they require during the stay ashore. For every seafarer the holiday is sacred, especially on the first days after signing of a ship. There are not many people who would be willing to sacrifice their time ashore for qualifying activity, especially if they have to take part in courses with themes that are already familiar.

The survey was performed in a period of two months. The participants were involved in short term qualification courses lasting from 2 to 5 days and long-term courses with duration over 5 days:

- Bridge team and resource management – 5 days;
- Operational use of ECDIS – 5 days;
- Ship handling and maneuvering – 3 days;
- Transport of dangerous, harmful and hazardous materials – 4 days;
- Maritime environment awareness – 2 days;
- Model course 7.01: Master and Chief Mate – 3 months.

The engineers were involved in only two courses - „Maritime environment awareness" and „Transport of dangerous, harmful and hazardous materials".

During the second decade of the 21st century, the Maritime Education System underwent major changes. Its development is obvious and undeniable. The industry, however, requires skills and knowledge, which much more practical from the maritime officers and engineers. This trend set on by the STCW and accepted by the 27th Assemble in Manila in 2010, shifted the educational focus towards acquiring and improvement of such knowledge, while theoretical training received a rather supportive role. In this regard, the first set of questions from the survey is directed to primary maritime education. In Table №1 questions are listed along with their response in the form of distractors. The respondents should choose only one answer.

Table 1 – First group of questions from the survey and the answers to them

FIRST GROUP OF QUESTIONS	№ 1	№ 2	№ 3	№ 4
1. Do you consider that the education in maritime university is quite enough as vocational education for future ship's officers?	Yes, completely.	Yes, but a periodical refresher training is necessary.	Rather enough.	No, it is definitely not enough.
2. Is there a need for further training for junior officers and management staff?	Yes, it is mandatory for ship officers to have further training	Yes, it would be good to organize such training.	Yes, but maybe without such training	No, it is not necessary.
3. During what period should refresher training be organized for seafarers?	Regularly every year	In the every change and amendment in the major conventions in the shipping	By decision of the shipowner	As long as possible.

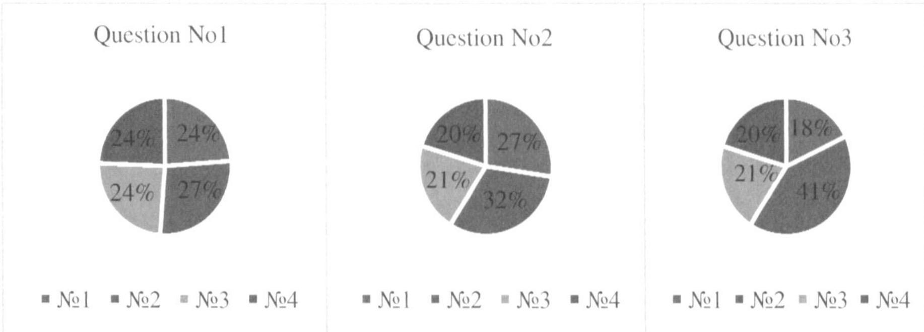

Figure 1 – Answers to questions № 1, 2 and 3 distributed in percentiles

The results from the first two questions showed a hesitant attitude towards maritime education and the possibility of further training. The answers to the third question showed that 82% of the respondents preferred to participate as little as possible in refresher training, or courses that would be organized in the event of change in some of the major conventions. Unfortunately, only 18% of the respondents wished to be involved in annual refresher training.

This contrast of opinions is largely the result of a lack of experienced young officers and a lack of sufficient maritime experience in establishing clear criteria for professionalism. It is logical for young officers who have recently been in an academic environment not to want to return to it again or at least not so soon. On the other hand, senior officers are clearly aware of the knowledge and skills required at sea and of whether they can acquire them at the university or while working aboard. We cannot ignore the differences between the generations that represent these two groups.

Nearly half of the respondents agreed that all innovations and changes to the rules of maritime conventions must be presented to seafarers in some form, but there is no unanimous opinion as to what period this should implement. Today's refresher training practice is three or five years [2, 5]. However, it does not exhaust the opportunities for refresher training. Important and up-to-date information can be presented in other forms and with other modern means.

The second group of questions is focused towards the essence of the study - the need for refresher training and qualification, and their structure. The questions with numbers: 4, 8, 9, 11 and 14 allow more than one answer.

Table 2 – Second group of questions from the survey and the answers to them

SECOND GROUP OF QUESTIONS	№ 1	№ 2	№ 3	№ 4
4. Do you think that refresher courses are necessary in your training activities?	Yes, definitely.	Yes, but maybe without them	Yes, but not for all training courses.	No, these courses are a waste of time.
5. In what period specific courses refresher training should carried out.	Every year	Once every three years	Once every five years	Only if the seagoing service is interrupted.
6. What tropics should cover the syllabus of a refresher course?	The whole issue on a given topic.	The only news of the problem, but very detailed.	Balanced information – innovation and the existing rules and	Novelties in the shortest form.

			requirements	
7. Do you consider that refreshing should be carried out in the absence of changes in the subject matter of a qualification course?	Yes, the rules are forgotten over the years.	Yes, a Marine Officer should not break up the connection with training institution.	Rather not.	No, once there is not changes, there is no need to refresh the knowledge.
8. Do you consider that each company needs to introduce its own training topics to its officers?	Yes, it is due to specific organization and requirements of the company.	Yes, because it is useful for officers to maintain their level of preparation.	No, the requirements of international maritime legislation are the same for everyone.	No, it is a waste of time.
9. In what format it is appropriate to conduct if a refresher course is required for a particular training.	In specialized institution under the guidance of a lecturer.	On the ship during the contract (onboard training).	Online in internet.	Using literature available in a specialized bookstore.
11. Which topics in the shipping need refresher training?	Topics concerning human factor.	Engineering topics related to innovations in the bridge electronic equipment	Topics related to changes in legislation (conventions and resolutions).	Topics related to new regulations in the shipping.
12. Is it necessary to standardized refresher training?	Yes, according to flag administratio n requirements.	Yes, according to international regulations	Yes, according to requirements of the company initialized it.	No, the educational institutions determine themselves the way and the syllabus they run.
13. Refresher training should cover:	All ship's crew despite of the type of cargo they carries.	Only the most risky ships caring potentially dangerous goods – tankers, chemical	Passengers ships only.	Vessels for which the ship owner considers that the cargo is potentially dangerous to the crew and the environment.

		tankers, gas carriers		
14. If refresher training on human factor in the shipping is to be carried out emphasis should be placed on the syllabus (please indicate the most important skills in your opinion)	Teamwork.	Decision making skills.	Communication skills	Other (please write)

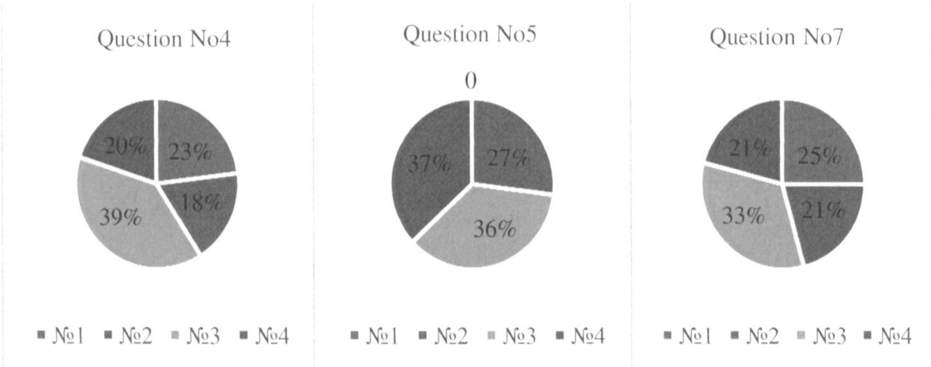

Figure 2 – Answers to questions 4, 5 and 7 distributed in percentiles

The analysis of questions 4 and 5 shows that 1/5 of the respondents indicated, "Answer №4" ("No, these courses are a waste of time" and "Only if the seagoing service is interrupted"), which is a direct refusal of periodic refreshing knowledge. Here the tendency for young officers to declare a strong denial to post-university education is also retained. It is interesting to note that 78% of those respondents sail aboard bulk carriers, general cargo vessels, yachts and pleasure craft where the risks in the working environment ~~is~~ are relatively small. Officers and engineers who sail ~~to~~ aboard tankers, offshore vessels and passenger ships indicate " Answer №1" and " Answer №3" to Question 4 and " Answer №2" and " Answer №3" to Question №5. Such a positive attitude to post-university studies is probably due to the continuous inspections to which these types of ships are subjected and the raised requirements for the seafarers' competence.

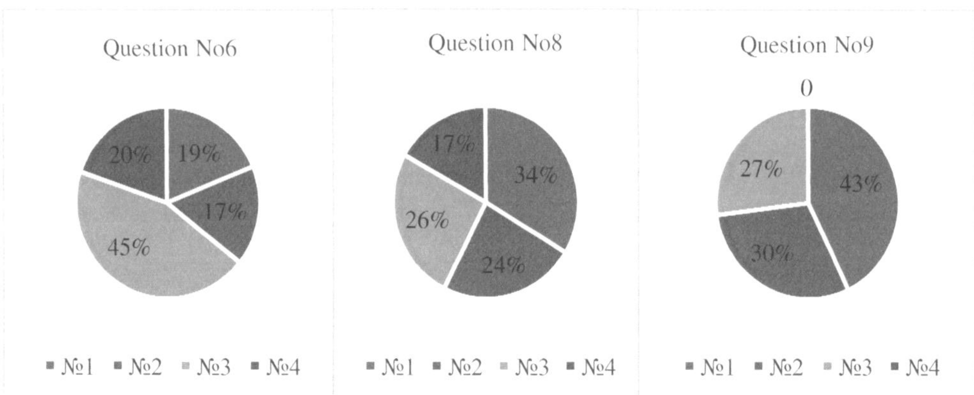

Figure 3 – Answers to questions 6, 8 and 9 distributed in percentiles

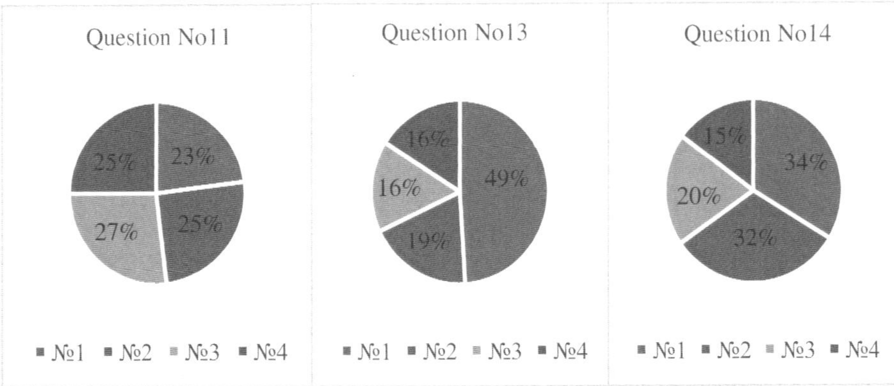

Figure 4 – Answers to questions 11, 13 and 14 distributed in percentiles

Questions № 6, 7, 8, 9, 11, 13, and 14 cover the topic of refresher trainings, which would be of interest for the process of education throughout the entire life. Almost half of the interviewed, 44.64%, would like better to receive balanced information – innovations and existing rules. About 1/5 of the interviewed prefer receiving the entire information of the topic, and 1/3 of the interviewed would demand only the innovations. Here the profile looks like this:

- The senior officers – 62% of those who chose Answer №3 of question 6, require balanced information.
- The innovations of a topic – junior officers mostly, 68% of those who chose answers №2 and №4 of question 6.

The analysis of the answers of question 9 gave an interesting result. Most of the interviewed chose answer №1. Their opinion is that qualified lectors should conduct the refresher trainings in a specialized training center. We can reach the conclusion that the live contact and the possibility for a dialogue are the major motives for the participation in refresher training. Sharing experience in solving specific practical issues is an irreplaceable and extremely useful part of the class form of training. This type of training is preferred even to the modern means of training, no matter the fact that in order to attend this type of qualification training sailors have to sacrifice part of their time between contracts, sometimes even away from their families.

Those who preferred the modern means of training – on board and online – are much less. As expected answers № 2 and № 3 were chosen by the junior officers and engineers – 77% of those who support these two types of training. The most common comments of the opponents of the modern means of training are:

- Unclear questions in the examination part of the training.
- Low level of the presented information for the topic.
- Lack of time for further insight of the issues of the training.
- Absence of live contact with qualified lector or instructor.

Table 3 – Third group of questions from the survey and the answers to them

THIRD GROUP OF QUESTIONS	№1	№2	№3	№4
10. Should state control over the refresher course be implemented?	Yes, as other STCW courses	No, until refresher courses becomes mandatory	-	-

15. Who should pay the participation of an officer in a refresher course?	The shipowner.	The account must be shared between shipowner and the officer.	Depending on the purpose of the course and what document required it – STCW or company rules.	Officer wholly

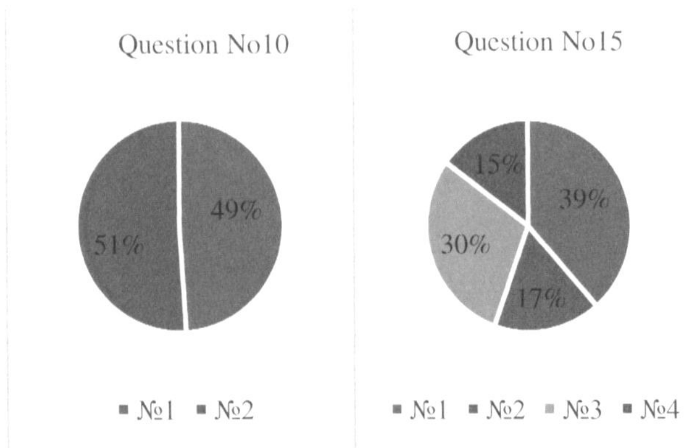

Figure 5 – Answers to questions 10 and 15 distributed in percentiles

In the seafarer`s comments regarding the type of training, the question "Who pays for the course?" is present. The worldwide practice for solving this issue is different from ship owners or seafarers or ship owners completely covering the expenses to implementing a voucher system and sharing the cost [5]. This issue is not major one but plays a part in forming the mood and attitude to the refresher training. Besides, providing the charge for training required by the ship-owner is an element of the social policy of the company, and an attainment that stimulates loyalty and dedication to work by the seafarers.

4. CONCLUSION

After the amendments in the STCW in 1995, the maritime community began paying more attention to the training and qualification of the Officers and Engineers on the ships. The Ship-owners are responsible for the hiring of qualified crew. The controlling organizations such as Flag Administrations or Port State Control via their inspections also provide assistance in ensuring that the crew`s qualification complies with IMO`s requirements. The continued accidents and loss of life at sea are the accelerator of the process of changes in the Maritime Conventions and therefore a stimulus for refresher training for all levels of maritime qualification.

The questionnaire`s goal is to express the opinions of the maritime Officers and Engineers about refresher training, its place and role in improving their qualification in the areas of work environment as well as in their social life ashore. The summary of the results shows us that most of the interviewed are positive about the general process of training throughout the entire life, part of which is the refresher training. It is clear that the Officers and Engineers, which work on ships with increased risk, are more likely to spend some of their free time for training and improving their qualification. The questionnaire also shows that when the training is not mandatory by the main regulations of IMO, the participation motivation is harder to reach.

References

[1] Green Ships, Eco Shippng, Clean Seas, IAMU AGA 12, Conference proceedings, 2011.
[2] International Convention on Standard of Training, Certification and Watchkeeping, STCW-78/95, IMO, London, 2011.

IOP Conf. Series: Earth and Environmental Science **172** (2018) 012022 doi:10.1088/1755-1315/172/1/012022

[3] Directive 2008/106/EC on the minimum level of training of seafarers (recast), Official Journal of the European Communities L323 of 3.12.2008 p.33-61.
[4] Directive 2012/35/EU of the European Parliament and of the Council of 21 November 2012 amending Directive 2008/106/EC on the minimum level of training of seafarers. L343 of 14.12.2012 p.78-105.
[5] Code of Federal Regulations, 46 CFR Ch. I (10–1–15 Edition), [Electronic version], Access through URL: https://www.gpo.gov/fdsys/pkg/CFR-2015-title46-vol1/pdf/CFR-2015-title46-vol1-sec12-609.pdf (Date of using: 06.09.2016).

Development of an Algorithm for Calculation and Application of Conformal Mapping Methods on the Calculation of Hydrodynamic Coefficients

S. Ferreira *Escola Naval, CINAV, Almada, Portugal*

P. Pires da Silva *Escola Naval, CINAV, Almada, Portugal, and Centre for Marine Technology and Ocean Engineering (CENTEC), Instituto Superior Técnico, Universidade de Lisboa, Lisbon, Portugal*

M. Moreira *Escola Naval, CINAV, Almada, Portugal, and Centre for Marine Technology and Ocean Engineering (CENTEC), Instituto Superior Técnico, Universidade de Lisboa, Lisbon, Portugal*

Escola Naval, *CINAV*, Base Naval de Lisboa, Alfeite, 2810-001 Almada-Portugal

hugo.silva.ferreira@marinha.pt

Keywords: *hydrodynamic coefficients, strip theory, conformal mapping, Lewis forms.*

Abstract. The multipole method was first developed by Ursell [1]. His method consists of the superposition of potential functions that satisfy the Laplace equation, the free-surface boundary condition, and the condition at infinity. The potential functions represent a source and horizontal dipole at the origin, which give the radiated waves at infinity, and a series of multipoles that die off rapidly as one moves away from the origin. The strengths of the source, dipole and multipoles are all determined so that the body boundary condition is met. Ursell used this method to solve the problem of a heaving circular cylinder. For sections that are not circular in shape, conformal mapping is used.

In the multipole method, the mapping function that transforms the ship section into a semi-circle is found. The mapping function can then be used in conjunction with Ursell's known solution for a circular cylinder to find the solution for the actual ship section.

The difficulty in the technique is to determine the proper mapping function for each cross section. Various methods have been proposed to find this mapping function. The most common mapping uses the so-called Lewis-forms [2], [3] and [4].

It is recognized that this method gives smooth solutions over all frequency range (no irregular frequencies). On the other hand, sharp corners are not well represented, and sections with very low sectional area coefficient may not be well represented as well. Even though, for first rough estimates in initial stages of ship design, this method may give results that agree reasonably well with the other more computational demanding methods, in terms of order of magnitude and trend.

Considering the difficulty to find good charts of the Lewis forms data – to the best of our knowledge, the best known are those published by Bhattacharyya [5] - and even more difficult to find the data in digital form. The main purpose of this work is to present a computational method of computing the data given through the *Lewis forms* and apply it to naval ship sections in order to find rough estimates of the hydrodynamic coefficients in heave.

1. Introduction

One of the areas of Ocean Engineering and Naval Architecture of extreme importance is *seakeeping*, as well as its ability to maneuver in the aquatic environment, allowing us to obtain a forecast of the ship's hull behavior in waves in its coupled six degrees of freedom. As a consequence of this, the study of ship dynamics has been traditionally separated into two main areas [6]:

- *Manoeuvring* or controllability in calm water and
- *Seakeeping* or vessel motion in a seaway.

With regard to *seakeeping*, there are a huge number of methods that can be applied, ranging from simple two-dimensional linear numerical methods in frequency or time domain, such as *Strip Theory*, to advanced nonlinear methods such as *Computational Fluid Dynamics* (*CFD*). Both methods have advantages and disadvantages. Although the *Strip Theory* is simpler and considers certain assumptions (e.g., the flow is *2D* and the fluid is inviscid), it continues to have results with good precision enough for initial stage design and in which the computational effort is not so demanding, being possible to obtain them with computers with less effort and faster [7].

The *2D* methods using a *Strip Theory* to find the ship's response in seaway need as one of the inputs the *hydrodynamic coefficients* to solve the equations of motion. From the several methods available today, the most used is the panel method *2D/3D*. However, other less computational, time a cost demanding methods are still possible to use, at least for first rough estimates.

The multipole method was first developed by Ursell [1]. His method consists of the superposition of potential functions that satisfy the *Laplace equation*, the free-surface boundary condition, and the boundary condition at infinity, and the body boundary condition. The potential functions represent a source and horizontal dipole at the origin, which give the radiated waves at infinity, and a series of multipoles that die off rapidly as one moves away from the origin. The strengths of the source, dipole and multipoles are all determined so that the body boundary condition is met. Ursell used this method to solve the problem of finding the *hydrodynamic coefficients* for a *heaving* circular cylinder. For sections that are not circular, *conformal mapping* has been used.

In Ursell's multipole method, a mapping function that transforms the ship section into a semi-circle is found, which can then be used in conjunction with Ursell*'s* known solution for a circular cylinder to find the solution for the actual ship section.

The difficulty in the technique is to determine the proper mapping function for each cross section. Various methods have been proposed to find this mapping function. The most common mapping uses the so-called *Lewis forms* [2], [3] and [4].

It is recognized that this method gives smooth behaviour of the solutions over all frequency range. On the other hand, sharp corners are not well represented, and sections with very low sectional area coefficient may not be well represented as well. Even though, for first rough estimates in initial stages of ship design, this method may give results that agree reasonably well with the other more computational demanding methods, in terms of order of magnitude and trend.

Nowadays it is difficult to find good published charts with Lewis form's data. The best known are those published by Bhattacharyya [5], which are difficult to read and convert reliably.

2. Ship Dynamics

Regarding the dynamic analysis of *seakeeping*, certain simplifications need to be accounted: the ship will be considered a rigid body with small amplitudes motions.

2.1. Motions and Reference Frame

It is necessary to predict the vessel´s translational (*surge*, *sway* and *heave*) and rotational (*roll*, *pitch* and *yaw*) motions. These motions are considered as being six degrees of freedom, as is shown in Figure 1.

IOP Conf. Series: Earth and Environmental Science **172** (2018) 012023 doi:10.1088/1755-1315/172/1/012023

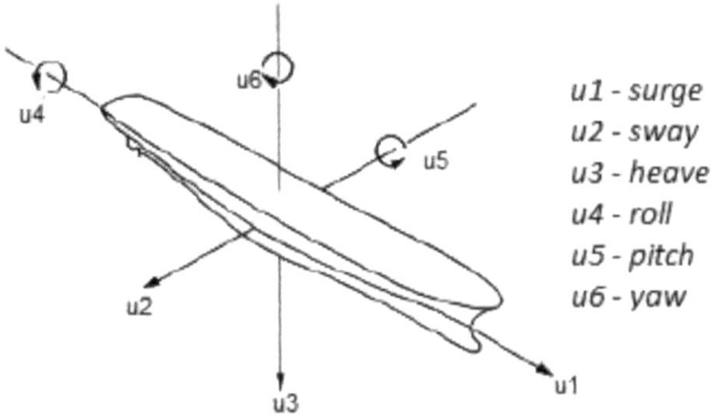

Figure 1. Six degrees of freedom of movement of the ship [6].

Unfortunately, there is no universal coordinate system accepted in the literature of the behavior of the ship at sea. Thus, taking into account the main linear numerical method in the frequency domain applied to linear waves (*Strip Theory*), two coordinate systems are normally used [8]:

- The *ship-fixed system* (*non-inertial system*) x, y, z, with axis pointing from amidships forwards, to starboard and towards the keel. In this system, the center of gravity of the ship is independent of the time x_g, y_g, z_g;
- The *Earth-fixed system* (*inertial system*) ξ, η, ζ, which follows the constant movement of the vessel with velocity $V = \sqrt{u^2 + v^2}$, dependent on the quasi-velocities in surge – u, and sway – v.

It should be noted that, in *seakeeping*, the coordinate system used is usually the *inertial system*.

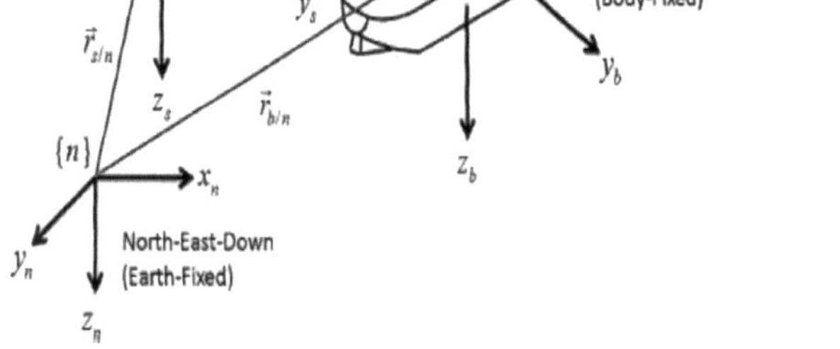

Figure 2. Seakeeping coordinate systems [9].

2.2. Linear Equations of Ship Dynamics in Regular Waves

The prediction of a ship's response in seaway, the *seakeeping*, is a complex process, involving the interactions between the ship's own dynamics and the surrounding hydrodynamic forces. Knowing the ship's responses in regular waves for different frequencies, we can predict its behavior for several sea states.

The general form of the linearized equations of ship dynamics in the six degrees of freedom, or in other words, the *Euler equations* of ship motion used in the literature devoted to *seakeeping*, using the fixed axes on the ship can be described as follows [9]:

$$\sum_{k=1}^{6} \Delta_{jk}\, \ddot{\eta}_k(t) = F_j(t) \quad j = 1,2 \ldots 6 \tag{1}$$

where:
Δ_{jk} – the inertia matrix components of the ship, such as mass and moment of inertia;
$\ddot{\eta}_k$ – the accelerations in mode k;
F_j – the sum of the forces and moments acting on the body in the direction j;
and F_j are harmonic functions in the time base.

Linearizing equation (1), certain terms in Δ_{jk} may be considered zero, as shown by [10], which for a ship with lateral symmetry [9], this equation can be reduced to the following six equations relating to the six degrees of freedom:

$$\begin{gathered}
\Delta(\ddot{\eta}_1 + z_c\ddot{\eta}_5) = F_1 \\
\Delta(\ddot{\eta}_2 - z_c\ddot{\eta}_4 + x_c\ddot{\eta}_6) = F_2 \\
\Delta(\ddot{\eta}_3 - x_c\ddot{\eta}_5) = F_3 \\
I_{44}\ddot{\eta}_4 - I_{46}\ddot{\eta}_6 - \Delta z_c\ddot{\eta}_2 = F_4 \\
I_{55}\ddot{\eta}_5 + \Delta(z_c\ddot{\eta}_1 - x_c\ddot{\eta}_3) = F_5 \\
I_{66}\ddot{\eta}_6 - I_{64}\ddot{\eta}_4 + \Delta x_c\ddot{\eta}_2 = F_6
\end{gathered} \tag{2}$$

where:
$F_j(t), j = 1,2,3$ - the sum of the forces in the directions x, y, z respectively;
$F_j(t), j = 4,5,6$ – the sum of the moments acting on the x,y and z axes, with the positive moment following the right hand rule;
Δ - the total mass of the ship;
$I_{jj}, j = 4,5,6$ – the moments of inertia about the x,y and z axes, respectively;
I_{46} - the product of inertia between the degrees of freedom *roll-yaw*= I_{64};
$(x_c, 0, z_c)$ – the coordinates of the center of gravity of the ship in the non-inertial system x, y, z;
$\ddot{\eta}_j(t)$ – the acceleration in the degree of freedom j, in the system with the fixed axes in the ship, referring to $j = 1, 2, 3, 4, 5, 6$ the *surge*, *sway*, *heave*, *roll*, *pitch* and *yaw*, respectively.

Comparing the equations (1) and (2) it is possible to write the inertia matrix Δ_{jk} as:

IOP Conf. Series: Earth and Environmental Science **172** (2018) 012023 doi:10.1088/1755-1315/172/1/012023

$$\begin{bmatrix} \Delta & 0 & 0 & 0 & +\Delta z_c & 0 \\ 0 & \Delta & 0 & -\Delta z_c & 0 & +\Delta x_c \\ 0 & 0 & \Delta & 0 & -\Delta x_c & 0 \\ 0 & -\Delta z_c & 0 & I_{44} & 0 & -I_{46} \\ +\Delta z_c & 0 & -\Delta x_c & 0 & I_{55} & 0 \\ 0 & +\Delta x_c & 0 & -\Delta x_c & 0 & I_{66} \end{bmatrix} \quad (3)$$

Taking into account the following assumptions [9]:

- Considering only the gravitational and fluid forces acting on the ship;
- Taking into account the linear theory, the ship's responses will be directly proportional to the wave amplitude, occurring at the frequency at which the ship suffers the incident waves;
- Considering only the ship's response in sinusoidal waves, the time-dependent responses of the vessel $\eta_j(t)$ will be sinusoidal at a given encounter frequency ω_e being represented by: $\eta_j(t) = \bar{\eta}_j e^{i\omega_e t}$ $j = 1,2 \dots 6$;
    - where $\bar{\eta}_j$ is the amplitude of the response of the ship in the direction j.
- The pressure around the hull can be obtained from the *Bernoulli* equation;
- The hydrostatic and hydrodynamic forces acting on the ship are obtained by integrating the fluid pressure along the surface of the submerged surface of the hull of the ship S (assuming the inviscid and irrotational flow, which allows the linear theory to be applied);
- The hydrodynamic forces resulting from the radiation problem involve the *added mass* and *damping* coefficients.
- F_j^I is the complex amplitude of the excitation force component due to the incident waves, commonly called the *Froude-Krylov* excitation force;
- F_j^D is the complex amplitude of the excitation force component due to the diffracted waves, called the diffraction excitation force.

Then, equation (1) results in the following expression:

$$\sum_{k=1}^{6} \left[-\omega_e^2 \left(\Delta_{jk} + A_{jk} \right) + i\omega_e B_{jk} + C_{jk} \right] \bar{\eta}_k = F_j(t) = F_j^I + F_j^D \quad j = 1, 2 \dots 6 \quad (4)$$

where:
A_{jk} - the *added mass coefficients* in dof *j* due to motion in dof *k*;
B_{jk} - the *damping coefficients* in dof *j* due to motion in dof *k*;
C_{jk} - the *hydrostatic restoring force coefficients* in dof *j* due to motion in dof *k*;
$F_j^I + F_j^D$ - the two components of the amplitude of the excitation forces acting on the ship.

2.3. Hydrodynamic Loads

Applying *Strip Theory*, and to simplify the concepts discussed so far, the three-dimensional (*3D*) problem is reduced to a two-dimensional (*2D*) problem by dividing the hull into several two-dimensional vertical sections along the length of the ship, each strip having a constant cross-section [7] and a flow that do not interferes longitudinally with the adjacent strip flow. Subsequently, some restrictions will be presented that need to be considered in applying the *Strip Theory*.

A common approach in the calculation of hydrodynamic loads can be made by dividing the hydrodynamic problem into two sub-problems [7]:

- *Sub-problem A*: The movement of the ship when exposed to incoming waves is predicted. In this sub-problem, the *Froude-Krylov* and the diffraction forces and moments of the wave excitation are computed.
- *Sub-problem B*: The incoming waves are not considered. In this sub-problem, we postulate that the ship is moving in its six degrees of freedom at the matching frequency corresponding to the wave frequency of *sub-problem A*. Here, the *added mass coefficients* A_{jk}, the *damping coefficients* B_{jk}, and the *hydrostatic restoring force coefficients* C_{jk} are calculated.

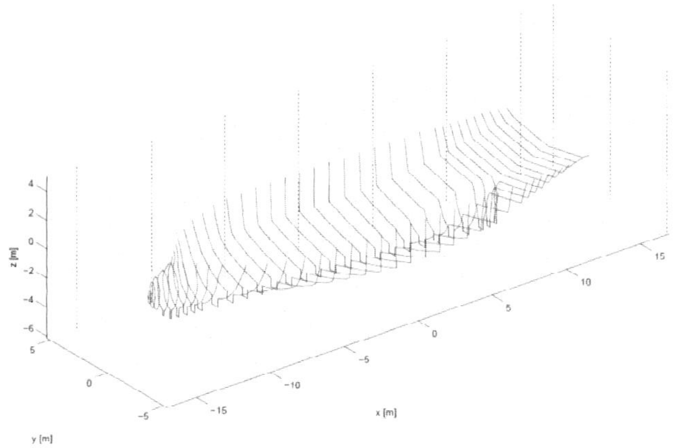

Figure 3. Strip Model of a ship. Each strip has constant cross section [5].

The decoupled equation of the ship's motion can be given by:

$$(M + A)\ddot{\eta} + B\dot{\eta} + C\eta = Fe^{i\omega_e t}$$
$$Sub - problem\ A \rightarrow\ Fe^{i\omega_e t} \quad (5)$$
$$Sub - problem\ B\ \rightarrow\ A\ddot{\eta} + B\dot{\eta} + C\eta$$

This work addresses the radiation problem – *sub-problem B*, that is the calculation of the *hydrodynamic coefficients* A_{jk} and B_{jk}.

3. Conformal Mapping

As started above the main principle in *Strip Theory* involves dividing the submerged part of the ship into a finite number of strips. Hence, *2D hydrodynamic coefficients* for *added mass* a_{jk} and *damping* b_{jk} can be computed for each strip and then be summed over the length of the body to yield the *3D* coefficients [11].

The *2D dynamic coefficients* can be calculated from boundary element methods or via *conformal mapping* [11].

Conformal mapping is used to transform the section into a circle, for which the form of the multipole potential is known. This representation is then transformed back into the physical plane using the derived mapping function [12].

The problem on *conformal mapping* now becomes one of determining the parameters in the transformation which map the arbitrary section to a unit circle.

The earliest two-parameter mapping technique was due to [2]. This method produces reasonable representations of conventional sections [12].

The determination of the local *2D added mass* and *damping coefficients* were being studied by many researchers, such as Ursell [1], [13], [14], [15] and [16], using the so-called *Lewis Forms* [17].

However, nowadays most of the computing tools available make use of more complex panel methods.

3.1. Computation of the Hydrodynamic Coefficients using a two-parameter conformal method

The algorithm developed computes the local *heave added mass* a_{33} and *heave damping coefficients* b_{33} on each section of the ship (the global ship *added mass* and *damping coefficients* in *heave* using *Strip Theory* being: $A_{33} = \int a_{33}dx$ and $B_{33} = \int b_{33}dx$ respectively).

Our work was based on the formulation developed by [18]. This scientific document gives the expressions for the *added mass* and *damping coefficients* for conventional hull cross-sections related with the *Lewis forms* approximations.

3.1.1. Lewis Transformation Method

Due to the fact that ship´s hulls do not have semi-circular cross-sections, *Lewis Transformation Method* is used to extend the results for the semi-circle into solutions for more realistic hull shapes. In this way, it must be considered that [17]:

- Small motion amplitudes are assumed.
- The *hydrodynamic coefficients* are calculated with the usual potential flow assumptions of:
  - Negligible viscosity;
  - Negligible compressibility;
  - No flow separation;
  - No skin friction.
- The transformation relates only the half cylinder below the free surface.
- The mapping relates only to the underwater shape of the hull cross section.

In this technique, the circle and the flow around it (stream and potential functions) are calculated in the complex z plane where:

$$z = x + iy = ire^{-i\theta}\ [m] \tag{6}$$

Then, these results are mapped into the flow around a hull section in the complex ζ plane (the hull cross section plane) defined as:

$$\zeta = x_{B2} + ix_{B3}\ [m] \tag{7}$$

These two complex planes can be related by the following transformation:

$$\zeta = f(z)[m] \tag{8}$$

It´s important to understand that for each size and shape of the section of the ship in the ζ plane, the functional form of the transformation equations must be determined for every individual case.

Therefore, the transformation that will map any point on a semicircle of radius a meters in the z plane into a corresponding point on a given shape in the ζ plane (if appropriate values of the coefficients a_0, a_1, a_3 are chosen) can be referred as [17]:

IOP Conf. Series: Earth and Environmental Science **172** (2018) 012023 doi:10.1088/1755-1315/172/1/012023

$$\zeta = f(z) = a_0\, a \left(\frac{z}{a} + \frac{a\, a_1}{z} + \frac{a^3\, a_3}{a^3} \right) \; [m] \tag{9}$$

Lewis forms are defined by the values of:

- The section area coefficient:

$$\sigma = \frac{A}{BD} \tag{10}$$

- The beam/draft ratio:

$$H = \frac{B}{D} \tag{11}$$

where:
A – the underwater sectional area;
B – the underwater sectional beam;
D – the sectional draft.

Taking into account that the section of the ship has radius $r = a\ [m]$, substituting equations (6) and (7) into equation (9), and separating real and imaginary parts, we can obtain a pair of parametric equations in θ (from $\theta = \pi/2$ to $\theta = 0$) describing the shape of the *Lewis form* in the ζ plane:

$$x_{B2} = a_0 a[(1 + a_1) sin\theta - a_3\, sin(3\theta)]\ [m] \tag{12}$$

$$x_{B3} = a_0 a[(1 - a_1) cos\theta - a_3\, cos(3\theta)]\ [m] \tag{13}$$

The coefficients a_1 and a_3 are obtained with these equations:

$$c = 3 + \frac{4\sigma}{\pi} + \left(1 - \frac{4\sigma}{\pi}\right)\left(\frac{H-2}{H+2}\right)^2 \tag{14}$$

$$a_3 = \frac{3 - c + \sqrt{9 - 2c}}{c} \tag{15}$$

$$a_1 = (1 + a_3)\left(\frac{H-2}{H+2}\right) \tag{16}$$

It should be noted that a_0 is a scale factor governing the overall size of the *Lewis form* [12]:

$$a_0 = \frac{B_0}{1 + a_1 + a_3} \tag{17}$$

Being B_0 the half underwater sectional beam. Some examples of *Lewis forms* are presented in Figure 4.

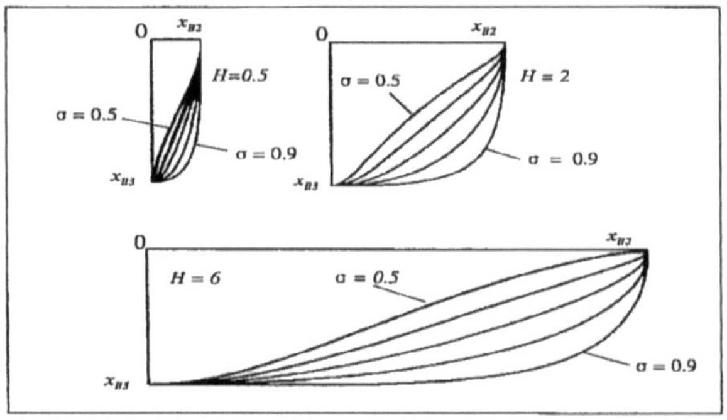

Figure 4. Examples of *Lewis forms* [11].

3.1.2. Added Mass and Damping Coefficients for a heaving Lewis Form – frigate cross section

Considering a cross section from a *Portuguese Navy frigate* with the following dimensions:

- $B = 13.2\ [m]$;
- $D = 4 \quad [m]$;
- $A = 22.4\ [m^2]$.

It is possible to obtain the sectional area coefficient and beam/draft ratio:

- $\sigma = \frac{\mathrm{A}}{\mathrm{BD}} = 0.4242;$
- $H = \frac{\mathrm{B}}{\mathrm{D}} = 3.3000.$

By the fact that only the half section of the ship is being considered on the *Lewis Transformation Method*, then in the ζ plane x_{B2} and x_{B3} it will be:

- $x_{B2} = \frac{13.2}{2} = 6.600\ [m]$;
- $x_{B3} = 4.000 \qquad [m]$.

The studied *frigate* cross section *Lewis form* is presented on Figure 5.

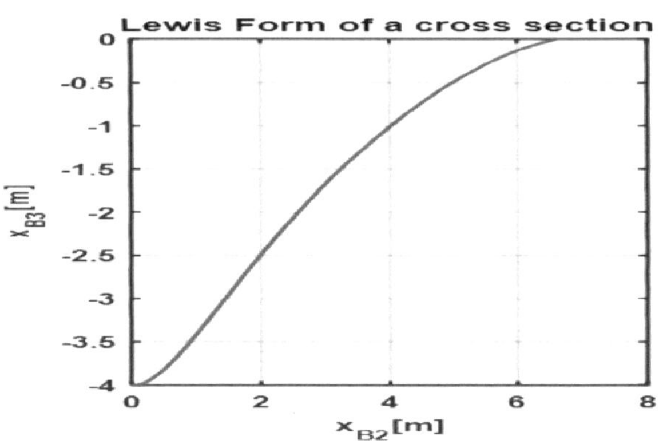

Figure 5. *Lewis form* of a cross section from *Portuguese Navy frigate*.

IOP Conf. Series: Earth and Environmental Science **172** (2018) 012023 doi:10.1088/1755-1315/172/1/012023

Considering the studies made by [18] the expressions for *added mass* and *damping coefficients* in *heave* for each section of the ship are:

$$a_{33} = \frac{\rho B^2 (A_* N_0 + B_* M_0)}{2({A_*}^2 + {B_*}^2)} \; [t] \tag{18}$$

$$b_{33} = \frac{\rho B^2 \omega \pi^2}{4({A_*}^2 + {B_*}^2)} \; [kN/(m/sec)] \tag{19}$$

An exhaustive description for both equations can be found in Appendix [17].

Comparing the *non-dimensional* curves of *hydrodynamic coefficients* in *heave* (*added mass and damping*) presented on page 120 [17] with the curves obtained using the algorithm written in *Matlab®*, it is possible to see that the results are similar with an approximate error of:

- $H = 2$, $\sigma = 1$; *2%* for *added mass* and *17%* for *damping;*
- $H = 4$, $\sigma = 1$; *4%* for *added mass* and *8%* for *damping*;
- $H = 8$, $\sigma = 1$; *8%* for *added mass* and *11%* for *damping*;
- $H = 1.155$, $\sigma = 0.5$; *5%* for *added mass* and *14%* for *damping*.

The respective curves are presented in Figures 6 and 7. This comparison serves the purpose of validation before making the calculations for the real cross section of the naval *frigate*.

The legend for the curves given by [17] and *Matlab® code*:

Form	*Key* *H*	σ	Symbol
Rectangle D	2	1.0	□
Rectangle E	4	1.0	▭
Rectangle F	8	1.0	▭
Triangle G	1.155	0.5	▽

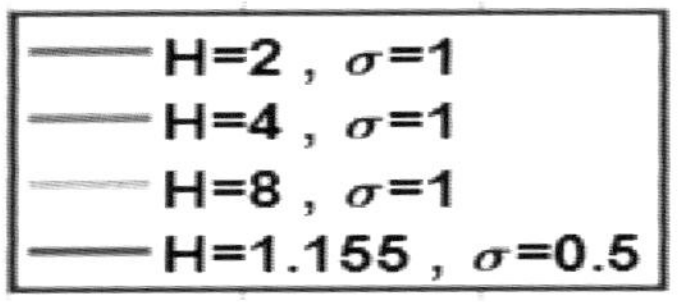

1. *Added mass coefficient in heave:*

a. Results from [17]: b. Results from *Matlab® code*:

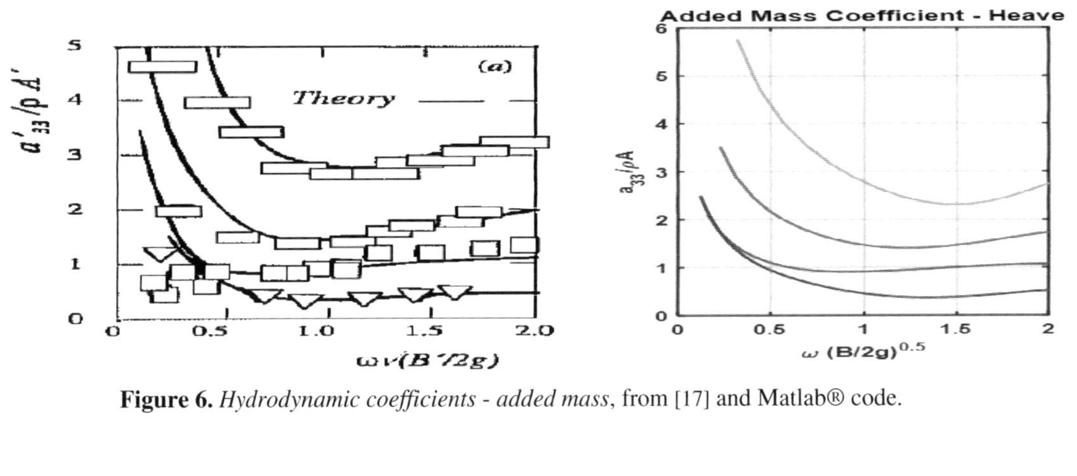

Figure 6. *Hydrodynamic coefficients - added mass*, from [17] and Matlab® code.

2. *Damping coefficient in heave:*
    a. Results from [17]:
    b. Results from *Matlab® code*:

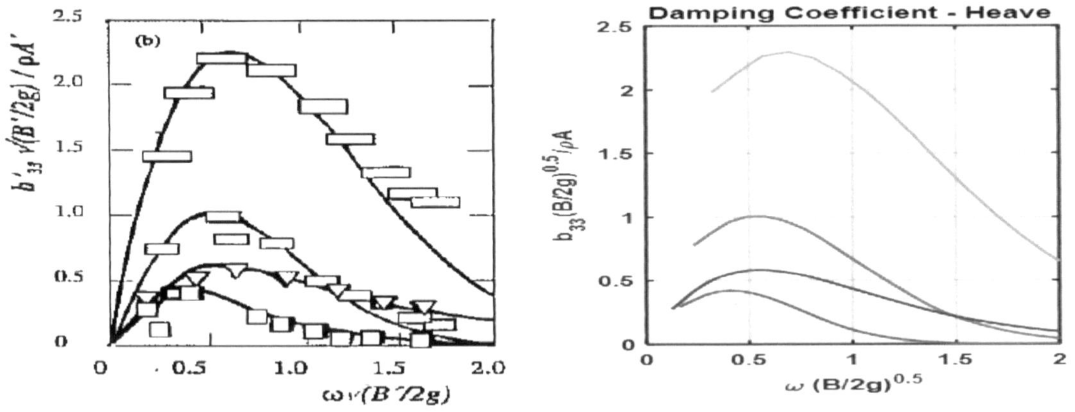

Figure 7. *Hydrodynamic coefficients – damping*, from [17] and Matlab® code.

Then the cross-sectional *non-dimensional added mass and damping hydrodynamic coefficients* for the cross section from *Portuguese Navy frigate* in *heave* were calculated and the results can be seen in Figures 8 and 9.

While the *damping* curve follows a expected tend, the *added mass* presents oscillations at the high frequencies that where not expected and need further study and validation.

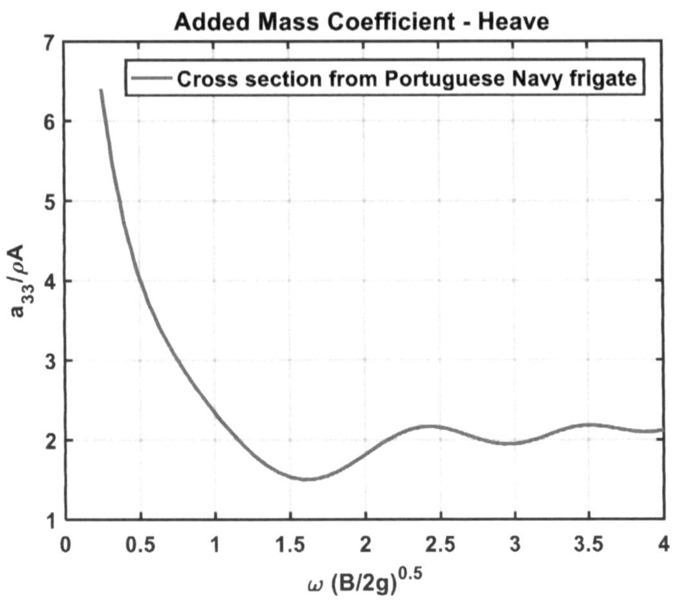

Figure 8. *Added mass coefficient* of a cross section from *Portuguese Navy frigate* in *heave*.

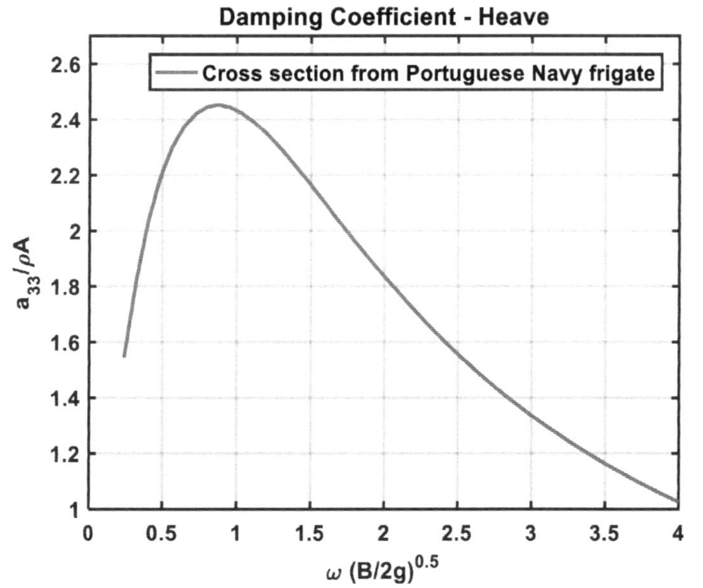

Figure 9. *Damping coefficient* of a cross section from *Portuguese Navy frigate* in *heave*.

4. Conclusions

This work tried to address the radiation problem in what respects the calculation of the *hydrodynamic coefficients* A_{jk} and B_{jk} in *heave*, A_{33} and B_{33}.

A computing code was developed to estimate the local *heave added mass* a_{33} and *heave damping coefficients* b_{33} using a two parameter *conformal mapping* method usually known as *Lewis forms*.

The code was based on the equations for the radiation problem presented by Lloyd [17].

The results in the form of *non-dimensional heave added mass* and *heave damping coefficients* were compared with those presented by de Jong [18] for several sectional forms and seem to agree fairly well, making a first validation of the code.

The code was applied to the cross section of a navy frigate and the results obtained for the *Lewis form* seem correct. The results obtained for a_{33} and b_{33} show that:

- for the *added mass* there are oscillations in higher frequencies region that need further study since they are not expected using this method which is known to be stable through all range of frequencies;
- the *damping* follows a expectable trend, however there are calculation instabilities to be solved in the very low frequency range ($\omega(B/2g)^{0.5} < 0.25$);
- for both *hydrodynamic coefficients* a validation needs to be made through comparison with other method for the same section.

References

[1] F. Ursell, "On the Heaving Motion of a Circular Cylinder on the Surface of a Fluid," *Quarterly Journal of Mechanics and Applied Mathematics,* vol. II, 1949.

[2] F. Lewis, "The Inertia of the Water Surrounding a Vibrating Ship," *Transactions, Society of Naval Architects and Marine Engineers,* vol. 37, pp. 1-20, 1929.

[3] L. Landweber and Macagno, "Added Mass of Two Dimensional Forms Oscillating in a Free Surface," Journal of Ship Research, Vol. 1, No. 3, 1957.

[4] L. Lanweber and Macagno, "Added Mass of a Three Parameter Family of Two-Dimensional Forces Oscillating in a Free Surface," Journal of Ship Research, Vol. 2, No. 4, 1959.

[5] R. Bhattacharyya, Dynamics of Marine Vehicles, Annapolis, Maryland: John Wiley & Sons, 1978.

[6] T. Perez, T. I. Fossen and A. Sorensen, A Discussion about Seakeeping and Manoeuvering Models for Surface Vessels, Norwegian University of Science and Technology (NTNU), 2004.

[7] P. M. Martinussen, "Parametric Roll Ressonance of a Fishing Vessel as Function of Forward Speed and Sea State," 2011.

[8] V. Bertram, Pratical Ship Hydrodynamics, Butterworth-Heinemann, 2012.

[9] E. V. Lewis, Principles of Naval Architecture (Second Revision), Volume III - Motions in Waves and Controllability, Society of Naval Architects and Marine Engineers, 1989.

[10] M. A. Abkowitz, Stability and Motion Control of Ocean Vehicles, The MIT Press, 1969.

[11] T. I. Fossen, Handbook of Marine Craft Hydrodynamics and Motion Control, John Wiley & Sons, Ltd, 2011.

[12] W. P. A. Westlake P. C., A New Conformal Mapping Technique for Ship Sections, International Shipbuilding Progress, 2000.

[13] F. Tasai, On the Damping Force and Added Mass of Ships Heaving and Pitching, Journal of Naval Architects of Japan, 1959.

[14] F. Tasai, "An Approximate Calculation of Hydrodynamic Pressure on the Midship Section Contour of a Ship Heaving and Pitching in Regular Head Waves," Reports of Research Institute for Applied Mechanics, Kyushu University, Japan, 1966.

[15] Grim, "Die Schwingungen von Schwimmenden, Zweidimenionalen Korpern," Hamburg Model basin Report 1171, 1959.

[16] Porter, "Pressure Distributions, Added Mass and Damping Coefficients for Cylinders Oscillating in a Free Surface," University of California, Berkeley, report 82-16, 1960.

[17] A. R. J. M. Lloyd, Seakeeping: Ship Behaviour in Rough Weather, Hampshire: Ellis Horwood, 1998.

[18] B. de Jong, Computation of the Hydrodynamic Coefficients of Oscillating Cylinders, Netherlands Ship Research Centre, 1969.

[19] H. T. d. S. Larsen, "Analise do comportamento de um navio utilizando a teoria das faixas," FCT/UNL, 2012.

[20] R. E. Randall, Elements of Ocean Engineering, Society of Naval Architects and Marine Engineers (SNAME), 2010.

[21] J. N. Newman, Marine Hydrodynamics, The MIT Press, 1999.

[22] M. E. McCormick, Ocean Engineering Mechanics, Cambridge University Press, 2010.

[23] E. M. Lewandowski, The Dynamics of Marine Craft: Maneuvering and Seakeeping, World Scientific Publishing Company, 2004.

[24] I. Erselcan, "A Frequency Domain Strip Theory Applied to the Seakeeping of the Zumwalt-Class Destroyer," Turkish Naval Academy, 2010.

[25] P. Bandyk, "A Body-Exact Strip Theory Approach to Ship Motion Computations," The University of Michigan - Naval Architecture and Marine Engineering, 2009.

[26] B. R. Clayton and R. E. D. Bishop, Mechanics of Marine Vehicles, Gulf Publishing Company, 1982.

[27] V. Wehausen and E. V. Laitone, Surface Waves, Springer, 1960.

IOP Conf. Series: Earth and Environmental Science **172** (2018) 012023 doi:10.1088/1755-1315/172/1/012023

Appendix

$$a_1 = (1 + a_3)\left(\frac{H-2}{H+2}\right)$$

$$a_3 = \frac{3 - c + \sqrt{9 - 2c}}{c}$$

$$c = 3 + \frac{4\sigma}{\pi} + \left(1 - \frac{4\sigma}{\pi}\right)\left(\frac{H-2}{H+2}\right)^2$$

$$x_{B2} = a_0 a\ \left[(1 + a_1) sin\theta - a_3\, sin(3\theta)\right]$$

$$x_{B3} = a_0 a\ \left[(1 - a_1) cos\theta - a_3 cos(3\theta)\right]$$

$$a_{33} = \frac{\rho B^2 (A_* N_0 + B_* M_0)}{2({A_*}^2 + {B_*}^2)}\ [t]$$

$$b_{33} = \frac{\rho B^2 \omega \pi^2}{4({A_*}^2 + {B_*}^2)}\ [kN/(m/sec)]$$

$$A_* = \Psi_c\left(1, \frac{\pi}{2}\right) + \sum_{m=1}^{\infty}\left[p_{2m}(-1)^{m-1}\frac{kBQ_1}{2Q_2}\right][m^2/sec]$$

$$B_* = \Psi_S\left(1, \frac{\pi}{2}\right) + \sum_{m=1}^{\infty}\left[q_{2m}(-1)^{m-1}\frac{kBQ_1}{2Q_2}\right][m^2/sec]$$

$$M_0 = \int_0^{\frac{\pi}{2}} \Phi_S(1,\theta)\frac{Q_3}{Q_2}d\theta + \frac{1}{Q_2}\left(\sum_{m=1}^{\infty}\left[q_{2m}(-1)^{m-1}Q_4\right] + \frac{\pi kBQ_{5_{q2m}}}{8Q_2}\right)[m^2/sec]$$

$$N_0 = \int_0^{\frac{\pi}{2}} \Phi_c(1,\theta)\frac{Q_3}{Q_2}d\theta + \frac{1}{Q_2}\left(\sum_{m=1}^{\infty}\left[p_{2m}(-1)^{m-1}Q_4\right] + \frac{\pi kBQ_{5_{p2m}}}{8Q_2}\right)[m^2/sec]$$

$$\Psi_c = \pi\, exp(-kx_{B3})\, sin(k|x_{B3}|)\ [m^2/sec]$$

$$\Phi_c = \pi\, exp(-kx_{B3})\, cos(kx_{B2})\ [m^2/sec]$$

$$\Psi_S = -\pi\, exp(-k\, x_{B3})\, cos(k\, x_{B2}) + \int_0^{\infty}\frac{exp(-v|x_{B2}|)}{v^2 + k^2}\left[v\, sin(v\, x_{B3}) + k\, cos(v\, x_{B3})\right] dv\ \ [m^2/sec]$$

$$\Phi_S = \pi\, exp(-kx_{B3})\, sin(k|x_{B2}|) - \int_0^{\infty}\frac{exp(-v|x_{B2}|)}{v^2 + k^2}\left[v\, cos(v\, x_{B3}) + k\, sin(v\, x_{B3})\right] dv\ \ [m^2/sec]$$

$$\Psi_c(1,\theta) - \frac{Q_6}{Q_2}\Psi_c\left(1, \frac{\pi}{2}\right) = \sum_{m=1}^{N} p_{2m} f_{2m}\ [m^2/sec]$$

$$\Psi_S(1,\theta) - \frac{Q_6}{Q_2}\Psi_S\left(1,\frac{\pi}{2}\right) = \sum_{m=1}^{N} q_{2m} f_{2m}\ [m^2/sec]$$

$$f_{2m} = -\left(sin(2m\,\theta) + \frac{kB}{2Q_2}Q_7 + \frac{kB}{2{Q_2}^2}(-1)^m Q_1 Q_6\right)[m^2/sec]$$

$$Q_1 = \frac{1}{2m-1} - \frac{a_1}{2m+1} - \frac{3a_3}{2m+3}$$

$$Q_2 = 1 + a_1 + a_3$$

$$Q_3 = (1 + a_1)cos\theta - 3a_3 cos(3\theta)$$

$$Q_4 = \frac{1 + a_1}{4m^2 - 1} + \frac{9a_3}{4m^2 - 9}$$

$$Q_{5_{p2m}} = (1 + a_1 - a_1 a_3)p_2 - a_3 p_4$$

$$Q_{5_{q2m}} = (1 + a_1 - a_1 a_3)q_2 - a_3 q_4$$

$$Q_6 = (1 + a_1)sin\theta - a_3 sin(3\theta)$$

$$Q_7 = \frac{sin((2m-1)\theta)}{2m-1} + \frac{a_1 sin((2m+1)\theta)}{2m+1} - \frac{3a_3 sin((2m+3)\theta)}{2m+3}$$

IOP Conf. Series: Earth and Environmental Science **172** (2018) 012024 doi:10.1088/1755-1315/172/1/012024

DESIGN CHALLENGES OF A MARITIME MULTIPURPOSE UNMANNED VEHICLE

P. Lopes[1], **P. Pires da Silva**[1,2], **M. Moreira**[1,2]
[1] Escola Naval, CINAV, Base Naval de Lisboa, Alfeite, 2810-001 Almada-Portugal
[2] Centre for Marine Technology and Ocean Engineering (CENTEC), Instituto Superior Técnico, Universidade de Lisboa, Lisbon, Portugal

E-mail: vasco.pereira.lopes@marinha.pt

Abstract. Military Navies have several missions in which due to their complex, dangerous, or disaster scenarios, unmanned vehicles are suitable to perform the necessary tasks. Besides other issues it is a current subject of research the trade-off between using many mission oriented unmanned vehicles or the use of less number of more robust and versatile multimission vehicles. Challenged by the technological advances achieved in mobile robotics and unmanned vehicles, and the need to develop in house know-how, this project aimed to put into practice the development of a concept design of a multimission unmanned surface vehicle, that can operate individually or cooperate with other systems whose development will be done in parallel on other projects.
This unmanned vehicle, designated Unmanned Surface Vehicle for the Portuguese Navy - USV – PoN is part of project based oriented learning strategy in development in the Portuguese Naval Academy. It poses particular challenges related to the broad spectrum of the concept of operations, such as mine warfare operations, port surveillance and protection operations, search and rescue operations, hydrographic data collection operations. All of these operations demand different and challenging missions and tasks, to which the concept design of the USV-PoN will answer in terms of feasibility studies and possible configurations.
The purpose of this work is to address the challenges that the concept of operations and corresponding technical specifications put into the design of the USV-PoN and show how to overcome them making the study and analysis necessary for the production of a prototype based on the concept design.
Out of the scope of this work were aspects related to sensor development and integration as well as the command and control of the vehicle.

1. Introduction

Due to the different missions and tasks undertaken by the military Navies, in complex, dangerous and disaster scenarios, it is necessary to reduce the risk to its military and increase responsiveness to different threats, in any type of scenario.
As part of a line of research with interest to the CINAV (Naval Research Center) and the strategic plan of the Portuguese Navy, it was started in 2017 the development of the conceptual design of an Unmanned Surface Vehicle (USV).
This vehicle is possibly intended to be an ocean-going platform. As such its design as any novel design is a complex and challenging activity. It demands a large spectrum of knowledge on diverse scientific areas combined with an iteration process that takes the designer through the same design detail several times, as needed to improve the object of design up to the point where the constraints of adequacy, feasibility and acceptability are met. Other way to put it is to apply an iteration process that improves slightly the design up to obtain an optimum, as Evans H. explains in his well-known Design Spiral [1].

IOP Conf. Series: Earth and Environmental Science **172** (2018) 012024 doi:10.1088/1755-1315/172/1/012024

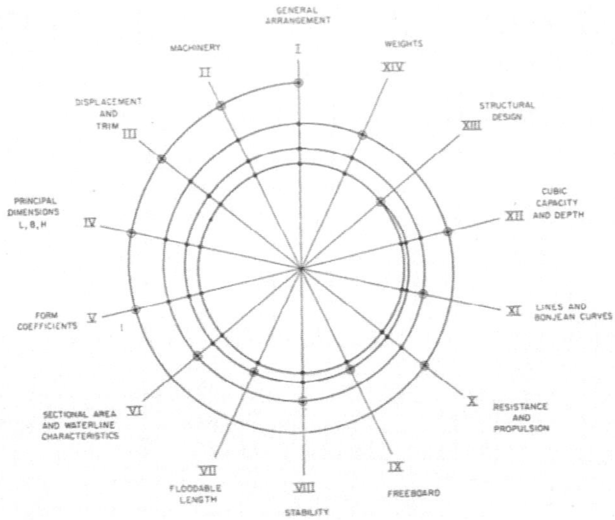

Figure 1 - Design Spiral from Harvey Evans [1].

As far as concerns to the acquisition process of naval means, it was adopted as a reference the publication NATO AAP-20: PAPS (Handbook on the Phase Armaments Programming System) [2], now known as NATO AAP-20: NATO Programme Management Framework (NATO Life Cycle Model) [3]. This publication is a manual which describes a phased programming system for the design and procurement of military assets. For the design of this platform the relevant phases of the NATO framework are the phases 3 to 6, which correspond to the tasks and deliverables of ship design [4].

Phase 3, the *Pre-feasibility Studies*, is when the design of a naval platform begins and a preliminary evaluation of technical concepts which can meet the objectives defined in phases 1 and 2, is performed. This phase ends with the preparation of the *Consolidated Operational Objectives*.

Phase 4, *Feasibility Studies*, involves the more detailed elaboration and evaluation of the most promising technical concepts that meet the *Consolidated Operational Objectives*, in order to identify the most advantageous alternative that is feasible from the technical point of view and from the point of view of the cost and duration constraints of the project. This phase concludes with the drafting of the *Operational Requirements* document, which reflects the selection of the most advantageous workable alternative.

It is only in phase 5, *Project Definition*, that the "*design*" of a ship takes shape and size. It is a phase whose objectives are the detailed specification of the ship and the development of the initial specifications of the sub-systems. This phase concludes with the elaboration of the *Project Development Objectives*.

The ship design is completed in phase 6, *Project Development*. It is at this stage that various engineering, detailed ship design and prototype manufacturing processes are developed, which will validate the project in order to allow the production contract to be advanced.

This work is focused on the practical application of the NATO framework, with the necessary adaptations to the nature and dimension of the asset, within which a first iteration of the design spiral is to be detailed.

For pursuing the previously described acquisition and design processes, it is necessary to reflect, analyze and list what are the tasks and activities to develop for the present platform.

Then it is necessary to work on what materials and equipment are required to perform the tasks.

Being known the first estimation of weight and volume of all equipment, it is possible to start the initial sizing studies, through empirical-analytical methods adapted to platforms with small dimensions.

The initial sizing gives first estimates of the main dimensions of the platform and the weight and volume of the structure of the platform, as well as the height of the centre of gravity.

IOP Conf. Series: Earth and Environmental Science **172** (2018) 012024 doi:10.1088/1755-1315/172/1/012024

Once the initial dimensions are known, 3D modelling of the platform can be carried out using a CAD/CAE software, such as *SolidWorks*®.
We conclude the study for this work, modelling the platform in *AutoHydro*® software where it is possible to obtain the stability curves, which allows to revise previous decisions and redesign the platform.

2. Concept of Operations

In the initial phase of the design, what is inserted in phases *Pre-feasibility* and *Feasibility Studies*, it is necessary to think "What is the platform for?", "What will the platform do?". It is at this stage that the Operational Requirements are defined, where the missions and tasks to be carried out by the platform are chosen.

2.1. Operational Requirements

Before defining any mission to be developed by the platform, it was necessary to define the area of operations. Then it was defined a broad range of operations, e.g. search and rescue, hydrographic survey, etc.
Due to the wide variety of operations required at this phase, and the predictable unfeasibility and acceptability of this spectrum of operations in a single platform as well as existence of overlap between them, it was necessary for a first iteration of the design to prioritize the most important and necessary operations. After some reflection and rational analysis based on the main needs of the organization and the maximization of the number of missions with common equipment it was concluded that a single platform will have to be able to carry out the following operations:

- maritime search and rescue operations;
- force protection operations and port protection;
- environmental data collection operations;
- hydrographic surveys operations.

Once defined the operations to be developed by the platform, it will be necessary to define the tasks to be performed in each mission/operation, e.g.:

- maritime search and rescue operations:
  - Detection of contacts (people and boats);
  - Automatic contact tracking;
  - Transport and release of rescue equipment.
- environmental data collection operations:
  - Water and air temperature;
  - Absolute and relative humidity;
  - Direction and intensity of the wind;
  - Atmospheric pressure.

2.2. Technical Requirements

In parallel (involving interaction between end-user and designer) or after defining the operational requirements, it is also imperative to define the technical requirements or technical specifications. These define the limitations/obligations of the technical characteristics of the platform, such as autonomy, maximum weight, maximum speed, etc that can fulfil de operational requirements within existing constraints. Some constraints are:
- the platform should be prepared to be stowed either on board of the Navy ships or in a trailer to be defined in its own specs. Therefore, it was necessary to verify the limiting working load that the cranes of the different ships, as well as the berthing. Thus limiting the vehicle maximum displacement and dimensions.

With the demanding maritime agitation and high intensity of wind present on the Portuguese coast, it is desirable that the vehicle is able to operate in rough sea conditions, defined as sea state 5 on Douglas scale - up to 4 meters significant wave height -, and wind force 6 on the Beaufort scale.
For the hull to withstand these conditions, it is necessary to be constructed with materials in agreement with. It is also necessary that the materials used have high resistance to corrosion.
Due to its maritime search and rescue mission, this vehicle should be designed to carry a liferaft for four shipwrecked. In addition to this requirement, the platform should provide for 2 shipwrecked to easily rise on board and to remain on board for short periods of time, namely from the place of the shipwreck to the mother ship, or to the nearest port.
In order to navigate areas common to other vessels, the platform must have the lighting provided in COLREGS (International Regulations for Preventing Collisions at Sea) for boats of its characteristics, or approximate, and thus be able to navigate safely.
Among many other technical requirements, it should be noted that the platform should have self-righting characteristics.

3. Preliminary Studies

The preliminary studies are considered in the *Project Definition* phase. Here we will present some studies that have been carried out to determine which type of hull is most suitable to be used on this vehicle and also what equipment will be needed to equip the platform in order to be able to perform the operations defined above.

3.1. Hull Type

An multicriteria initial evaluation was developed in order to obtain the types of hulls most used in USV's, from which the following list was obtained:

- Planning;
- Semi-planning;
- Hydrofoil;
- Multi-hull;
- SWATH.

Some general characteristics were defined that may characterize the different types of hull, and they are:

- survivability;
- seakeeping;
- hull resistance to advance;
- deck floor area;
- flexibility in load arrangement;
- ease of design/production;
- launch and recovery;
- transportability.

Once the different hull types were chosen and the general characteristics defined, a matrix was constructed which is represented in figure 2, where a weight of importance is assigned to the general characteristics, where the value 1 represents the least important characteristic and 8 the most important characteristic. For the different types of hull, a value of 1 to 5 has been assigned depending on whether the type of hull is "*well*", to "*not appropriate*" for a given characteristic, where the value 1 means that it is not appropriate and 5 well appropriate.

IOP Conf. Series: Earth and Environmental Science **172** (2018) 012024 doi:10.1088/1755-1315/172/1/012024

		Hull Types				
Importance	**General Features**	**Planing**	**Semi-Planing**	**Hydrofoiled**	**Multihull**	**SWATH**
1	Resistance to sinking	3	3	3	5	5
2	Seakeeping	2	3	1	4	5
3	Resistance to advance	3	2	5	4	1
4	Deck floor area	2	3	1	4	5
5	Flexibility in load arrangement	4	5	4	3	3
6	Ease of design/production	4	5	1	3	2
7	Launch and Recovery	5	4	1	2	2
8	Transportability	5	5	1	2	1
		109	111	42	82	67
36		3.0	3.1	1.2	2.3	1.9

Figure 2 - Decision matrix for hull type selection

The best type of hull will be the one that will present the highest value when applied to equation (1) to the different types of hull.

$$\frac{\sum_{i=1}^{8}(importance_i \times Hull\ Type_i)}{\sum_{i=1}^{8} Importance_i} \tag{1}$$

In this way it is concluded that the best type of hull to apply on the platform could be a semi-planning hull, but a planning hull is also an adequate choice. A planning hull was chosen to develop due to easiness of modelling and time constraints.

3.2. Equipment

To know the payload, it is necessary to choose which equipment to put on the platform. The equipment is chosen so that the platform can carry out the different operations defined previously.
As an example, for search and rescue operations at sea, the platform should be equipped with:

- Sidescan sonar;
- Sonar;
- IR Camera;
- Life raft.

For the other missions the same or other equipment is needed.
In addition, the vehicle will require equipment that allows it navigate safely, such as:

- GPS;
- GNSS;
- RADAR;
- Batteries;
- Propulsion system.

In this work the technical questions related with autonomous operation were not addressed yet.
With the equipment defined, a search was made in the market with the purpose of obtaining the weight and volume of each equipment, and adding up all the weights and volumes of the equipment, up to obtain the payload.

Table 1 - Payload example

	Weight [kg]	Volume [m^3]
RADAR	5,60	0,193
GPS	0,24	0,001
GNSS	0,50	0,005
BATTERIES	21,50	0,011
SIDESCAN SONAR	1,10	0,001
SONAR	27,60	0,035
IR CAMERA	27,20	0,369
LIFE RAFT	28,00	0,091

The first estimate for total payload weight was about 230 kg and 1 m^3 of volume. A margin of 5% was taken for cable weight, platform fastening and other unknown payload variables.

4. Sizing and Geometric Modelling

In the initial sizing of the USV-PoN it is necessary to identify the main characteristics of the hull such as length (L), beam (B), draft (T) and depth (D). This process is still inserted in the *Project Definition* phase.
To calculate these characteristics, one made use of analytic-empirical methods adapted for small-scale platforms.
To start the calculations, a list of Unmanned Surface Vehicles (USV's) with characteristics similar to those intended for the study of this platform were chosen, in order to obtain an average of the characteristic form coefficients for this type of vehicle. The list of vehicles studied and their main characteristics are shown in table 2.

Table 2 - Database of USV's characteristics.

	C-Sweep	C-Worker 5	C-Worker 6	C-Worker 7	C-Worker 8	C-Target 3	C-Target 6	C-Target 9	C-Target 13
Length [m] L	10,8	5,5	5,8	7,2	7,7	3,5	6,5	9,6	12,9
Beam [m] B	3,5	1,7	2,2	2,3	2,1	1,4	2,1	2,4	2,7
Depth [m] D	2,9	1,8	2,2	2,1	2,1	1,3	1,0	1,4	2,0
Draft [m] T	0,7	0,9	0,9	0,9	1,0	0,6	0,3	0,4	0,5
Displacement [ton] Δ	9,0	1,9	13,5	5,3	4,8	0,3	1,2	2,8	4,2
Volume of Displacement [m^3] ∇	8,8	1,9	13,2	5,2	4,7	0,3	1,2	2,7	4,1

With the main characteristics it is possible to calculate the form coefficients of each hull in order to establish an average of these form coefficients, which in turn allows for the calculation of the initial dimensions of the USV-PoN [5]. The form coefficients considered were:

IOP Conf. Series: Earth and Environmental Science **172** (2018) 012024 doi:10.1088/1755-1315/172/1/012024

- Circular M:

$$M = \frac{L}{\nabla^{1/3}} \tag{2}$$

- The Beam/Draft ratio (K_B):

$$K_B = \frac{B}{T} \tag{3}$$

- Block coefficient (C_B):

$$C_B = \frac{\nabla}{L \times B \times T} \tag{4}$$

It was also calculated the following ratios:

$$\frac{L}{B} \tag{5}$$

$$\frac{T}{D} \tag{6}$$

The form coefficients are shown in table 3:

Table 3 - Hull Coefficients

	C-Sweep	C-Worker 5	C-Worker 6	C-Worker 7	C-Worker 8	C-Target 3	C-Target 6	C-Target 9	C-Target 13	Mean
M	5,2	4,5	2,5	4,2	4,6	5,1	6,2	6,9	8,1	5,24
K_B	5,0	1,9	2,4	2,6	2,2	2,3	7,0	6,0	5,4	3,86
C_B	0,3	0,2	1,1	0,3	0,3	0,1	0,3	0,3	0,2	0,36
L/B	3,09	3,24	2,64	3,13	3,60	2,50	3,10	4,00	4,78	3,34
T/D	0,24	0,50	0,41	0,43	0,47	0,46	0,30	0,29	0,25	0,37

4.1. Sizing

To begin the sizing process, it is necessary to start with the payload already calculated previously:

- pv=payload volume≈1 m^3
- pw=payload weight≈0,23 ton

To arrive at an initial estimate of the platform dimensions it is necessary to follow several equations and perform several iterations such that at the end the platform is balanced. Some of these equations are:

Volume of Displacement:

$$\nabla = \frac{\Delta}{\rho_w} \tag{7}$$

$\rho_w = Density\ of\ Seawater = 1{,}025\ ton/m^3$

Immersed Hull Dimensions:

$$\mathrm{L} = \mathrm{M}\nabla^{1/3} \tag{8}$$

$$\mathrm{T} = \frac{\nabla^{1/3}}{[MK_B C_B]^{1/2}} \tag{9}$$

$$\mathrm{B} = K_B \mathrm{T} \tag{10}$$

When the entire sequence of iterations is completed, we have as a result an estimate of weight (W) and required volume (V_{req}) of the platform.

$$W = \sum_i w_i = \Delta \tag{11}$$

where:
W – weight estimate [kg]
w_i – is the ith platform mass, [kg]
Δ - displaced mass, [kg]

$$V_{req} = \sum_i v_i = \mathrm{V} \tag{12}$$

where:
V_{req} – required volume $[m^3]$
v_i – is the ith platform volume, $[m^3]$
V - volume, $[m^3]$

If $W = \Delta\ and\ Vreq = V$, then we have a balanced design. Otherwise, if $W \neq \Delta\ or\ Vreq > V$ the design is not balanced, and we should iterate using new values of W and V_{req} to replace the initial estimation [6].
After several iterations performed, the desired result was reached, that is $W = \Delta$ and $Vreq = V$. The main initial dimensions of the USN-PoN are presented in table 4.

Table 4 - Main dimensions of the platform.

L [m]	7,83
T [m]	0,55
B [m]	2,13
D [m]	1,36
W [ton]	3,42

In order to determine the impact of each mission on the dimensions of the platform, a study has been developed where a mission and its equipment (a configuration) is be removed, varying the payload (weight and volume).
Starting by removing the maritime search and rescue operations and repeating the entire sizing process. Then performing the same for the other configurations, we obtained some platform variations as presented in figure 3.

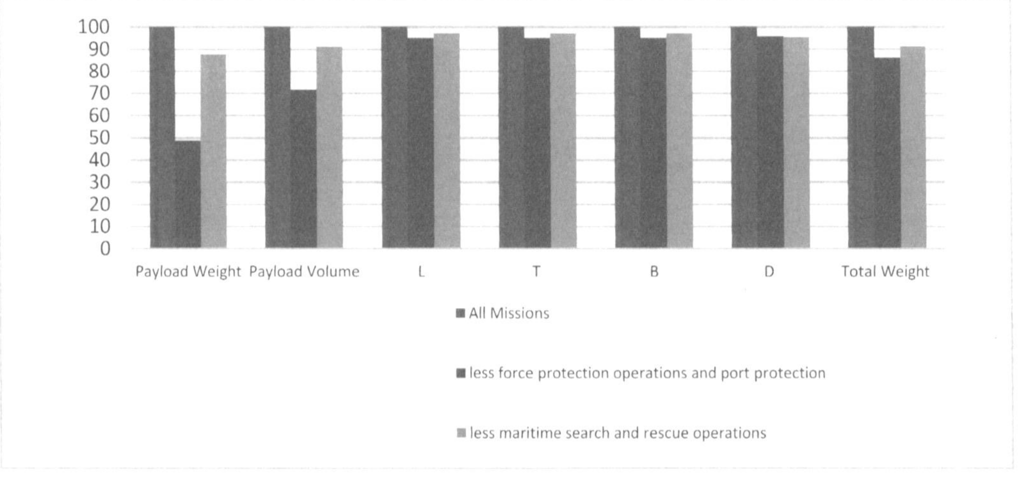

Figure 3 - Effects of configuration changings on vehicle sizing

Removing environmental data collection operations and hydrographic surveys operations, the payload will not vary, therefore the dimensions of the platform will not vary. As removing maritime search and rescue operations and force protection operations and port protection do not cause significant changes in the dimensions of the platform, it was decided to maintain the initially calculated dimensions.
In order to have an initial estimate of the hull thickness a simple barge model was conceptualized where the applied load corresponds to its displacement, and it has been defined that the material to be used will be fiberglass-reinforced polymer as a low-cost and technologically accessible material.

4.2. Geometric Modelling
Still in the *Project Definition* phase, we proceeded to model the platform using the *SolidWorks*® software.

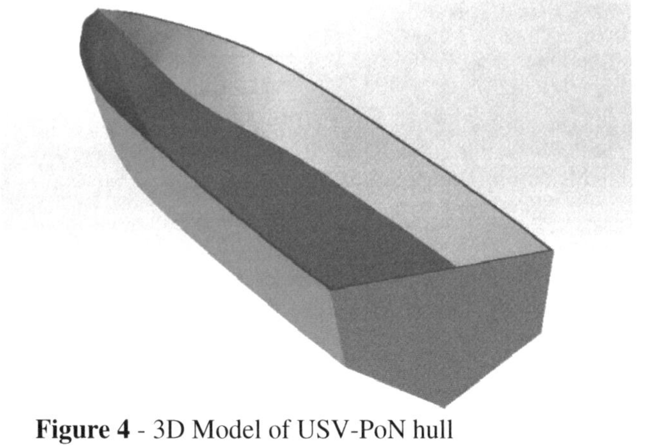

Figure 4 - 3D Model of USV-PoN hull

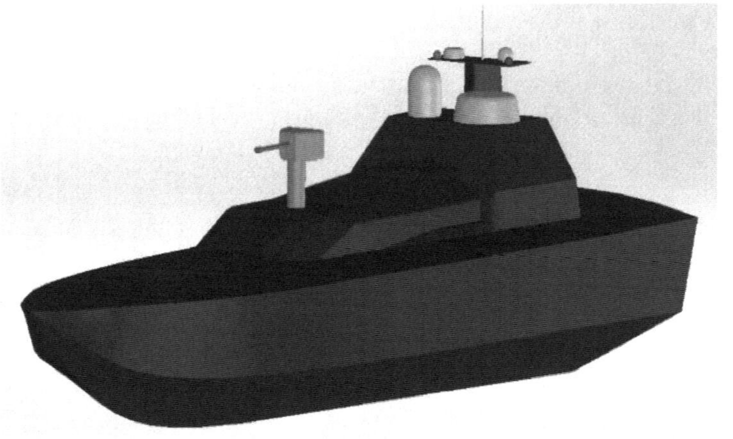

Figure 5 . Initial 3D Model of USV-PoN.

5. Stability

It will be important to study the stability of the USV-PoN in order to ensure that the platform can be used at sea as required and being self-righting. This study is inserted in the *Project Development* phase because it is a validation process.
To begin the study of stability, the platform was modelled in the *Modelmaker®* with the same coordinates used in *SolidWorks®*.

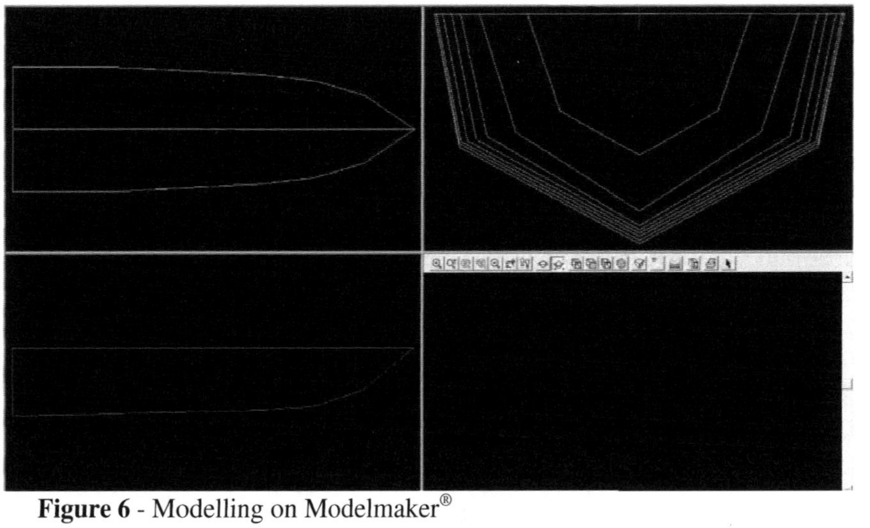

Figure 6 - Modelling on Modelmaker®

After *Modelmaker®* modelling, it is possible to export the platform to *AutoHydro®* to perform stability calculations. The calculated righting arm is represented in figure 7.

IOP Conf. Series: Earth and Environmental Science **172** (2018) 012024 doi:10.1088/1755-1315/172/1/012024

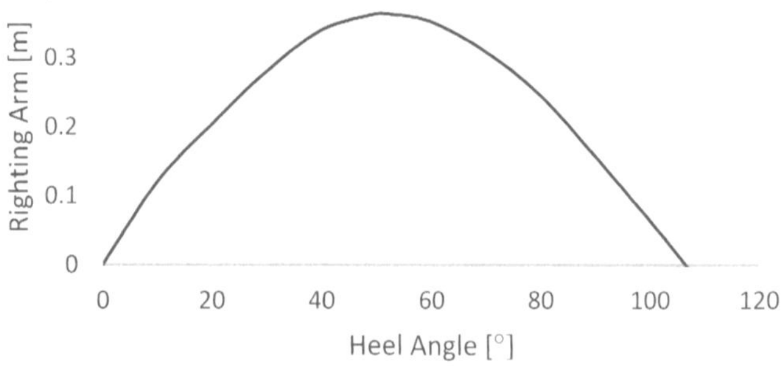

Figure 7 – Initial righting arm curve of the USV-PoN.

As from the study of the resulting curve in figure 7 it can be seen that the USV-PoN doesn´t have self-righting characteristics, a further detailed stability study had to be developed to ensure the that characteristic.

After several iterations altering the shape platform, a model of the self-righting vehicle was reached. The curve of righting arm is represented in figure 8.

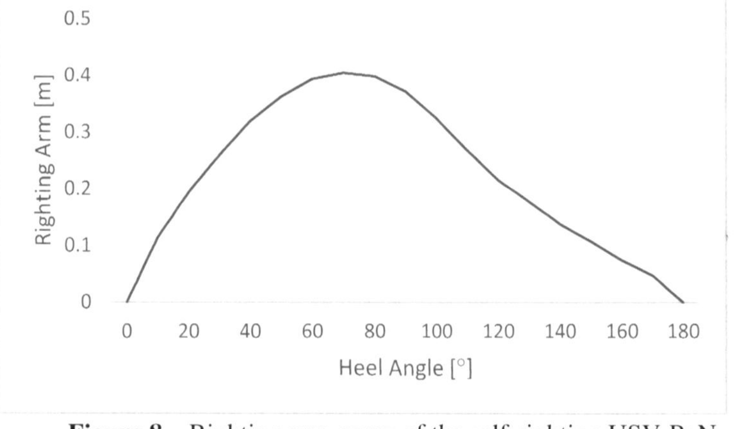

Figure 8 – Righting arm curve of the self-righting USV-PoN

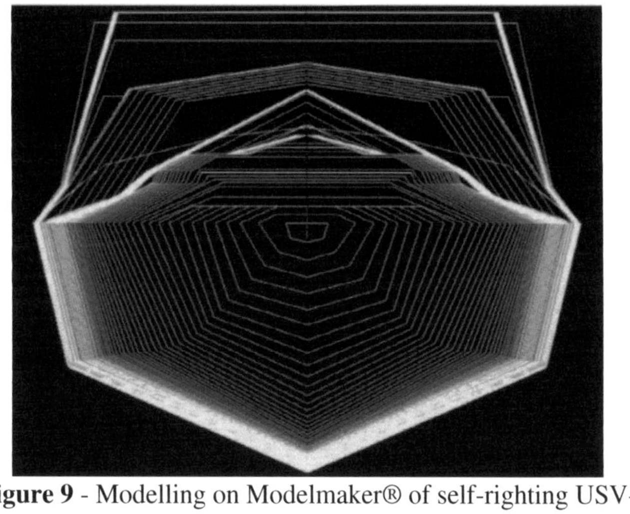

Figure 9 - Modelling on Modelmaker® of self-righting USV-PoN

IOP Conf. Series: Earth and Environmental Science **172** (2018) 012024 doi:10.1088/1755-1315/172/1/012024

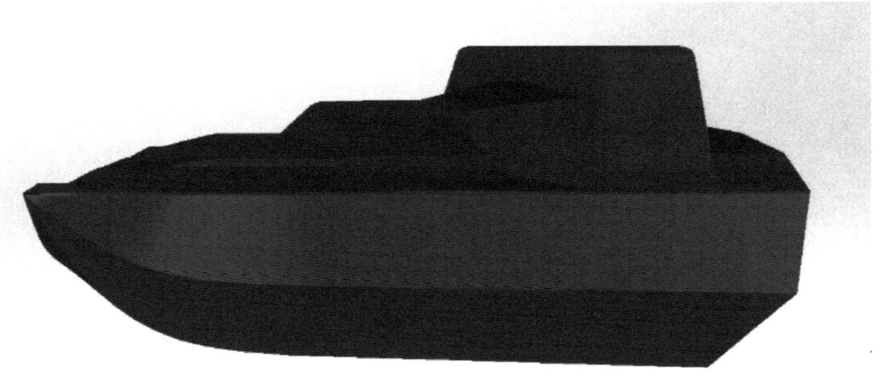

Figure 10 - 3D Model of self-righting USV-PoN.

6. Conclusions

In this paper the conceptual design of an USV to be applied in the Portuguese Navy was developed, following a framework of acquisition and construction of military means. Inserted in this framework, several analytical-empirical methods were used in order to obtain the desired objectives.

With a wide variety of missions to perform by the platform, which leads to a high number of equipment and sensors, it was a challenge to make the design comply with the sizing and weight constraints.

The objective of designing a self-righting platform was achieved, however, much further work will be needed. It will be necessary to perform the following studies:

- resistance and propulsion, with associated maximum speed levels, propulsion selection, and vehicle endurance,
- cost estimates,
- weapon and sensors arrangement and energy demands,
- seakeeping,
- command and control and associated levels of artificial intelligence.

After all these studies it will be necessary make a new iteration with the inputs of the end user.

IOP Conf. Series: Earth and Environmental Science **172** (2018) 012024 doi:10.1088/1755-1315/172/1/012024

References

[1] Evans, H. (1959). Basic Design Procedure.

[2] North Atlantic Treaty Organization. (2010). AAP20 - Phased Armaments Programming System (PAPS).

[3] North Atlantic Treaty Organization. (2015), AAP20 - NATO Programme Management Framework (NATO Life Cycle Model).

[4] Lérias, R. (2005). Crónica de Arquitetura Naval. Anais do Clube Militar Naval, 2005.

[5] Rawson, K. J. & Tupper, E. C. (2001). Basic Ship Theory, Volume 1.

[6] Andrews, D. (1985). An Integrated Approach to Ship Synthesis. The Royal Institute of Naval Achitects, London.

[7] Yeh, E. (2004). Design, Build and Trial of 19 meters self-righting SAR boat. The Royal Institution of Naval Architects, London.

[8] Rawson, K. J. & Tupper, E. C. (2001). Basic Ship Theory, Volume 2.

[9] Lewis, E. V. (1988). Principles of Naval Architecture, Vol. I: Stability and Strength.

[10] Moore, C. (2010). The Principles of Naval Architecture Series: Intact Stability.

[11] Molland, A. (2008). The Maritime Engineering Reference Book - A guide to ship design, construction and operation.

[12] Blomberg, K. & Gross, J. (2014). Conceptual Study of a USV for the Swedish Navy.

[13] d'Oliveira, R. (1964). Elementos de Arquitetura Naval, Volume I. Serviço de Publicações Escolares da Escola Naval.

[14] Letcher, J. (2010). The Principles of Naval Architecture Series: The Geometry of Ships.

IOP Conf. Series: Earth and Environmental Science **172** (2018) 012025 doi:10.1088/1755-1315/172/1/012025

NONSYMMETRICAL SINUSOIDAL THREE-PHASE SYSTEMS

E DRAGOMIR[1], O N VOLINTIRU[2], T M ŞTEFĂNESCU[2], A POCORA[3]

[1]NAVAL ACADEMY „Mircea cel Bătrân", CONSTANŢA
[2]MILITARY TECHNICAL ACADEMY, BUCUREŞTI
[3]POLITEHNICA UNIVERSITY OF BUCHAREST
edy_catalin27@yahoo.com

Abstract. This article deals with the theoretical aspects of the three-phase systems sinus lop-sided, especially the calculation of active, reactive power and the unbalanced and oscillating power systems. The unbalanced power or improperly called "lop-sided" introduced for pragmatic reasons to explain the loss of power caused by the unbalanced powers replaces the deficit of the three-phase system cumulated as power of the phases which are partial compensated under the three-phase systems. The unbalanced power called the "lop-sided power" should not be confused as it is sometimes with the oscillating (fluctuating) powers from which it differs by the sign of nonorthogonallity component.

1. Introduction

For this analysis, symmetrical components phasor [1] and representative vectors (space-temporally phasor [2-4] are used sometimes improperly called symmetrical components [5-6] also.

Symmetrical components phasors, homopolars, direct and inverted (with 0, 1 respectively 2 indexes) have been defined for the three-phase system of the tension phasor (index a, b, c) thus:

$$\begin{aligned} \underline{u}_0 &= \frac{1}{3}(\underline{u}_a + \underline{u}_b + \underline{u}_c) \\ \underline{u}_1 &= \frac{1}{3}(\underline{u}_a + a\underline{u}_b + a^2\underline{u}_c) \\ \underline{u}_2 &= \frac{1}{3}(\underline{u}_a + a^2\underline{u}_b + a\underline{u}_c) \end{aligned} \tag{1}$$

In these expressions $a = \varepsilon^{j2\pi/3}$ and $a^2 = \varepsilon^{-j2\pi/3}$ are rotation operators of phasors multiplied with $\frac{2\pi}{3}$ respective $-\frac{2\pi}{3}$.

Similar relations can be written for current phasors. The spatial phasors associated to rotating magnetic field of three-phase machines and induced tension by this, has similar expressions with direct and inverted symmetrical components but are functions of instantaneous values of phases tensions for unlike the symmetrical components phasors which are functions of phasors which are associated to phases tensions.

$$\begin{aligned} \underline{u}^+ &= \frac{2}{3}\left(u_a + au_b + a^2u_c\right) \equiv \left(\underline{u}^-\right)^* \\ \underline{u}^- &= \frac{2}{3}\left(u_a + a^2u_b + au_c\right) \equiv \left(\underline{u}^+\right)^* \end{aligned} \tag{2}$$

IOP Conf. Series: Earth and Environmental Science **172** (2018) 012025 doi:10.1088/1755-1315/172/1/012025

Operators a and a^2 have here the signification ***of directional versor*** [4]. From these result the relations with the symmetrical components phasors with which they must not be confused.

$$\underline{u}^+ = \underline{u}_1 + \underline{u}_2^*$$

$$\underline{u}^- = \underline{u}_1^* + \underline{u}_2$$

The same relations are obtained from phasors associated to currents.

2. Active and reactive powers of the sinus lop-sided three-phase system

The instantaneous complex power of three-phase systems without homopolar component of tension ($u_0 = 0$) equal with the sum of instantaneous values of direct and inverted complex powers of the phases has the following mathematical expression:

$$\underline{s} = [\underline{u}_a \quad \underline{u}_b \quad \underline{u}_c] \begin{bmatrix} i_a \\ i_b \\ i_c \end{bmatrix} \equiv \underline{u}_1 [1 \quad a^2 \quad a] \begin{bmatrix} i_a \\ i_b \\ i_c \end{bmatrix} + \underline{u}_2 [1 \quad a \quad a^2] \begin{bmatrix} i_a \\ i_b \\ i_c \end{bmatrix} \tag{3}$$

$$\begin{aligned} \underline{s} &= \underline{u}_1 (i_a + a^2 i_b + a i_c) + \underline{u}_2 (i_a + a i_b + a^2 i_c) \equiv \frac{3}{2} \underline{u}_1 \underline{i}^- + \frac{3}{2} \underline{u}_2 \underline{i}^+ \equiv \\ &\equiv \frac{3}{2} \underline{u}_1 (\underline{i}_1^* + \underline{i}_2) + \frac{3}{2} \underline{u}_2 (\underline{i}_2^* + \underline{i}_1) \equiv 3 (\underline{U}_1 \underline{I}_1^* + \underline{U}_1 \underline{I}_2 \varepsilon^{2j\omega t} + \underline{U}_2 \underline{I}_2^* + \underline{U}_2 \underline{I}_1 \varepsilon^{2j\omega t}) \equiv \\ &\equiv 3 [\underline{U}_1 \underline{I}_1^* + \underline{U}_2 \underline{I}_2^* + (\underline{U}_1 \underline{I}_2 + \underline{U}_2 \underline{I}_1) \varepsilon^{2j\omega t}] \equiv \underline{S}_1 + \underline{S}_2 + \underline{\tilde{s}} \end{aligned} \tag{4}$$

The tensions $\underline{U}_{1,2}$ and currents $\underline{I}_{1,2}$ are represented in simplified complex of the respective phasors

$$\underline{u}_{1,2} = \sqrt{2} \underline{U}_{1,2} \varepsilon^{j\omega t}, \qquad \underline{i}_{1,2} = \sqrt{2} \underline{I}_{1,2} \varepsilon^{j\omega t}$$

The two real components (three-phase active power) are thus obtained in one vectorial

$$P = 3U_1 I_1 \cos\varphi_1 + 3U_2 I_2 \cos\varphi_2 \tag{5}$$

and the "imaginary" reactive power

$$Q = 3U_1 I_1 \sin\varphi_1 + 3U_2 I_2 \sin\varphi_2 \tag{6}$$

as well as the oscillating power of three-phase system

$$\tilde{s} = \mathrm{Re}\, \underline{\tilde{s}} = 3U_1 I_2 \cos(2\omega t + \alpha_1 + \alpha_2 - \varphi_2) + 3U_2 I_1 \cos(2\omega t + \alpha_2 + \alpha_1 - \varphi_1) \tag{7}$$

The amplitude of these oscillating or fluctuant power is

$$S_{osc} = 3\sqrt{(U_1 I_2)^2 + (U_2 I_1)^2 + 2U_1 U_2 I_1 I_2 \cos(\varphi_1 - \varphi_2)} \tag{8}$$

Reported to the algebraic sum of direct and inverted apparent powers, it was considered an indicator of "unbalanced" system [7].

Other expressions of sinusoidal non-symmetrical power system which use Park spatial vectors or biphasic components of this [8-17] offer one exact mathematical expression of the active power.

$$p=[u_a u_b u_c]\begin{bmatrix}i_a\\ i_b\\ i_c\end{bmatrix}=\frac{1}{2}\left[\underline{u}^+\underline{u}^-\right]\begin{bmatrix}1 & a^2 & a\\ 1 & a & a^2\end{bmatrix}\cdot\frac{1}{2}\begin{bmatrix}1 & 1\\ a^2 & a\\ a & a^2\end{bmatrix}\begin{bmatrix}\underline{i}^+\\ \underline{i}^-\end{bmatrix}=\frac{3}{4}\left[\underline{u}^+\underline{u}^-\right]\begin{bmatrix}0 & 1\\ 1 & 0\end{bmatrix}\begin{bmatrix}\underline{i}^+\\ \underline{i}^-\end{bmatrix}=$$

$$=\frac{3}{4}\left(\underline{u}^+\underline{i}^-+\underline{u}^-\underline{i}^+\right)=\frac{3}{4}\left[\underline{u}^+\underline{i}^-+\left(\underline{u}^+\underline{i}^-\right)^*\right]=\frac{3}{2}\mathrm{Re}\,\underline{u}^+\underline{i}^-=\frac{3}{2}\mathrm{Re}\,\underline{u}^-\underline{i}^+$$

$$p=\frac{3}{2}\mathrm{Re}\,\underline{u}^+\underline{i}^-=\frac{3}{2}\mathrm{Re}\left(\underline{u}_1+\underline{u}_2^*\right)\left(\underline{i}_1^*+\underline{i}_2\right)=$$

$$=3\,\mathrm{Re}\left(\underline{U}_1\underline{I}_1^*+\underline{U}_2^*\underline{I}_2+\underline{U}_1\underline{I}_2\varepsilon^{j2\omega t}+\underline{U}_2^*\underline{I}_1^*\varepsilon^{-j2\omega t}\right)$$

respectively

$$p=\frac{3}{2}\mathrm{Re}\,\underline{u}^-\underline{i}^+=3\,\mathrm{Re}\left(\underline{u}_1^*+\underline{u}_2\right)\left(\underline{i}_1+\underline{i}_2^*\right)=$$

$$=3\,\mathrm{Re}\left(\underline{U}_1^*\underline{I}_1+\underline{U}_2\underline{I}_2^*+\underline{U}_1^*\underline{I}_2^*\varepsilon^{-j2\omega t}+\underline{U}_2\underline{I}_1\varepsilon^{j2\omega t}\right)$$

but arbitrary, incorrect for reactive

$$Q=\mathrm{Im}\,3\left(\underline{U}_1\underline{I}_1^*+\underline{U}_2^*\underline{I}_2\right)=3U_1I_1\sin\varphi_1-3U_2I_2\sin\varphi_2=Q_1-Q_2 \tag{9}$$

respectively

$$Q=\mathrm{Im}\,3\left(\underline{U}_1^*\underline{I}_1+\underline{U}_2\underline{I}_2^*\right)=-3U_1I_1\sin\varphi_1+3U_2I_2\sin\varphi_2=-Q_1+Q_2 \tag{10}$$

as well as for complex instantaneous power

$$\begin{aligned}\underline{s}&=\frac{3}{2}\underline{u}^-\underline{i}^+=\frac{3}{2}\left(\underline{u}_1^*+\underline{u}_2\right)\left(\underline{i}_1+\underline{i}_2^*\right)\\ \underline{s}&=\frac{3}{2}\underline{u}^+\underline{i}^-=\frac{3}{2}\left(\underline{u}_1+\underline{u}_2^*\right)\left(\underline{i}_1^*+\underline{i}_2\right)\end{aligned} \tag{11}$$

The persistent error in numerous theses in specialized literature of the expression that contains spatial vectors exclusive ("Park") are explained through the lock of originality for the solution for instantaneous active power

$$p=\frac{3}{4}\left(\underline{u}^+\underline{i}^-+\underline{u}^-\underline{i}^+\right)=\frac{3}{2}\mathrm{Re}\,\underline{u}^+\underline{i}^-=\frac{3}{2}\mathrm{Re}\,\underline{u}^-\underline{i}^+ \tag{12}$$

Any of these two expressions offer concrete solutions for active power but different, incorrect and ambiguous for reactive power.

A formula considered "general" for reactive power from three-phase systems [18] that cures the deficiency of the expression $\underline{s}=3/2\,\underline{u}^+\underline{i}^-$ to reproduce incorrectly the three-phase reactive power of non-symmetrical systems uses one representative tension vector rotated with $-\pi/2$ in rapport with direct succession vector and with $+\pi/2$ in rapport with the inverted conjugated succession

$$\underline{u}'=-j(\underline{u}_1-\underline{u}_2^*) \tag{13}$$

thus obtain the following expression

$$\underline{s}'=\frac{3}{2}\underline{u}'\underline{i}^*=\frac{3}{2}\underline{u}'\underline{i}^-=\frac{3}{2}(-j)(\underline{u}_1\underline{i}_1^*-\underline{u}_2^*\underline{i}_2+\underline{u}_1\underline{i}_2-\underline{u}_2^*\underline{i}_1^*) \tag{14}$$

The three-phase complex power (average in one period) results in:

$$\underline{S}=3(-j)\left(\underline{U}_1\underline{I}_1^*-\underline{U}_2^*\underline{I}_2\right)=3(Q_1+Q_2)-3j(P_1-P_2) \tag{15}$$

In consequence, reactive power $Q = \mathrm{Re}\underline{S}$ is reproduced correctly while active power $P = \mathrm{Im}\underline{S}$ is wrong.

In conclusion, the only one correct expression at three-phase complex power for unbalanced system, single and complete [19-22] is the following

$$\underline{s} = \frac{3}{2}[\underline{u}_1\underline{i}^- + \underline{u}_2\underline{i}^+] = \frac{3}{2}[\underline{u}_1\underline{i}_1^* + \underline{u}_2\underline{i}_2^* + \underline{u}_1\underline{i}_2 + \underline{u}_2\underline{i}_1]$$

or

$$p = \frac{3}{4}(\underline{u}^+\underline{i}^- + \underline{u}^-\underline{i}^+) \tag{16}$$

Following a correct grouping of the pairs of terms complexly conjugated

$$p = \frac{3}{4}[(\underline{u}_1\underline{i}^- + \underline{u}_2\underline{i}^+ + \underline{u}_1^*\underline{i}^+ + \underline{u}_2\underline{i}^-] \tag{17}$$

With the election of the first two terms of this expression

$$p = \frac{3}{2}\mathrm{Re}[(\underline{u}_1\underline{i}^- + \underline{u}_2\underline{i}^+] = \mathrm{Re}\,\underline{s} \tag{18}$$

results in a correct solution for complex power $\underline{s}$ and therefore for the imaginary parts (reactive power) and for real parts (active power) of it.

As in case with the three-phase systems symmetrical voltages and currents where the total instantaneous power does not have the characteristic reactive power of each phase (because of the compensation between system phases), similarly to systems with non-symmetrical three-phase voltages and currents, the total instantaneous power does not contain the other specific power of phases.

Therefore, before analyzing the various proposals for the definition of effective power of three-phases systems, we express the instantaneous physical powers of phases using symmetrical components of voltages and currents and not representative vectors (Park or Clark) as is usually done incorrectly [8-17].

In the absence of homopolar components, with relations between the tensions and currents phasors of phases and symmetrical components

$$\begin{bmatrix}\underline{u}_a\\ \underline{u}_b\\ \underline{u}_c\end{bmatrix} = \begin{bmatrix}1 & 1\\ a^2 & a\\ a & a^2\end{bmatrix}\begin{bmatrix}\underline{u}_1\\ \underline{u}_2\end{bmatrix} \qquad \begin{bmatrix}\underline{i}_a\\ \underline{i}_b\\ \underline{i}_c\end{bmatrix} = \begin{bmatrix}1 & 1\\ a^2 & a\\ a & a^2\end{bmatrix}\begin{bmatrix}\underline{i}_1\\ \underline{i}_2\end{bmatrix}$$

obtained [19-21] following the expression of complex instantaneous power for each phases (a, b, c) of the system.

$$\underline{s}_{a,b,c} = (\underline{u}\cdot i)_{a,b,c} = (\underline{u}\cdot\frac{\underline{i}+\underline{i}^*}{2})_{a,b,c} =$$

$$= \frac{1}{2}[\underline{u}_1(\underline{i}_1^* + \underline{i}_2) + \underline{u}_2(\underline{i}_2^* + \underline{i}_1) + \underline{u}_1(\underline{i}_1 + \underline{i}_2^*) + \underline{u}_2(\underline{i}_2 + \underline{i}_1^*)]_{a,b,c}$$

or

$$\underline{s}_a = \frac{1}{2}\left[\underline{u}_1\underline{i}_1^* + \underline{u}_2\underline{i}_2^* + \underline{u}_1\underline{i}_2 + \underline{u}_2\underline{i}_1 + \underline{u}_1\underline{i}_2^* + \underline{u}_2\underline{i}_1^* + \underline{u}_1\underline{i}_1 + \underline{u}_2\underline{i}_2\right]$$

$$= \frac{\underline{s}}{3} + \frac{\underline{u}_1\underline{i}_2^* + \underline{u}_2\underline{i}_1^* + \underline{u}_1\underline{i}_1 + \underline{u}_2\underline{i}_2}{2} \quad (19)$$

$$\underline{s}_b = \frac{\underline{s}}{3} + \frac{a\underline{u}_1 \cdot \underline{i}_2^* + a^2\underline{u}_2\underline{i}_1^* + a\underline{u}_1\underline{i}_1 + a^2\underline{u}_2\underline{i}_2}{2} \quad (20)$$

$$\underline{s}_c = \frac{\underline{s}}{3} + \frac{a^2\underline{u}_1 \cdot \underline{i}_2^* + a\underline{u}_2\underline{i}_1^* + a^2\underline{u}_1\underline{i}_1 + a\underline{u}_2\underline{i}_2}{2} \quad (21)$$

where $\underline{s} = \frac{3}{2}\left(\underline{u}_1 \cdot \underline{i}_1^* + \underline{u}_2\underline{i}_2^* + \underline{u}_1\underline{i}_2 + \underline{u}_2\underline{i}_1\right)$ is the complex instantaneous power of three-phase system, common to the three phases.

If the phasors of symmetrical components of tensions and currents are explained the following expressions of instantaneous complex power are obtained

$$\underline{s} = 3\left[\underline{U}_1\underline{I}_1^* + \underline{U}_2\underline{I}_2^* + \left(\underline{U}_1\underline{I}_2 + \underline{U}_2\underline{I}_1\right)\varepsilon^{j2\omega t}\right] \quad (22)$$

$$\underline{s}_a = \frac{\underline{s}}{3} + \underline{U}_1\underline{I}_2^* + \underline{U}_2\underline{I}_1^* + \left(\underline{U}_1\underline{I}_1 + \underline{U}_2\underline{I}_2\right)\varepsilon^{j2\omega t} \quad (23)$$

$$\underline{s}_b = \frac{\underline{s}}{3} + a\underline{U}_1\underline{I}_2^* + a^2\underline{U}_2\underline{I}_1^* + \left(a\underline{U}_1\underline{I}_1 + a^2\underline{U}_2\underline{I}_2\right)\varepsilon^{j2\omega t} \quad (24)$$

$$\underline{s}_c = \frac{\underline{s}}{3} + a^2\underline{U}_1\underline{I}_2^* + a\underline{U}_2\underline{I}_1^* + \left(a^2\underline{U}_1\underline{I}_1 + a\underline{U}_2\underline{I}_2\right)\varepsilon^{j2\omega t} \quad (25)$$

The expressions of medium powers (measured) components of classical vectorial complex power of non-symmetrical three-phase system are resulted

$$\underline{S} = 3(P_1 + P_2) + 3j(Q_1 + Q_2) = 3\left(\underline{U}_1\underline{I}_1^* + \underline{U}_2\underline{I}_2^*\right) =$$
$$= 3\left(U_1 I_1\cos\varphi_1 + U_2 I_2\cos\varphi_2\right) + j3\left(U_1 I_1\sin\varphi_1 + U_2 I_2\sin\varphi_2\right)$$

as well as the (measured) medium powers, components of complex power for each phase.

$$\underline{S}_a = \frac{\underline{S}}{3} + U_1 I_2\cos\varphi_{12} + U_2 I_1\cos\varphi_{21} + j\left(U_1 I_2\sin\varphi_{12} + U_2 I_1\sin\varphi_{21}\right)$$

$$\underline{S}_b = \frac{\underline{S}}{3} + U_1 I_2\cos\left(\varphi_{12} + \frac{2\pi}{3}\right) + U_2 I_1\cos\left(\varphi_{21} - \frac{2\pi}{3}\right) +$$
$$+ j\left[U_1 I_2\sin\left(\varphi_{12} + \frac{2\pi}{3}\right) + U_2 I_1\sin\left(\varphi_{21} - \frac{2\pi}{3}\right)\right] \quad (26)$$

$$\underline{S}_c = \frac{\underline{S}}{3} + U_1 I_2\cos\left(\varphi_{12} - \frac{2\pi}{3}\right) + U_2 I_1\cos\left(\varphi_{21} + \frac{2\pi}{3}\right) +$$
$$+ j\left[U_1 I_2\sin\left(\varphi_{12} - \frac{2\pi}{3}\right) + U_2 I_1\sin\left(\varphi_{21} + \frac{2\pi}{3}\right)\right]$$

It can be observed that the active powers associated to the phases

$$P_a = U_1 I_1 \cos\varphi_1 + U_2 I_2 \cos\varphi_2 + U_1 I_2 \cos\varphi_{12} + U_2 I_1 \cos\varphi_{21}$$
$$P_b = U_1 I_1 \cos\varphi_1 + U_2 I_2 \cos\varphi_2 + U_1 I_2 \cos(\varphi_{12} + \frac{2\pi}{3}) + U_2 I_1 \cos(\varphi_{21} - \frac{2\pi}{3}) \quad (27)$$
$$P_c = U_1 I_1 \cos\varphi_1 + U_2 I_2 \cos\varphi_2 + U_1 I_2 \cos(\varphi_{12} - \frac{2\pi}{3}) + U_2 I_1 \cos(\varphi_{21} + \frac{2\pi}{3})$$

contain components of mutual unbalanced which is compensated within of total (sum) power. So the total three-phase power is deficient because it does not contain all powers available for each phase.

Because of mutual unbalanced components, the powers associated phases may have different values, some phases with greater or smaller powers, positive or negative in relation to other phases.

The same remark resulted for reactive powers (measured by wattmeter) through the use of some artificial tensions dephased with $\pi/2$ in relation with real tensions as well as for "vectorial" complex power.

$$Q_a = U_1 I_1 \sin\varphi_1 + U_2 I_2 \sin\varphi_2 + U_1 I_2 \sin\varphi_{12} + U_2 I_1 \sin\varphi_{21}$$
$$Q_b = U_1 I_1 \sin\varphi_1 + U_2 I_2 \sin\varphi_2 + U_1 I_2 \sin(\varphi_{12} + \frac{2\pi}{3}) + U_2 I_1 \sin(\varphi_{21} - \frac{2\pi}{3}) \quad (28)$$
$$Q_c = U_1 I_1 \sin\varphi_1 + U_2 I_2 \sin\varphi_2 + U_1 I_2 \sin(\varphi_{12} - \frac{2\pi}{3}) + U_2 I_1 \sin(\varphi_{21} + \frac{2\pi}{3})$$

Therefore they sought other practical expressions of apparent powers, active and reactive and for "unbalanced" to correct this deficiency.

As the sinusoidal monophase system, the movement of power does not have a special physical significance because the power components do not have the property of identifiable substances whose "flowing" could be tracked [23]. However, it stressed the usefulness of conservation property of some power.

Decomposition of the unbalanced regime in balanced components as well as a non-sinusoidal periodic in harmonic components has an important role regarding the methological point of view for observing the different effects in relation to symmetrical sinusoidal reference and for achievement of compensating filters.

3. Power of the unbalanced sinus lop-sided three-phase system

„The apparent effective power'' is suggested by the expressions of actual values tensions and currents as functions of the harmonic expressions of the non-sinusoidal single phase systems, similarly for expressions corresponding to the phases of unbalanced three-phase systems.

Defined [25] and adopted [24][26] an expression of "actual apparent" power

$$S_e = 3U_e I_e \quad (29)$$

with

$$3U_e^2 = U_a^2 + U_b^2 + U_c^2 = 3(U_1^2 + U_2^2 + U_0^2)$$
$$3I_e^2 = I_a^2 + I_b^2 + I_c^2 = 3(I_1^2 + I_2^2 + I_0^2)$$

To simplify, in the absence of homopolar components, if using Lagrange identity similar to that used by C Budeanu at non-sinusoidal monophase systems by explaining the active and reactive components of currents on orthogonal relationship of power is obtained.

$$\left(\frac{S_e}{3}\right)^2 = (U_1{}^2 + U_2{}^2)(\mathrm{I}_1{}^2 + \mathrm{I}_2{}^2) \equiv$$
$$\equiv (U_1{}^2 + U_2{}^2)(\mathrm{I}_1^2 \cos^2 \varphi_1 + \mathrm{I}_1^2 \sin^2 \varphi_1 + \mathrm{I}_2^2 \cos^2 \varphi_2 + \mathrm{I}_2^2 \sin^2 \varphi_2) \equiv$$
$$\equiv (U_1 \mathrm{I}_1 \cos\varphi_1 + U_2 \mathrm{I}_2 \cos\varphi_2)^2 + (U_1 \mathrm{I}_1 \sin\varphi_1 + U_2 \mathrm{I}_2 \sin\varphi_2)^2 +$$
$$+ (U_1 \mathrm{I}_2 \cos\varphi_2 - U_2 \mathrm{I}_1 \cos\varphi_1)^2 + (U_1 \mathrm{I}_2 \sin\varphi_2 - U_2 \mathrm{I}_1 \sin\varphi_1)^2$$

or

$$S_e^2 = P^2 + Q^2 + N_p^2 + N_q^2 \qquad (30)$$

In addition to the powers

active $P = (U_1 \mathrm{I}_1 \cos\varphi_1 + U_2 \mathrm{I}_2 \cos\varphi_2)$

reactive $Q = 3(U_1 \mathrm{I}_1 \sin\varphi + U_2 \mathrm{I}_2 \sin\varphi_2)$

that exist in the complex classic expression (vectorial) with symmetrical components, new orthogonal "unbalanced" components appear

active $N_p = 3(U_1 \mathrm{I}_2 \cos\varphi_2 - U_2 \mathrm{I}_1 \cos\varphi_1)$

(31)

reactive $N_q = 3(U_1 \mathrm{I}_2 \sin\varphi_2 - U_2 \mathrm{I}_1 \sin\varphi_1)$

that forming apparent unbalanced power similar to deforming power

$$N^2 = N_p^2 + N_q^2 = 9\left[(U_1 \mathrm{I}_2)^2 + (U_2 \mathrm{I}_1)^2 - 2U_1 U_2 \mathrm{I}_1 \mathrm{I}_2 \cos(\varphi_1 - \varphi_2)\right] \qquad (32)$$

For a balanced three-phase system with equal impedances of phases

$$\underline{Z} \equiv \frac{\underline{U}_1}{\underline{I}_1} = \frac{\underline{U}_2}{\underline{\mathrm{I}}_2}, \qquad \varphi_1 = \varphi_2$$

the unbalanced power is cancelled (*N=0*) resulting a decrease of total losses of power by the interconnection line.

$$\Delta P = 3r(\mathrm{I}_a{}^2 + \mathrm{I}_b{}^2 + \mathrm{I}_c{}^2) = 9r\mathrm{I}_e{}^2 = r\frac{Se^2}{U_e^2} = r\frac{P^2 + Q^2 + N^2}{U_e^2} \qquad (33)$$

The "unbalanced" power or improperly called "non-symmetrical" [24][27] introduced for pragmatic reasons to explain the loss of power caused by the unbalanced regime replaces the deficit of the cumulative three-phase powers like powers of phases which are partially compensated under the three-phase systems.

Even if in calculations and judgment the powers of unbalanced three-phase systems (and not algebraic sum of them like three-phase) the utility of effective three-phase power is justified because, "unbalanced power" will be cancelled simultaneously with cancellation of the unbalanced (the identity of ***impedances*** phases system).

The power of "unbalanced N" called the "lop-sided power" [24][28] should not be confused as it is sometimes [28-29] with the oscillating (fluctuating) powers S_{osc} [7] from which it differs by the sign of non-orthogonality components.

$$\left(\frac{N}{3}\right)^2 = U_1{}^2 \mathrm{I}_2{}^2 + U_2{}^2 \mathrm{I}_1{}^2 - 2U_1 U_2 \mathrm{I}_1 \mathrm{I}_2 \cos(\varphi_1 - \varphi_2)$$
$$\left(\frac{S_{osc}}{3}\right)^2 = U_1^2 \mathrm{I}_2^2 + U_2^2 \mathrm{I}_1^2 + 2U_1 U_2 \mathrm{I}_1 \mathrm{I}_2 \cos(\varphi_1 - \varphi_2)$$

Fluctuating power S_{osc} (8) will be cancelled $U_1/I_1 = U_2/I_2$ only if $\varphi_1 - \varphi_2 = \pi$ theoretical condition cannot be realized physically with passive circuits when the unbalanced power N is cancelled for purely balanced receiver, with $\varphi_1 = \varphi_2$.

Expressed in relative units by reporting at direct apparent power

$$\left(\frac{S_1}{3}\right)^2 = \left(\frac{U_1 I_1}{3}\right)^2$$

a function of factors of non-symmetrical tension and circuits (factors as the functioning of the unbalanced) is obtained.

$$\left(\frac{N}{S_1}\right)^2 = \left(\frac{I_2}{I_1}\right)^2 + \left(\frac{U_2}{U_1}\right)^2 - 2\frac{I_2}{I_1}\frac{U_2}{U_1}\cos(\varphi_1 - \varphi_2)$$

$$\left(\frac{S_{osc}}{S_1}\right)^2 = \left(\frac{I_2}{I_1}\right)^2 + \left(\frac{U_2}{U_1}\right)^2 + 2\frac{I_2}{I_1}\frac{U_2}{U_1}\cos(\varphi_1 - \varphi_2)$$

The effective apparent power with his active, reactive, deforming and unbalanced components was previously deduced [24] as the maximum power transmitted for loss of power by transmission line (proportional to the squares effective currents) data. The expressions obtained include both distorting as the unbalanced power.

A simplified expression of the effective apparent power S_e to highlight the quality factors of unbalanced and distorting functioning is obtained in relative units by reporting at direct fundamental apparent power $3U_1 I_1$.

$$S_e^2 = 3U_e{}^2 \cdot 3I_e{}^2 = 3\left[U_1{}^2 + U_2{}^2 + U_H{}^2\right] \cdot 3\left[I_1{}^2 + I_2{}^2 + I_H{}^2\right] \quad (34)$$

$$\left(\frac{S_e}{3U_1 I_1}\right)^2 = \left[1 + \left(\frac{U_2}{U_1}\right)^2 + \left(\frac{U_H}{U_1}\right)^2\right]\left[1 + \left(\frac{I_2}{I_1}\right)^2 + \left(\frac{I_H}{I_1}\right)^2\right] \quad (35)$$

U_H and I_H are the effective value of all voltage respectively current harmonics direct and inverted components.

4. Conclusions

Using formal representative spatial vectors lead to erroneous expressions of reactive power from the three-phase lop-sided. Using rotating phasors (Fresnel) leads to the correct expressions.

Correct expression of complex three-phase power for the unbalanced, single and complete system is the following $p = \frac{3}{4}(\underline{u}^+ \underline{i}^- + \underline{u}^- \underline{i}^+) \equiv \frac{3}{4}\left[(\underline{u}_1 + \underline{u}_2^*)i^- + (\underline{u}_1^* + \underline{u}_2)i^+\right]$ with $p = \frac{3}{2}\mathrm{Re}\left(\underline{u}_1 \underline{i}^- + \underline{u}_2 \underline{i}^+\right) = \mathrm{Re}\,\underline{s}$ and $q = \frac{3}{2}\mathrm{Im}\,\underline{s} = 3\left(U_1 I_1 \sin\varphi_1 + U_2 I_2 \sin\varphi_2\right)$ which provides a fair solution for complex power and as well as for the real (active power) and for the imaginary (reactive power) parts.

The unbalanced power N called the "lop-sided power" should not be confused as it is sometimes with the oscillating (fluctuating) power S_{osc} from which it differs by the sign of non-orthogonality.

Only the analysis of quantities (current, voltage, and power) characteristic to phases of the three-phase systems presents theoretical and practical importance to achieve balance of the unbalanced systems as well as to analyze the current and powers that load the phases of the system.

The analysis of the entire three-phase system provides little indications due to the compensation between the phases of active and reactive powers of the phases.

References

[1] Fortesque C *1918 Method of symmetrical co-ordinates applied to the solution of polyphases networks* AIEEE vol 37 p 1027-1115
[2] Manea F 1957 Regimuri electromagnetice tranzitorii simetrice ale maşinii asincrone *Electrotehnica* 5 nr 6 p 179-183
[3] Manea F 1959 *Studiul regimurilor reţelelor trifazate cu vectori reprezentativi şi simetrici* Electrotehnica 7 nr 5 p 155-158
[4] Manea F 1966 *Unitatea ecuaţiilor şi transformărilor de coordonate în teoria maşinilor electrice* Electrotehnica 14 nr 3 p 79-83
[5] Ku Y 1929 *Transient Analysis of A-C Machinery* TAIEE 48 p 707-715
[6] Ku Y 1951*Transient Analysis of Rotating Machines and Stationary Networks by Means of Rotating Reference Frames* TAIEE 70 p 343-956
[7] Crug C 1953 *Bazele electrotehnicii* vol II Ed Energetica de Stat p 299
[43] Ferrero A, Furga G 1991 *A New Approach to the Definition of Power Components in three Phase Systems under Nonsinusoidal Conditions* IEEE No 3
[8] Cristaldi L, Ferrero A, Furga G 1994 *Current Decomposition in Asymmetrical Unbalanced Three-Phase Systems under Nonsinusoidal Conditions* IEEE Trans On Instru And Meas vol 43 No 1 February,
[9] Marshall D, Van Wyk J 1991 *An Evaluation of the real-time compensation of fictions power in electric energy* IEEE Trans. On Power Delivery vol 8 No 2 October
[10] Watanabe E, Stephan R, Aredes M 1993 *New concepts of instantaneous active and reactive powers in electrical systems with generic loads* IEEE Trans On Power Delivery vol 8 No 2 April
[11] Willems J 1994 *Instantaneous Sinusoidal and Harmonic Active and Reactive Currents in Three-Phase power System* ETEP vol 4 No 5 September/October
[12] Willems J 1996 *Mathematical Foundations of the Instantaneous Power Concepts* ETEP vol 6 No 5 September/October p 299-305
[13] Cristaldi L, Ferrero A 1996 A *Mathematical Foundations of the Instantaneous Power Concepts an Algebric Approach* ETEP vol 6 No 5 September/October, p 305-311
[14] Salvatore L, Savino M 1994*Time-Frequency Representations of Instantaneous Complex Power Voltage and Currant Space Vectors* ETEP vol 4 No 5 September/October
[15] Ferrero A 1998 *Considerations about the Different Possible Approaches to the Study of the Electrical Power Systems under Nonsinusoidal Conditions* 8th ICHQP (International Conference of Harmonics and Quality of Power) October 14-16 Athens Greece
[16] Segura A, Hernandez M 1998 *Instantaneous Active and Reactive Current Compensation for AC/DC Three-Phase Harmonic Generating Loads* 8th ICHQP (International Conference of Harmonics and Quality of Power) October 14-16 p 521-526 Athens Greece
[17] Nabae A, Tanaka T 1998 *A New Approach to Individual Phase Reactive Power Compensation for Non-sinusoidal and Unbalanced Three-Phase Systems* 8th ICHQP (International Conference of Harmonics and Quality of Power) October 14-16 Athena Greece
[18] Nedelcu V 1967 *O nouă formulă generală pentru puterea reactivă în sistemele trifazate* Electrotehnica Nr 12 p 447-449
[19] Manea F, Samoilescu G, Dragomir E 2002 *Aspecte vechi şi noi ale puterilor sistemelor periodice nesimetrice şi nesinusoidale* Sesiunea jubiliară de comunicări ştiinţifice Academia Navală „Mircea Cel Bătrân Constanţa 14-16 noiembrie
[20 Manea F, Samoilescu G, Dragomir E 2004 *Puteri instantanee, medii şi aparente ale sistemelor sinusoidale* Sesiunea de comunicări ştiinţifice Academia Navală Mircea Cel Bătrân Constanţa

[21] Manea F 2002 *The instantaneous power of three-phase non-symmetric systems* Rev Roum Sci Techn Energ 47 2 p 135-140 Bucureşti
[22] Manea F 2005 *The measurable reactive instantaneous power of three-phase non-symmetric sinusoidal system* Sci Techn Energ 50 1 p 11-18 Bucureşti
[23] Tugulea A 1985*Considération concernant la définition et le rôle de la puissance réactive en régime déforment* Rev Roum Sci Techn et Energ 30 4 p 361-365 Bucureşti
[24] Manea F 1960 *Puterile sistemelor trifazate nesimetrice şi nesinusoidale* Studii şi cercetări de energetică şi electrotehnică vol 10 No 4 p 771-777
[25] Emanuel A 1998 *The Bucholtz-Goodhne apparent power definition: The practical approach for nonsinusoidal and unbalanced systems* IEEE Trans On Power Delivery vol 13 No 2 April p 344-349
[26] Emanuel A 1993*On the definition of power factor and apparent power in unbalanced polyphase circuits with sinusoidal voltage and currents* IEEE Trans On Power Delivery vol 8 No 3 July p 841-847
[27] Nedelcu V 1968 *Regimurile de funcţionare ale maşinilor de curent alternativ* Editura tehnică p 60-63
[28] Ghiţă C1993 *Asupra puterii fluctuante în sistemele trifazate* EEA Electrotehnica 41 vol I p 6-9
[29] Ghiţă C 2003 *Fluctuating power within polyphase systems without homopolars components* Rev Roum Sci Techn El Et Energ 48 2-3 p 283-289

Professional Adjustment Variables Onboard the Seagoing Ships

Carmen Luminita COJOCARU[1], Catalin POPA[2], Taner ALBAYRAK[3], Alecu TOMA[4]
[1] *Mircea cel Batran Naval Academy, Romania, carmen.cojocaru@anmb.ro*
[2] *Mircea cel Batran Naval Academy, Romania, catalin.popa@anmb.ro*
[3] *Piri Reis University, Turkey, talbayrak@pirireis.edu.tr*
[4] *Mircea cel Batran Naval Academy, Romania, alecu.toma@anmb.ro*

Abstract. *The professional activity carried in the maritime sector requires not only a broadly educated and intensively trained human resources, but also a very strong psychosocial profile on individual and crew level, for onboard team members, for the sake of economic efficiency and effectiveness of transportation services. In this complex environment, the paperwork is depicting a statistical study on the major factors influencing the professional adjustment and adaptation variables to be considered on a professional level for seagoing personnel as following: the "human factor", the "working conditions onboard" and "cultural diversity". Using a questionnaire applied on 58 subjects, men and women gender seafarers, the authors had carrier a statistical analysis of the chosen item impact on both genders, reflecting the major differences, in any, in between. Conclusions have shown that both genders will perform properly onboard, with no significant difference, but assuming the assigned onboard role in the team, transcending from an individual perception to the collective rationales and imperatives. The present research has been carried out by the authors within MENTORESS project - Maritime Education Network to Orient and Retain Women for Efficient Seagoing Services.*

1. Introduction

The professional activity carried in the maritime sector requires not only broadly educated and intensively trained human resources, but also a strong psychosocial profile on individual and crew level. Onboard the ships, in the center of this psychosocial entity, can be found specific mixtures among social multicultural micro-groups, defined in particular by diversity. The crew diversity is not only a determined by culture, religion, customs, traditions or gender but also by the educational, training or experience background of the individuals, aggregated on a complex interactive manner as crew, on different ship services (A. Tabachiu, 2003, Gh. Iosif, 2001).

On the other hand, the professional activity in the naval sector can be oriented toward efficiency and effectiveness taking into account both major aspects of the environment of activity, namely physical labour conditions and the psycho-organizational environment embedded by the professional and behaviour features of the professional activity onboard, carried out within a high risk and uncertainty climate. In regard of onboard particularities, the risk is essentially a subjective psychological resort that refers to the possibility of negative events to interfere in a certain situation, causing a certain loss. The way of how the crew members perceive this potential threat and how they would react to this threat is depending on several factors as following: personality structure, personal attitudes, actual disposition, context specificity or cognitive biases. (Hăvârneanu, 2015, pp. 16)

In this context, the work security and safety and the success of the carried mission become the major focusing objective to the crew management, all of other psychosocial variables being considered subordinated, including here the inventory of individual factors (i.e. as the inter-individual differences on gender) and situational factors (i.e. onboard working conditions, environmental risk, cultural diversity) that would affect behavioural adjustment and its effectiveness.

2. Research methodology and study hypothesis

The null hypothesis (H_0) of the study postulates that there are no statistically significant determinations caused by the gender issue against the factors that could influence the success and safety of the professional performance on board the vessels. Under the conditions of the rejection of the null hypothesis, the following alternative assumptions could become plausible:

• H_1: There are statistically significant differences between women and men in terms of the **human factor** involved in achieving the success and safety of the work mission on board ships;

• H_2: There are statistically significant differences between women and men in terms of **cultural diversity** as supporting factor in achieving the success and safety of the professional performance on board ships;

• H_3: There are statistically significant differences between women and men in terms of **working conditions** provided in order to achieve the success and safety of the mission on board ships;

The verification of these hypotheses is based on a simple factorial design, including the following variables:

• Biological gender (nominal independent variable with 2 levels: male and female);
• Human factor (scalar dependent variable);
• Cultural diversity (scalar dependent variable);
• Working conditions on the ship (scalar dependent variable).

The minimal size of the research group in order to meet the parametric analyzes assumptions should consist of 60 observations at least, equally divided into the two groups determined by the independent variable.

3. Research methodology

3.1. Collecting data instrument

The study is to be conducted using the questionnaire survey method, the research instrument having the following structure, drafted on Likert type of scale consisting 4 steps of qualitative responses measurement (1-Very important, 2-Important, 3-Less Important, 4-Not important):

• **Human factor** - contains 16 items, the theoretical amplitude of the scale being between 16 and 64 points maximum, with the average theoretical scale of 24 points;

• **Cultural Diversity** - contains 8 items, the theoretical amplitude of the scale being between 8 and 32 points, with the average theoretical scale of 12 points;

• **Working conditions** onboard the ship - contains 13 items, the theoretical amplitude of the scale being between minimum 13 and max 52 points , with the theoretical average of 19.5 points.

The research procedure involves undergoing the following steps:

• Analyzing the consistency of the measurement scales to decide about the relationships between the items and the measured factor;

• Total score calculation and descriptive analyzes to verify the basic assumptions;

• Analysis of the general tendency of the responses comparing to the theoretical mean of the scales as referential;

• Verification of postulated hypotheses by comparing the two groups determined by the independent variable.

3.2. Research batch of participants

Regarding the independent variable, the research group will comprise 58 subjects, cadets and seafarers (3rd deck officers), graduates of the "Mircea cel Batran" Naval Academy in Constanta of the last 5 years, 37 (63.8%) of them having as male biological gender, and 21 (36.2 %) female gender, assuming the volume not being procedurally fulfilled (figure no. 1). We consider in this context that, assuming the other assumptions about the distribution of the scalar variables will be met, we could use the parametric analysis in which the "biological gender" variable will determine the research groups.

3.3. Analysis of instrument quality

The instrument quality analysis was carried out by studying the internal consistency of the items with Cronbach's internal Alpha coherence coefficient.
The initial values of the internal consistency coefficients for each of the three factors were 0,686 (0,554 - 0,792) for the "Human factor" scale, 0,826 (0,748 - 0,886) for the Cultural "Diversity scale" and 0,750 (0,645 - 0,836) in the case of "Working conditions onboard ship" scale.

The last two factors have a suitable consistency of items for the present research, while for the first factor case the correlation of 4th item (Management of emotions) presents an extremely small total score (0,05), its elimination leading to an increase of the consistency of the scale "Human Factor" at 0.747. Therefore, the "Human Factor" scale will lose one item, the theoretical amplitude in this simplified case becoming on a range between 15 and 60 points, with a theoretical mean of 22.5 points.

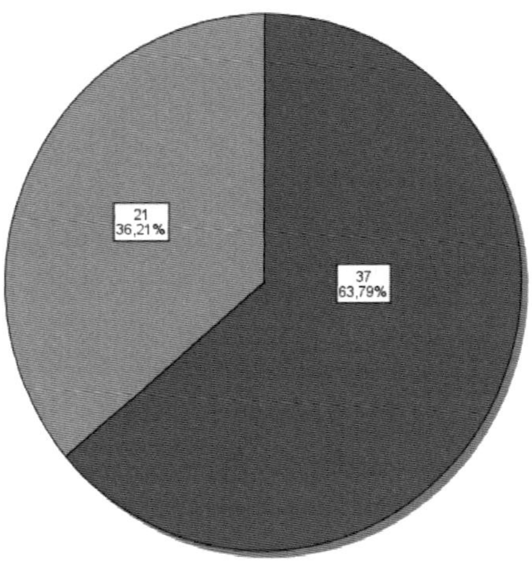

Figure no. 1: *The research group structure on gender criterion (63,79% males, 36,21%)*

4. Single varied descriptive analysis

As figured in the table no. 1 values, the "Human factor" presents a score distribution with the average of 23.76 points and a standard deviation of 4.31 points, the median being at the score of 24, but there are more modal values (20, 21 and 26). The empirical amplitude of the distribution is 20 points, ranging from a minimum of a theoretical minimum (15 points) to a maximum of 35 points (compared to a theoretical maximum of 60 points). However, the distribution is symmetric (Skewness = 0,388, Skewness error = 0,314) and normally arched (Kurtosis = 0,198, Kurtosis error = 0,618), the variable being suitable to be used in parametric analysis.

Table no. 1: *Unvaried descriptive values measurement*

Indicators		Human Factor	Cultural diversity	Working conditions onboard
N	Valid	58	58	58
	Absent	0	0	0
Mean		23,76	19,72	21,38
Median		24,00	19,50	21,00
Module (val ce mai mare)		20[a]	18[a]	20
Standard Deviation		4,314	4,340	4,167
Skewness (simtrie)		,388	-,378	,368
Skewness Standard Error		,314	,314	,314
Kurtosis		,198	,562	,074
Kurtosis Standard Error		,618	,618	,618
Amplitude		20	22	18
Minimum		15	8	13
Maximum		35	30	31

The "Cultural diversity" factor presents a distribution of scores averaging 19.72 points and a standard deviation of 4.34 points, the median being at 19.5, but there are more modal values (18 and 19). The empirical amplitude of the distribution is 22 points, ranging from a minimum equal to the theoretical minimum (8 points) and a maximum of 30 points (compared to a theoretical maximum of 32 points). However, the distribution is also symmetric (Skewness = -0,378, Skewness error = 0,314) and normally arched (Kurtosis = 0,562, Kurtosis error = 0,618), the variable being suitable to be used in parametric analysis.

The "Working conditions onboard the ship" factor shows a distribution of the scores averaging 21.38 points and a standard deviation of 4.16 points, the median being at 21, with the modal value 20. The empirical distribution amplitude is 18 points, a minimum equal to a theoretical minimum (13 points) and a maximum of 31 points (compared to a theoretical maximum of 52 points). However, the distribution is symmetric (Skewness = 0.368, Skewness error = 0.314) and normally arched (Kurtosis = 0.074, Kurtosis error = 0.618), the variable being also suitable to be used in parametric analysis.

5. Global analysis of trends by comparison with the mean of the scale

The global trend analysis involves comparison on the basis of the statistical values of the mean value resulted for each factor applied on the research group with the theoretical average of the respective scale. Following the analysis conclusion, statistical significant differences in all three factors analyzed can be observed. Thus, in the case of the "Human factor" variable, it was found that the average of the subjects' answers on the research group ($m_{(H1)} = 23,76$) is statistically significantly higher than the theoretical mean of the scale ($m_{(H0)} = 22,5$) in the sense that the subjects tend to give high overall values to this factor ($t_{(57)} = 2.22$; $p < 0.05$), but the significance of the high values indicates a rather unimportant appreciation of the human factor (d = 0.2923), the effect being medium to weak.

Regarding the "Cultural diversity" factor, it could be noticed the same overall trend. The average of the responses in the research group ($m_{(H1)} = 19.72$) is statistically significantly higher compared to the theoretical mean of the scale ($m_{(H0)} = 12.0$), the effect being much stronger (d = 1,77). Thus, we observe a general propensity to appreciate cultural diversity as not having an important role in the adaptation process ($t_{(57)} = 13.55$; $p < 0.05$).

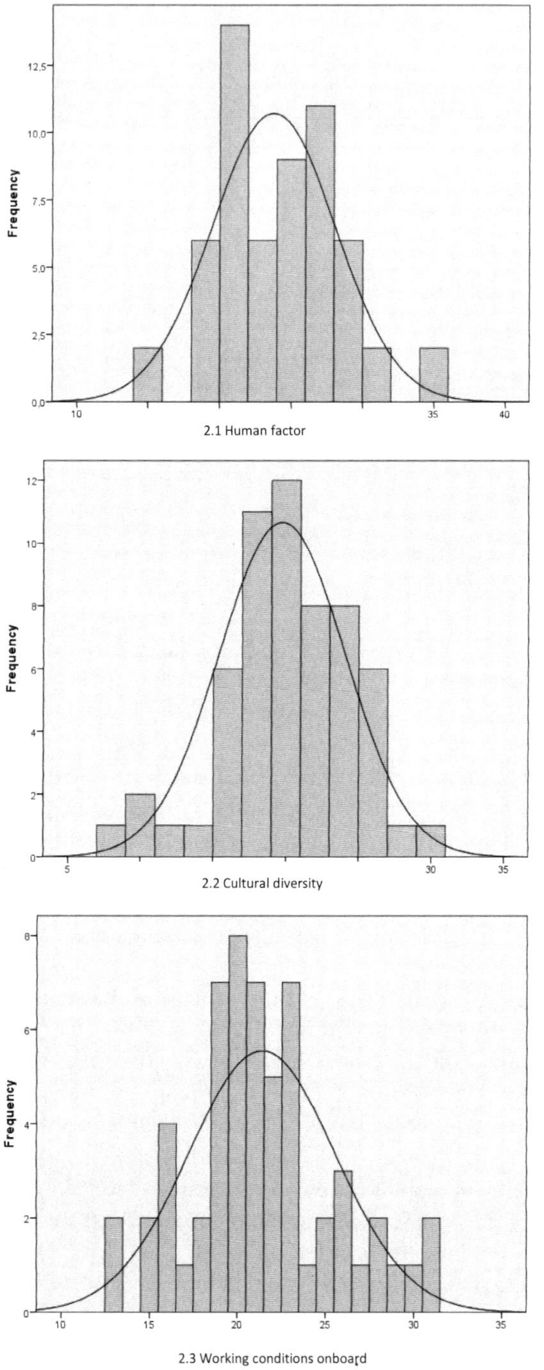

Figure no. 2: *Variable distributions on descriptive values*

In the case of "Working conditions onboard ships" factor, the distance between the theoretical mean of the scale ($m_{(H0)} = 19.5$) and the average of the subjects' answers ($m_{(H1)} = 21.38$) shows the same tendency to minimize the work conditions importance throughout the adjustment and adaptation processes onboard ($t_{(57)} = 3.43$; $p < 0.05$), this factor impact proving a medium-strong intensity ($d = 1$).

Comparing and sizing the differences in terms of impact magnitude, it can be asserted that alongside the process of professional adaptation onboard the ship, the strongest lack of valorisation is encountered in the "Cultural Diversity" factor case, followed by "Working Conditions onboard" and the "Human factor" respectively.

6. Verifying the research hypothesis

The analysis of the statistical average of male and female responses has reflected that is no significant changes in the role of the three studied factors in the adjustment and adaptation processes, with no statistically significant differences. Therefore, the null hypothesis cannot be rejected, the perceptions of adaptation following the general trend described above without being influenced by the biological gender becoming feasible.

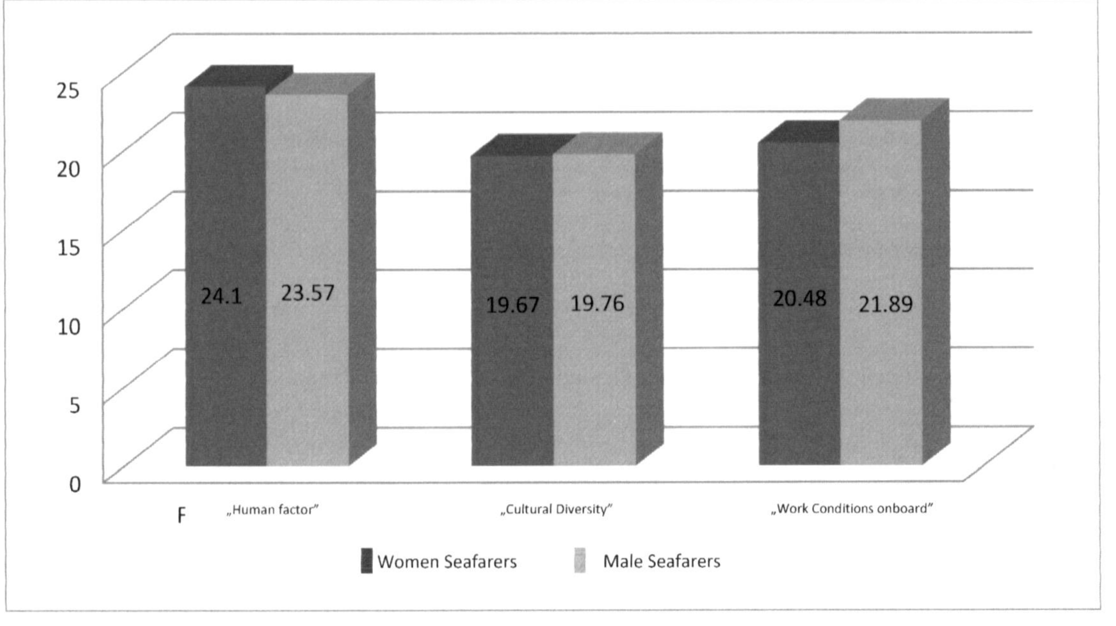

Figure no. 3: *Statistical average of male and female responses on each studied factor impact against the adjustment and adaptation onboard*

For a more detailed analysis, there were used non-parametric analyzes based on the U Mann-Whitney test at each item level. The results indicated significant statistical differences for items 5 ("Status awareness") and 12 ("Attitudes in interpersonal relationships"). Also, without being able to talk about statistical significance yet, there were noticed significant differences for Items 6 ("Decision making"), 26 ("Management") and 36 ("Environmental conditions").

In terms of "situation awareness", "decision making" and "management" factors, male seafarers had considered the variables to be significantly more important in adjusting and adapting processes onboard working conditions than women ($z_{(awareness)} = -2.05$, $p = 0{,}04$); ($z_{(decision\ making)} = -1.65$, $p = 0.098$); ($z_{(management)} = -1.83$, $p = 0.067$). The situation is reversed in regard of attitudes in "interpersonal relationships" and "environmental conditions". In this case, women consider the factors significantly more important in the process of professional adjusting and adaptation compared to men ($z_{(attitudes)} = -2{,}18$, $p = 0{,}02$); ($z_{(conditions)} = 1{,}79$, $p = 0{,}07$)

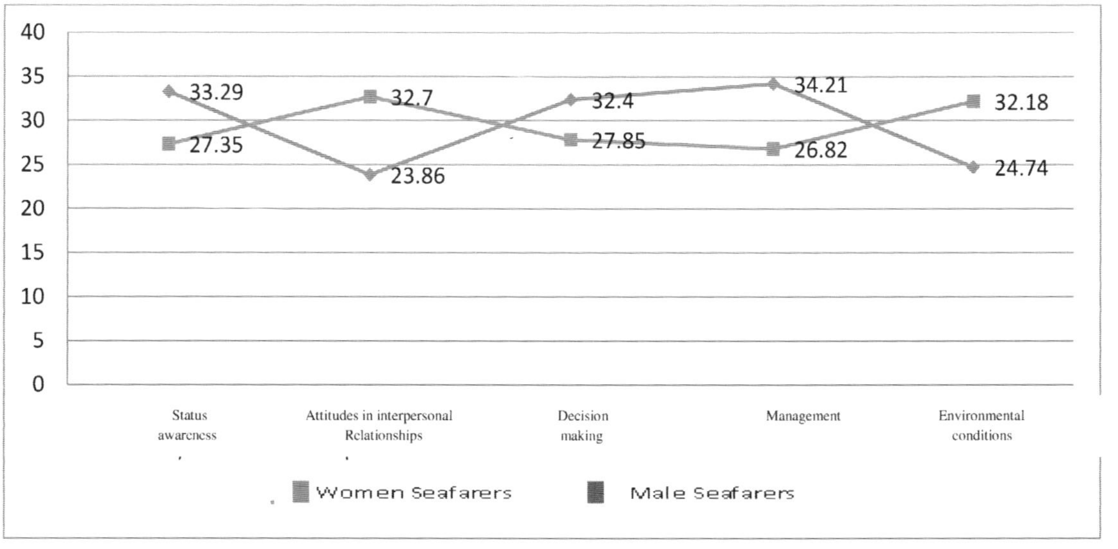

Figure no. 4: *Statistic differences on non-parametric analyzes*

7. Interpretation of results. Conclusions and remarks

The psychosocial interpretation of the results obtained by statistical processing of the data collected from the subjects undergoing the investigation - women and men - is not at all as simple as it might seem at first glance. That is why the authors aimed to seek for the most scientifically plausible interpretations drawn from the analysis carried against the specificity of onboard work, group dynamics and the particular characteristics of the onboard crew, considered as a multicultural social group on a smaller scale.

A first observation is that there are no significant differences, statistically supported, in the scale of the individual impact, depending on the gender, deducted from the assessment of the three dimensions studied based on the applied questionnaire, in case of onboard environment. In this sense, it can be sought for the similarities and consonance of male and female gender responses, once the seagoing crew looks more like a homogeneous structure focused on the assign tasks and responsibilities, as a unitary group, whose dynamics is influenced by the interaction of the two substructures: formal and informal.

From this perspective, within the seagoing crew the formal substructures dominates, which implies the mainly existence of procedural and routine activities, a hierarchical functionality, a well-articulated network of statuses and roles, clear and precise regulations to be followed up. Thus, a double coercion is exerted against the members of the naval group: on the one hand, the formal structure of the group requires the compliance with the rules, procedures stimulating the cold interpersonal relationships settlements, and, on the other hand, the informal organizational culture of the group of sailors will inquiry the adherence to traditions, maritime values, being defined by "face to face", affective-sympathetic and direct interpersonal relations.

In these specific conditions regarding the working environment onboard the ship, three distinct categories of psychosocial phenomena will be developed: a) conformism and non-conformism; b) the affective life of the group materialized in the tensions among the crew members and c) the cohesion and the dissociation of the group" (Zlate, 2004, pp. 466). Considering the compliance tendency of each individual crew member, by accepting the group rules as the essential factor for group survival, regardless of the compositional structure of the group related also to gender, as proved by the present study conclusions, we will assist to the unification and harmonization of the various psychological manifestations within the group, in regard of individual perceptions, affective states or behaviours. For the sake of a good functionality of the onboard crew, the individuals will seek for compliance to the rules and procedures, contributing to the consensus enforcement among the group members, adopting afterwards similar attitudes and behaviours, related to the same cause of object (Zlate, 2004, pp. 468). In the maritime seagoing context, where the normality and rules/roles conformity are praised, the preferential interpersonal relationships, seduction or attractiveness promoted among genders different ones are diluted and transformed progressively onboard, into working relationships whose dynamics describe much accurate, the interactions based on cooperation, mutuality, competition and conflict.

In this way, it can be found a proper explanation for the extremely low score obtained by the "Emotion Management" item, included applied questionnaire, in "Human factor", in regards of the gender issue assessment, where both women and men have intellectualized the emotions transforming them into a facilitator vector of self-control and self-discipline, useful for the sake of onboard activity efficiency and effectiveness.

Although the professional adjustment and adaptation of the crew members involves personality factors such as: emotional balance, motivational variables, aspiration level, need for self-reliance, risk attitudes or resilience to monotony, it could be noticed that within the dimensions of the "Human Factor", male seafarers are looking more to the "situation awareness" and "decision-making" variables, while the women use to attach a greater importance to "attitudes in interpersonal relationships". On the other hand, within the "Working conditions onboard" dimension, placed on the second place in test participants' opinion, men value more the item "management" while women are inclined to value the "environmental conditions" (Tabachiu, 2003, pp. 72).

The authors have observed that on the formal perspective of the onboard crew, where there is a higher need for observing and complying with the norms, codes or standards, more pragmatic males' personality structure, push them to believe that several items as "situation awareness", "decision making" and "management" are most likely to contribute to the proposed objectives achievement. The perceptive responses replied by women suggest that in the formal operation of the seagoing crew, there will be mechanisms interfering to compensate rigidity, formal relations, normality being represented by informal relations, subjective, affective, sociometric structures, where the female personality structure recognizes easily its reference frame. It can be stated that while men resonate more with the axiological-normative structures, close to active side and power dimension, women resonate more with the socio-affective and communicational structures, aiming the collective affective phenomenon, that energize the life of the group. These different affinities within the group will determine tensions, that can be defined as latent and collective emotional states that occur within the group, facilitating or disrupting their work and their harmony (Zlate, 2004, pp. 474).

These gender differences, as observed in the questionnaires' answers on different applied items, reflect in fact the differences between the personality attributes established between men and women as a result of the pressure towards conformism, from the primary socialization, by the membership groups, with the explicit purpose of developing socially desirable behaviours, according to the onboard standards.

Considering the human being as a complex biological, psychosocial and cultural entity, the recalled onboard adjustment and adaptation process come close as part of the "enculturation" process, which is defined as everything that individuals will acquire in a lifetime because it is available and socially valorised (Gavreliuc, 2011, pp. 53) or viewed as the process by which a cultural group incorporates to their descendants specific axiological elements for an optimal integration into community life (Cucos, 2000, pp. 118).

A surprising aspect residing from the investigated subjects' perspective is about the "Cultural Diversity" lower impact on individual adjustment and adaptation to onboard working conditions and to the life at sea hardships, when dealing with multinational and multicultural crews. This is due most probably, to the fact that the authentic cultural identity is changeable, structurally adaptive and, self-constructively, since the personality is organized around relatively stable axiological cores. Therefore on board ships, the seafarers will appreciate that there are no individual valuable judgments formulated as results of a specific culture, but there are only different cultures that must be respected as such. (Cucos, 2000, pp.147)

By carrying out the onboard activities in a restricted physical area, for a short time horizon (3 to 6 months), seagoing crew members learn to co-exist with the "others" by developing non-valid identity adaptation strategies by sharing the language used in communication, other than their native language, and also by values and work procedures, axiological neutral. This organizational culture of onboard work within a multicultural crew is defined by coherence, unity, continuity, safety because the multicultural environment describes the specificity of the interaction between the social actors living with each other (Gavreliuc, 2011, pp. 37).

In the multicultural seagoing crew, the process of integration into the particular life and activity conditions at sea, supports certain characteristics determined by successive adjustments and adaptations to new, unfamiliar stimulus, that confer vulnerability to human psychic. In order to restore the balance of the human psychic system and to reduce cognitive-emotional dissonance, the stages of socialization should be described by some defining patterns. First, in the tripartite "self ↔ working environment ↔ the others" interaction, due to cultural heterogeneity, there is a superficial assimilation, the individual preserving his original culture in order to avoid his identity disaggregation, on a surface conformism by adhering to rules and to the imposed language (Gavreliuc, 2011, pp. 57).

In other words, the personality supports permanent adjustments and readjustments between the contents of his own culture and the new stimulus, unfamiliar as issued from a different social and cultural environment, following to integrate with exigency into a socio-cultural environment with various new cultural references. These fine retouching are in line with the external requirements, producing what is so called "to cope with", implying the new situational stage of dealing with the uncertainties, situational risks and ambivalences. In this conjuncture, judging this specific context of work and life, it can be considered that the crew members will undergo a "re-socialization" phase passing through the onboard adjustment, whereby individuals, through a competition of circumstances, are in part or totally isolated from their previous social and cultural background. (Achim, 2008, pp. 215)

Far from being exhaustive, the psychosocial interpretations considered on his study as explanatory, in regard of how our male and female subjects had assessed the importance of human factors, working conditions and cultural diversity alongside the professional adjustment and adaptation onboard ships, incite to other questions, other answers, other more demanding scientific searches, more in-depth on this topic or other related topics.

References

[1] Achim, M. (2008). *Sociology.* Cluj-Napoca: Ed. EIKON.
[2] Cucoș, C.(2000). *Education. Cultural and Intercultural Dimensions.* Iași: Polirom Publishing House.
[3] Gavreliuc, A. (2011). *Intercultural Psychology*, Iași: Polirom Publishing House.
[4] Havârneanu, C., Havârneanu, G. (2015). *The Psihology of Risk*, Iași: Polirom Publishing House.
[5] Iosif, Gh. (2001). *Human Resources Management.* Bucharest: Ed. Victor Publishing House.
[6] Tabachiu, A. (2003). Working Psychology. Bucharest: Titu Maiorescu University Press.
[7] Zlate, M. (2004). *Managerial Psychology Manual, vol. 1.* Iași: Polirom Publishing House.

IOP Conf. Series: Earth and Environmental Science **172** (2018) 012027 doi:10.1088/1755-1315/172/1/012027

The application of new noise and vibration standards onboard ships

T Pazara[1], M Pricop[2], G Novac[3]and C Pricop[4]

[1]Professor Assistant, PhD, Naval Academy „Mircea celBătrân", Constanța, RO
[2]Associate Professor Eng., PhD, Naval Academy „Mircea cel Bătrân", Constanța, RO
[3] Professor Assistant, PhD, Naval Academy „Mircea cel Bătrân", Constanța, RO
[4]Associate Professor Eng., PhD, Maritime University, Constanța, RO
[1]tiberiu.pazara@anmb.ro
[2]mihail.pricop@anmb.ro
[3]george.novac@anmb.ro
[4]pricopcodruta@yahoo.com

Abstract. The ship represents a very complex system of noise and vibration sources determined by the operation of numerous onboard installations and specific activities of crew and passengers. Shipboard noise can cause discomfort to passengers and hearing damage to the crew and in this regard the need for a safe work environment is growing. Due to these effects, it is necessary to strictly observe the noise and vibration criteria by noise and vibration analysis in the design phase and their adequate control. In this paper are discussed the aspects regarding the latest standards of noise and vibration for ships. It is well known that changes made in regulations are not easy to be implemented on board old ships. The authors made a careful analysis of these changes and made a few recommendations for the shipbuilding companies and naval operators.

1. Introduction

The progress in naval industry is associated with changes in naval standards and one of these standards regards the comfort onboard ships. On commercial ships is important to assure a safe environment for the crew. A criterion is represented by the noise and vibration to which the crew is exposed Jegaden [1,2]. During the years, national and international organizations established limits for noise and vibration exposure. These limits changed over the years due to observations and studies made to improve the conditions of the crew onboard.

In 1981 the IMO organization set the standards for noise and vibration onboard ships. Since then, numerous studies and research made observations and conclusions that improved the conditions of navigator's onboard commercial and military ships. DNV, IMO, ABS and other organizations collect information and update the standards. Due to globalization, these standards cannot vary from country to country. The need to assure a uniform standard and in the same time to protect and preserve the safety and the health of the crew have determined a continuous upgrade of the noise and vibration limits.

In time, the noise and vibration criteria have been established for civil and military ships. For civil ships, there are standards for cruise ships and commercial ships. This means that nowadays three different sets of rules are in effect. Of course, the limits don't vary that much from one standard to another but it is important to apply these standards onboard every ship regardless of its type.

Biot and de Lorenzo [3] studied the applicability of noise and vibration criteria onboard cruise ships. They have pointed that the changes in ISO 6954:1984 that were made in 2000 (ISO 6954:2000) do not cover all the aspects of the problems. Their conclusion was that changes must be made carefully with consideration to human perception of noise and vibration.

In 2009, the IMO Sub-committee on Ship Design and Equipment made proposals to improve the noise criteria onboard ships [4]. Approximately, for each type of compartment these proposals suggest a 5dB reduction of noise limits. In the following years, specialists from different countries analyzed the proposals and concluded that these exposure limits must be applied Shuri 2010 [5] Kitamura 2010 [6], Yudong 2010 [7]. But in November 2012, the MSC-337(91) [11] was adopted without a part of specialists' amendments. Beltran et al [9] pointed that very small changes have been made from the old standard and the protection of the crew is not improved.

Constructors and shipyards must implement these standards onboard ships. S-H Kim [10] presented a few improvements that Hyundai Heavy Industries have introduced onboard their ships according to ISO standards. However, these improvements were made only on new ships and not on old ships. For ship, builders the new noise and vibration criteria can easily be applied onboard their new projects. The challenge for the ship owners is to implement the new limits to their current sailing ships.

2. Sources of noise and vibrations on-board ships

In this paragraph are presented some of noise and vibration sources and a few solutions to reduce their effects.

2.1. The main engine

Noises and vibrations due to the structure

Solutions:

- Main engine adjustments by minimizing imbalance to avoid any further noise transmission.
- Operation the main engine in the proper mode: whenever possible revolutions and propeller pitch must be adjusted in order to reduce noise and vibration.
- A correct alignment of main engine, gear and shaft is needed

Noises and vibrations due to the engine mounting

Solutions:

- Flexible mounting of main engine with: engine is placed on vibration absorbers designed against increased dynamic and axial load; between main engine and gear must be fitted a flex-coupling; It is necessary to ensure that the bed-plate is rigid as the engine is not a part of the rigidity.

Air borne noise radiation

Solutions:

- Sealing the engine: if the main engine is the main noise source it is necessary to seal the engine, a solution that requires a lot of space around the engine; this measure must be combined with flexible mounting of the engine and it's also important to ensure adequate cooling air presence with absorbing labyrinth openings; all lead-ins must be tightly and flexibly sealed.
- Sealing engine room from noise penetration: when sealing the engine room all surfaces must be insulated (rock wool and galvanized sheet) and lead-ins and component mountings also.
- Absorption panels for noise reduction in engine room: by hanging absorption systems for reduction of the reverberation (the system must not absorb oil fumes);
- Packings in doors and openings: by mounting close-fitting rubber profile packings in doors or in hatches.

2.2. The exhaust system

Noises and vibrations due to the structure

Solutions:

- General solutions: by removing all fixed spots between structure and exhaust system; the expansion joints must never be fully compressed or extracted; the exhaust system with absorber must not hang in or be supported by expansion joints;
- Mounting sound absorber: sound absorbers can be mounted on rubber or full-metal absorbing elements;
- Exhaust pipe with rain cap: by mounting a fixed rain cap with overflow and coming on top of funnel (the funnel top must not touch the exhaust pipe).

Air borne radiation
Solutions:

- Insulating sound absorber and pipes: by insulating and sealing the suspension, flanges and expansion joints of exhaust pipe and sound absorber;
- Providing new sound absorber: by placing a high efficient sound absorber with new absorbing features made specifically for individual vessel;
- Exhaust system tuning: avoid any resonance by tuning the entire exhaust system;
- Casing insulation: by insulating surfaces towards accommodation spaces.
- Mounting baffles in casing: mounting sound baffles or insulating of side sections will reduce standing sound waves and noise from the exhaust system (it is necessary to ensure free air flow).

2.3. The propeller

Effects on the structure due to the cavitation
Solutions:

- New or modified propellers: using wake field adapted new propellers or nozzle propellers;
- By changing blade tilt: blade tilt with low-noise shape (can affect on drawing power).
- Improve the area above propellers by:
  - Reinforcing the hull plate above the propellers by means of bracing and/or thicker plates;
  - Mounting vibration reduction system material on hull plate above propellers and on deck;
  - Using absorptive material in rooms with reverberation;
  - Placing cement in hull plate area above propellers.
- Insulating the engine room towards propellers: by Insulating hull plate and abaft bulkhead;
- Placing of zinc: in order to avoid unnecessary eddy currents around propellers zinc must be placed in current paths

Effects on the structure due to the noise from nozzle
Solutions:

- Nozzle shaft reinforcement: the hull nozzle shaft must be fixed and efficiently mounted and also the soft plate areas must be strengthened.

2.4. Auxiliary engines, pump and other equipments

Effects on the structure due to machinery mounting
Solutions:

- Ensuring a solid foundation: pumps and engines foundation must be rigid and thoroughly mounted to the construction;
- Using absorbers: machinery must be placed on rubber absorbers/mats.

Effects on the structure due to connections
Solutions:

- Using flexible connections: using whenever possible only flexible connections when mounting a component flexibly.

Air borne radiation from components
Solution:

- Better sealing: using sound absorbing case for components to mount in without ignoring heat release or cooling air;
- Shielding: if sealing is not possible the use absorptive panels.

2.5. The pipe system

Effects on the structure due to pipe diameter, suspension or rigidity, pipe penetration
Solutions:

- By changing the pipe dimensioning: enlarging the pipe diameter for high speeds or pressure ;
- Flexible mounting of pipes: using suspended flexible brackets or rubber lining brackets;
- Using u-bend mounting for Long rigid pipes; pipes must not be attached to two fixed points without the ability to move.
- Using waterproof lead-ins: If possible use flexible lead-ins or ensure that the pipe is not both sides fixed;
- Using open lead-ins: make sure that the pipe does not clank and is free from the opening.

2.6. The hydraulic system

Effects on the structure due to piping/suspension
Solutions:

- General solutions: avoid using sharp elbows and transitions and also avoid using straight pipes between two structure fixing points; suspension must be flexible whenever possible and flexible connections used at bulkhead lead-in, engine and pump.

Effects on the structure due to pressure deviation or speed
Solutions:

- Using sound absorber in the high pressure system;
- Using pressure accumulator in substantial pressure deviations systems;
- Increasing pipe diameter when using high or many pipe bending.

Effects on the structure due to hydraulic engine
Solutions:

- Using suspension: by suspending with absorbers and flexible connections (hose connections).

Air borne noise from hydraulic engine
Solutions:

- By sealing: installing components in an airproof box without ignoring heat release or cooling air;

2.7. The ventilation system

Effects on the structure due to air duct rumble
Solutions:

- Making reinforcement: using cross bracing of plate areas or bending the plates in pyramidal shape;
- By silencing mass: adding silencing mass on plate area (bitumen plates or coating);
- Lowering air speed by using larger duct;
- Sleek Transitions by avoiding sharp elbows and large changes in cross cut.

Effects on the structure and air borne noise due to ventilator
Solutions:

- Making modification: using flexible suspension of ventilator, changing number of revolutions or replacing the rotor.
- Using new ventilator with low noise level.

IOP Conf. Series: Earth and Environmental Science **172** (2018) 012027 doi:10.1088/1755-1315/172/1/012027

Air borne air noise in duct
Solutions:

- Avoid tearing by using absorptive material with thin plastic coating to;
- Mounting sound absorbers in the air ducts;
- Lowering air speed by enlarging the duct cross cut or increasing the air ducts number.

Air borne air noise in blow-off and air diffusion
Solutions:

- Using silenced blow-off valve with larger openings to reduce opening speed;
- Using baffles inside air intake grill;
- Making larger opening: a larger air intake grill with smooth transition to air duct will cause an intake speed reduction.

3. Case study: noise on-board training ship Mircea

Although is not a ship from one of the three categories mentioned in paragraph 1, training ship Mircea must comply with the standards regarding noise and vibration. We chose this example because we want to see if the new standards can be applied onboard an old ship. For this purpose, a set of measurements were taken during a short voyage in the Black Sea. The noise was measured with a sound level meter B&K type 2250. In the following figures (1-5) are presented the noise levels measured in several compartments.

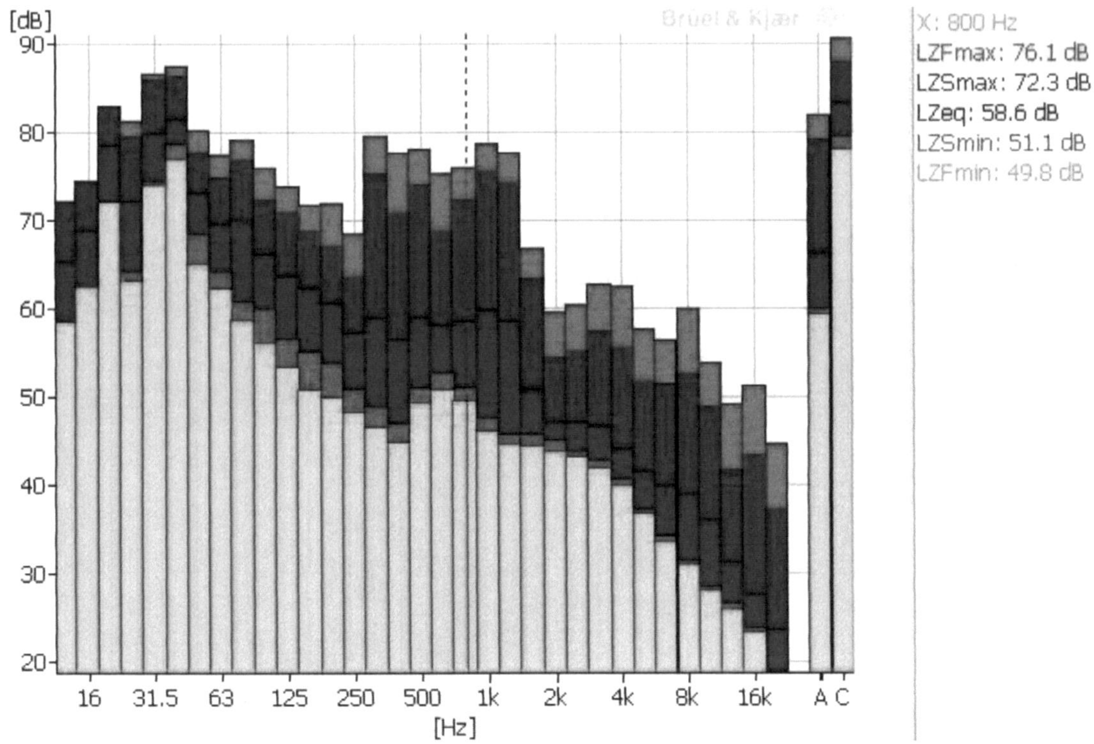

Figure 1 Second officer room

IOP Conf. Series: Earth and Environmental Science **172** (2018) 012027 doi:10.1088/1755-1315/172/1/012027

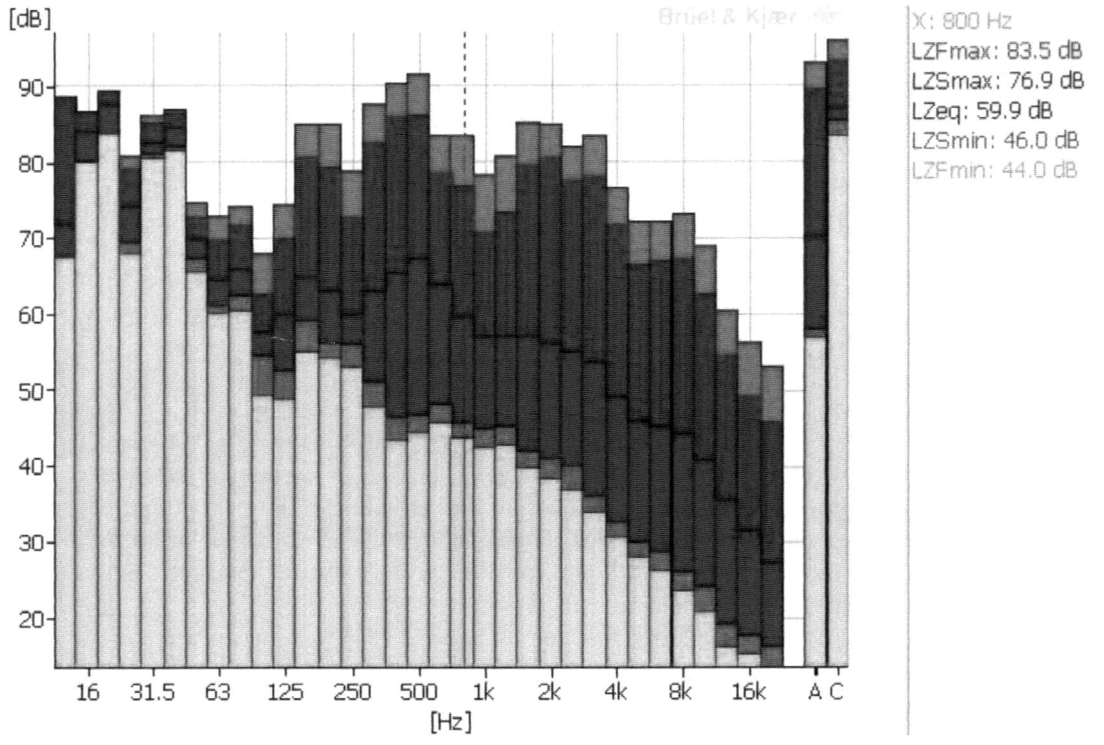

Figure 2 Command deck

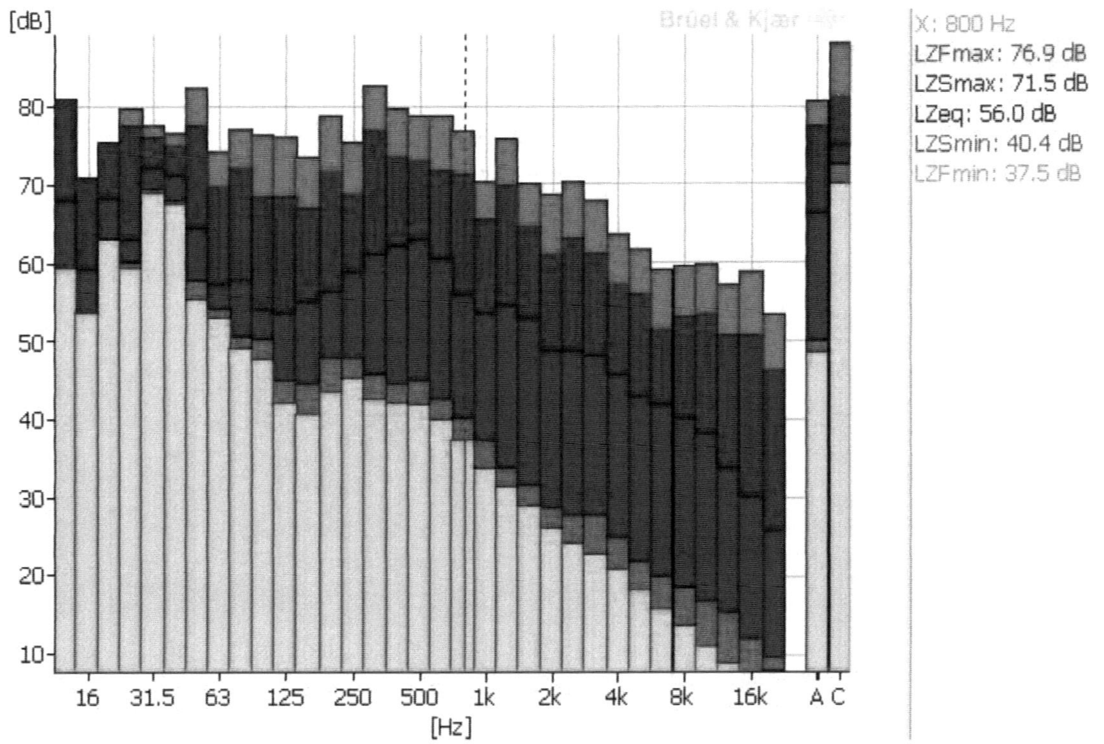

Figure 3 Storage compartment

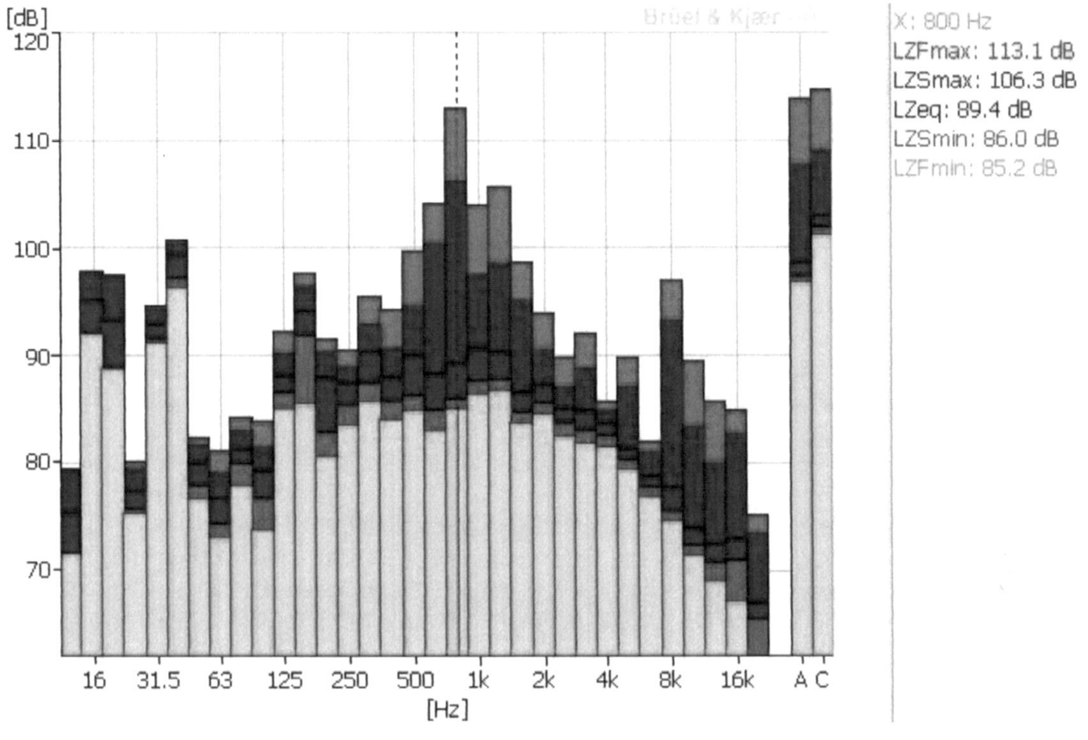

Figure 4 Machine room

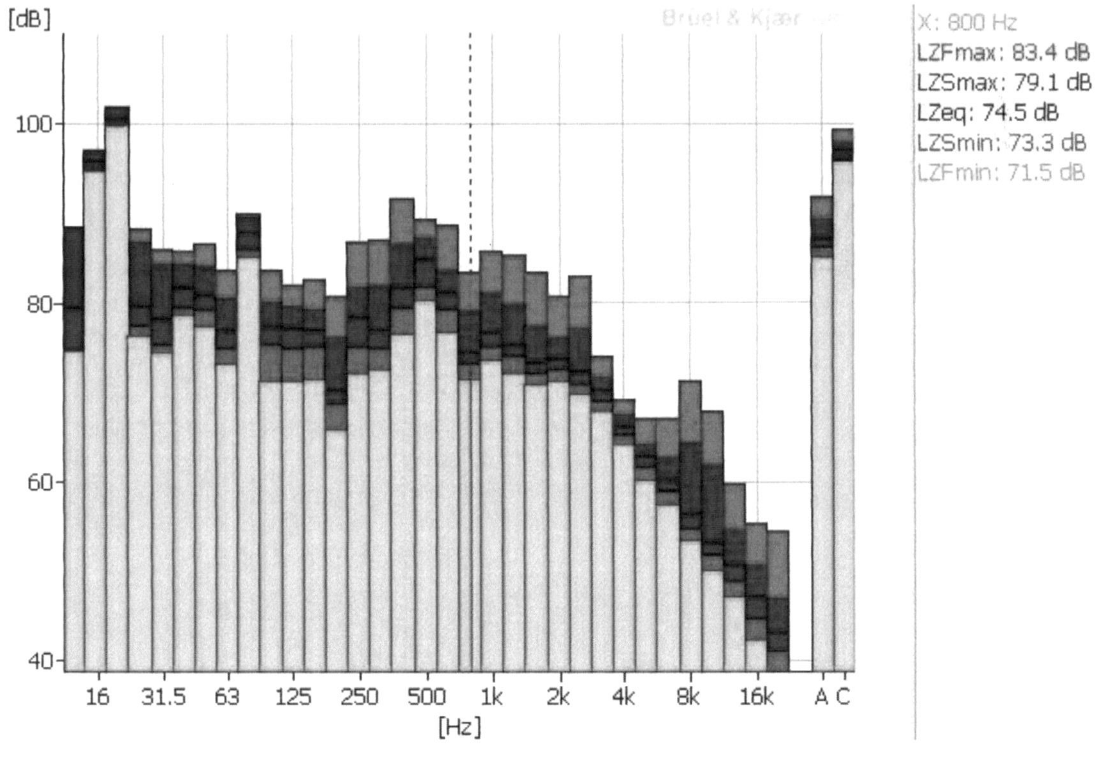

Figure 5 Open deck

By comparing these results with the noise limits imposed by [MSC-337(91)– 11] we can observe that these limits are exceeded. Of course, the differences are little, but due to extended time of exposure these noise levels are not acceptable onboard. Comparing the measured noise levels with the old standard [12] they comply only for the machine compartment.

4. Conclusions

From the design and construction phase of the ship's body the sources of noise and vibrations are analyzed in terms of energy level, frequency spectrum, their interaction with the structure of the ship, air and water transmissibility, as well as effect on receptors (crew members and passengers).Due to the increase in ship size and the power of propulsion installations for increasing comfort in passenger ships, the criteria of international standards on acoustics and vibrations have been introduced over time in the naval field, globally approved by the regulations for limiting the noise level and vibration on ships [1].

The ship is a complex construction in terms of both of the metal structure and numerous machines, equipment and installations used for good efficiency. All of these items are sources that emit noise and vibrations in air and water, with special effects on humans, fauna and environment.

The main engine and auxiliary engines generate noise and vibration in air and structure in which they are fixed. The frequencies of these signals are determined by the speed of rotation, the number of cylinders, the number of cycles, the crankshaft, the number of piston valves or the number of piston rings [1].

The propeller is the second source of noise and vibration, which propagates in water and inside the hull. The phenomena that produce these noises and vibrations are more complex than in the case of thermal machines and are determined by the cavitation, the variation of the pressure around the propeller due to suction and rotation of liquid inside the propeller [13, 14].

Also, the ship's wake has a very large influence and is determined by the unevenness of the current in which the propeller operates as direction and amplitude [15].

Air conditioning, ventilation and air conditioning are also sources of noise. The structure of the body (deck, walls, ceilings), pipes and tubes that vibrate, also become noise sources in the spaces occupied by passengers and crew [16].

Depending on the type of ship, mainly loading / unloading and handling equipment on board (grapples, conveyors, vehicle ramps, etc.) there are also other sources of noise and vibration [17].

Because of these noise and vibration sources, the air and water propagation pathways, the phenomena of reflection, diffraction, absorption, dispersion and subjectivity of the receptors (crew and passengers) at different times of co-existence on board the ship (in a state of recurrence / rest or work), it is difficult to overcome the same criteria and standardized measurement and analysis methods for all types of ships. At the same time, it is difficult to model in the initial project phase, by numerical methods, the sources of noise and vibration [9, 11, 12, 18, 19].

On the other hand, due to lack of standards, there are difficulties in determining SPL (Sound Power Level) for large-scale noise sources such as the ship. Difficulties are caused by methods of measuring tonal components and directivity, as well as interaction and overlapping phenomena, especially at low frequencies.

Vibration and noise measurement methods are based on the following principles:

- Identify spaces from the ship with high values;
- Noise level;
- Frequency band;
- Periodicity of occurrence;
- Identification of sources;
- Analysis of working conditions;
- Analysis of the propagation medium (air, structure and water) [20].

Measurements are performed on a real-time ship with multi-channel systems with microphones, accelerometers or hydrophones, and a dedicated signal processing software. Measurements can also be made on a scale model [9, 11, 12, 18, 19].

The acoustic mobility of the engine foundation structure is determined by modal analysis, theoretical or experimental methods, or Statistical Energy Analysis [21].

The modeling of the signal transmission paths from the source through air, water and structure is done by using the near-field acoustical holography (NAH) method for modeling acoustic properties of the source by modeling radiation, diffraction and transmission between source and receiver through the vibro-acoustic transfer matrix [22,23].

Although lately there have been many studies and research on noise generation, acoustic modeling of air and water propagation, and environmental management studies in this area, there is no global approach to vibration and airborne noise emissions in ships.

Noise and vibrations can be reduced by following the source-path transmission-receiver path.

A first way is to reduce noise and vibration at source.

The source isolation method is made from the design phase and constructive, each machine and equipment supplier must ensure isolation of the noise produced by the components or the moving agent.

The design of the propeller to eliminate or reduce cavitation is achieved by optimizing the propeller load, designing the optimal shape of the aft body for uniformity in the propeller disk, optimizing the propeller geometry by diameter, number of blades, pitch, skew angle, blade profile.

Another way is to isolate the source of the propagation medium (air, structure).

For structure, vibration reduction and thus noise propagation is achieved by increasing the structure's own frequency, which depends on the rigidity of the structure.

Last resort is the protection of the receiver.

Isolation of the receiver (space, ear protection of crew members) is a little expensive but does not solve the noise problem at the ship's level.

The use of the absorbent material generally does not alter the transmission of the noise from the source, but the reverberation time changes, creating a different sound image by diminishing the reflexive phenomenon.

5. References

[1] D. Jegaden, Vibration on-board ships and its effects

[2] D. Jegaden, Ship noise and its effects on seafarer's health

[3] M. Biot, F. de Lorenzo, Noise and vibration on-board cruise ships: are new standards effective?, ICMRT '07, June 28-30, 2007, Ischia (Naples), Italy

[4] IMO Sub-committee on ship design and equipment, DE 53/10, 18 December 2009

[5] H. Shuri, Protection against noise on-board ships, Tripartite 16th October 2010, Tokyo

[6] O. Kitamura, Mandatory control of noise on-board ships, The 4th ASEF in Kyoto, 17-18 November 2010

[7] S. Yudong, China shipbuilding and new request of protection against noise on-board ships, 4th Asian Shipbuilding Forum, 2010 Japan

[8] E. Brubakk, Comfort on-board ships, an important safety factor, Sea Asia 2009 Conference

[9] P. Beltran, R. Salinas, A. Moreno, The new IMO noise code: a lost technical opportunity. Irreversable and high cost consequences for fisherman and other seaman that will continue being deaf, ICSV 21, 13-17 July, 2014, Beijing, China

[10] S-H Kim, Technical practices for revision as mandatory rule of the IMO resolution A.468 for noise, 6th Asian shipbuilding experts' forum, 22nd November 2012

[11] IMO Resolution MSC.337(91), ''Code on noise levels on board ships'', 30 November 2012

[12] IMO Resolution A.468(XII) "Code on noise levels on board ships". International Maritime Organization, 4, Albert Embankment, London, UK, 1981

[13] J. S. Carlton. Marine Propellers and Propulsion. Butterworth- Heinemann, 2007

[14] Song Li, Desen Yang, Haoyang Zhang, Experimental study of underwater propeller low-frequency noise flied, Ocean Acoustics (COA), 2016 IEEE/OES China
[15] Sing-Kwan Lee, Min Liao, and Suqiu Wang, Propeller-Induced Hull Vibration – Analytical Methods, Proceedings of the 2nd International Ship Noise and Vibration Conference, London, UK, June 28, 2006
[16] Stephen Moore, A Review of Noise and Vibration in Fluid-Filled Pipe Systems, Proceedings of ACOUSTICS 2016 9-11 November 2016, Brisbane, Australia
[17] The Danish Fishermen's Occupational Health Counsil, How to handle noise and vibrations in ships, July 2000
[18] Davide Borelli, Tomaso Gaggero, Enrico Rizzuto, and Corrado Schenone, Control of Ship Noise Emissions, De Gruyter Open, Noise Mapp. 2016; 3:107–119
[19] ABS, Guidance Notes on Noise and Vibration Control for Inhabited Spaces, sept 2017
[20] Sang R. Kim, Jae S. Kim, Hyun S. Kim, and Hyun J. Kang, Method to Determine Structure-Borne Noise Level from Machinery in Sea Application, Fifth International Congress on Sound and Vibration, December 15-18, 1997 Adelaide, South Australia
[21] Mateusz Weryk, Ship on-Board Noise Propagation Analysis Methods, Hydroacoustic Journal, 2012, Vol 15, Pag. 187-194
[22] A. Derouiche, N. Hamzaoui, B. Bouzouane And A.H. Miloudi, Use of the Vibroacoustic Transfer Function Built for the Prediction of Noise Radiated to other Vibrational States, 22ème Congrès Français de Mécanique, Lyon, 24 - 28 Août 2015
[23] Jeong-Guon Ih., Acoustic Holography Based On The Inverse-Bem For The Source Identification Of Machinery Noise, ICSV 14, 9-12 July, 2007

Study the parameters of a petrol engine using analysis of variance

C O Ilie[1], M Marinescu[2], L Barothi[3], R Vilau[4] , I Lespezeanu[5]

[1,2,3,4,5] Military Automotive Engineering and Transportation Department, Military Technical Academy, 39-49 George Coşbuc Blvd, Bucharest, Romania

E-mail: ovidiuilie66@yahoo.com

Abstract. This paper presents o method to analyse the influence of various parameters on spark ignition engine operation. The mathematical tool it was used is analysis of variance. Based on statistical data, analysis of variance is a method of study the influence of factors acting simultaneously on any process. From a statistical point of view, it consists of comparing the average values of different collectives and checking of the statistical hypothesis on their homogeneity. Based on its results, the structure of the mathematical models obtained by regressions and predictions is established. Data were obtained by measuring the dynamic parameters of vehicles and engines. We have used 14th automobiles of the same type with different wages and mileages. Testing program aimed to capture a large range of operating regimes. Vehicles moved on different surfaces in different weather and traffic conditions. Some of the engine parameters like torque and power were measured on the test bench.

1. Introduction

This paper presents the analysis carried out in a phase of a vehicle dynamics research program based on statistical models. The first step was to design and apply a testing program to obtain a variety of data defining dynamic behavior of a gasoline engine vehicle. There were used 15 cars having 1.6 l engines and mileages ranging between 13500 to 115000 km. Most of the experimental data were acquired and stored in the tester intended to diagnose the cars used. There were capture with a frequency of 10 Hz. Same parameters like fuel consumption, power and engine's torque could not be measured using the onboard system. These were calculated using static engine characteristics obtained by measuring on the test bench. Eventually, we selected 64 tests, each of them having 15 strings with 256 values. Each string represents a measured or calculated size lasts approximately 26 s. The data were saved in a tensor. The tensor was used to issue mathematical models that describe vehicle's dynamics.[1]

2. Analysis of variance

Making statistical models, in this case regressions and predictions, requires a prior analysis of the data. [2] In other words, should be compared the samples of each parameters to see which one has a stronger influence to the studied process. The models will only be set based on the insightful series. An instrument for this is alongside the correlation and frequency analysis, ANalysis Of VAriance (ANOVA). The method consists in comparing the samples based on their means. The means of two or more groups are checked if they are significantly different from each other. Depending on the number of factors whose influence is studied there are one way, two-way ore multi-variate analyses of

IOP Conf. Series: Earth and Environmental Science **172** (2018) 012028 doi:10.1088/1755-1315/172/1/012028

variance. All of them can be performed either on samples of the same volume or using different volume samples.

3. One-Way ANOVA (Analysis of Variance)

The One-Way ANOVA compares the means of two or more series of the same parameter in order to determine whether there is statistical evidence that the associated population means are significantly different. Also, it is known as One-Factor ANOVA.

Samples of the same n volume as are the experimental data were used to perform One-Factor analysis of variance. In this respect, it is considered a number of k sub-collectivities that hypothetically have the same mean m of the general characteristic X. One random sample of the same volume n is drawn from each sub-collectivity, in order to find out if their averages differ significantly from m.

The initial data forms a matrix x_{ij} with n lines and k columns, so $i=1,n$ and $j=1,m$. Ramdom k sub-collectivities reflect the influence of a certain factor acting on the studied process considered by the X characteristic.

The foollwing are defined:

- means of sub-collectivities:

$$\overline{x}_j = \frac{\sum_{i=1}^{n} x_{ij}}{n} \tag{1}$$

- great mean (the mean of the means):

$$\overline{x} = \frac{\sum_{i=1}^{n}\sum_{j=1}^{k} x_{ij}}{nk} \tag{2}$$

The distance of each value of the characteristics from the great mean is divided in two components:

$$x_{ij} - \overline{x} = (x_{ij} - \overline{x}_j) + (\overline{x}_j - \overline{x}) \tag{3}$$

Summating the squares of expression (3) results:

$$\sum_{i=1}^{n}\sum_{j=1}^{k}\left(x_{ij} - \overline{x}\right)^2 = \\ = \sum_{i=1}^{n}\sum_{j=1}^{k}\left(x_{ij} - \overline{x}_j\right)^2 + 2\sum_{i=1}^{n}\sum_{j=1}^{k}(x_{ij} - \overline{x})(\overline{x}_j - \overline{x}) + \sum_{i=1}^{n}\sum_{j=1}^{k}\left(\overline{x}_j - \overline{x}\right)^2 \tag{4}$$

Expresion (4) becames:

$$\sum_{i=1}^{n}\sum_{j=1}^{k}\left(x_{ij} - \overline{x}\right)^2 = \sum_{i=1}^{n}\sum_{j=1}^{k}\left(x_{ij} - \overline{x}_j\right)^2 + n\sum_{j=1}^{k}\left(\overline{x}_j - \overline{x}\right)^2 \tag{5}$$

because

$$2\sum_{i=1}^{n}\sum_{j=1}^{k}(x_{ij}-\overline{x})(\overline{x}_j-\overline{x}) = \\ = 2\sum_{j=1}^{k}\left[\sum_{i=1}^{n}(x_{ij}-\overline{x}_j)(\overline{x}_j-\overline{x})\right] = 2\sum_{j=1}^{k}\left[(\overline{x}_j-\overline{x})\sum_{i=1}^{n}(x_{ij}-\overline{x}_j)\right] = 0 \tag{6}$$

Expression (5) shows that the total sum of squares noted SPA_t of the X characteristic from the k samples (sub-collectivities) of n volume was decomposed in two main components:

- sum of squares within noted SPA_r:

$$SPA_r = \sum_{i=1}^{n}\sum_{j=1}^{k}\left(x_{ij}-\overline{x}_j\right)^2 \tag{7}$$

- sum of squares between noted SPA_f:

$$SPA_f = n\sum_{j=1}^{k}\left(\overline{x}_j-\overline{x}\right)^2 \tag{8}$$

Thus expresion (5) becames:

$$SPA_t = SPA_f + SPA_r \tag{9}$$

Mathematical expresions of variances of the three sums, SPA_t, SPA_r, SPA_f, are obtained dividing them by number of degrees of freedom involved.[4] Result:

- factorial variance D_f

$$D_f = \frac{n\sum_{j=1}^{k}\left(\overline{x}_j-\overline{x}\right)^2}{k-1} \tag{10}$$

- residual variance D_r

$$D_r = \frac{\sum_{i=1}^{n}\sum_{j=1}^{k}\left(x_{ij}-\overline{x}_j\right)^2}{k(n-1)} \tag{11}$$

- total variance D_t

$$D_t = \frac{\sum_{i=1}^{n}\sum_{j=1}^{k}\left(x_{ij} - \overline{x}\right)^2}{nk-1} \tag{12}$$

Analysis of variance consists of comparing the two variances, factorial and residual variances, based on the null hypothesis H_0. The null hypothesis in ANOVA is valid when all the sample means are equal, or they don't have any significant difference. The alternate hypothesis H_1 is valid when at least one of the sample means is different from the rest of the sample means.[3]

If the two variances do not differ significantly (with a level of significance α), than the Fisher's computed coefficient noted F_c which is the ratio between them is close to the unit value.

If the k samples are extracted from collectivities with different means than the factorial variance D_f will be considerrabily higher than ten residual one D_r and the calculated Ficher's coefficient F_c will be greater than the table Fisher's coefficient F_t.

Table 1 presents all the calculus to be performed for One-Way Anova.

Table 1. One-Way Anova Calculus

Tipe of variance	Sum of squares *SPA*	Nuber of degrees of freedom *GL*	Variance *D*	Fisher's coefficient (computed) F_c	Fisher's coefficient (table) F_t
Factorial	$n\sum_{j=1}^{k}\left(\overline{x}_j - \overline{x}\right)^2$	$\upsilon_1 = k-1$	D_f	$F_c = \frac{D_f}{D_r}$	$F_{\alpha,\upsilon_1,\upsilon_2}$
Residual	$\sum_{i=1}^{n}\sum_{j=1}^{k}\left(x_{ij} - \overline{x}_j\right)^2$	$\upsilon_2 = k(n-1)$	D_r		
Total	$\sum_{i=1}^{n}\sum_{j=1}^{k}\left(x_{ij} - \overline{x}\right)^2$	$\upsilon = nk-1$	D_t		

The One-Way Anova has the following steps:

1. F_c is computed according with the formula given in Table 1;
2. It is adopted the level of significance α which usually is 0,05;
3. F_t is established on the basis of the tables with critical values of the distribution F depending on the level of significance α adopted and the values of the related degrees of freedom;
4. By comparing the two values of Fisher's coefficient, it follows:
   a. If $F_c < F_t$, the null hypothesis H_0 with the significance level α is accepted, which means that the grouping factor considered did not have a large influence on the means of the samples;
   b. If $F_c >= F_t$, the null hypothesis H_0 with the significance level α is rejected, , the alternative hypothesis H_1 is accepted, and thus the grouping factor considered has had a significant influence on the means of the samples.[1]

4. Study the parameters of petrol engine using One-Way ANOVA

ANOVA has been used in the presented research program to established the data strings that will be used to calculate mathematical models. Input (factorial) and output (end result) sizes for each model were determined. The series corresponding to the input for the 64 tests were compared using One-Way ANOVA. Based on the analysis, it was established for each model the series to be used.

An example is shown bellow. It was set the throttle ξ as an input and series corresponding test I1n and I2n to be compared. The results are presented in table 2 and figure 1.

Table 2. One-Way ANOVA for throttle ξ and tests I1n and I2n

Tipe of variance	Sum of squares *SPA*	Nuber of degrees of freedom *GL*	Variance *D*	Fisher's coefficient (computed) F_c	Fisher's coefficient (table) F_t
Factorial	50474.6	1	50474.6	148.446	0
Residual	173410.6	510	340		
Total	223885.1	511			

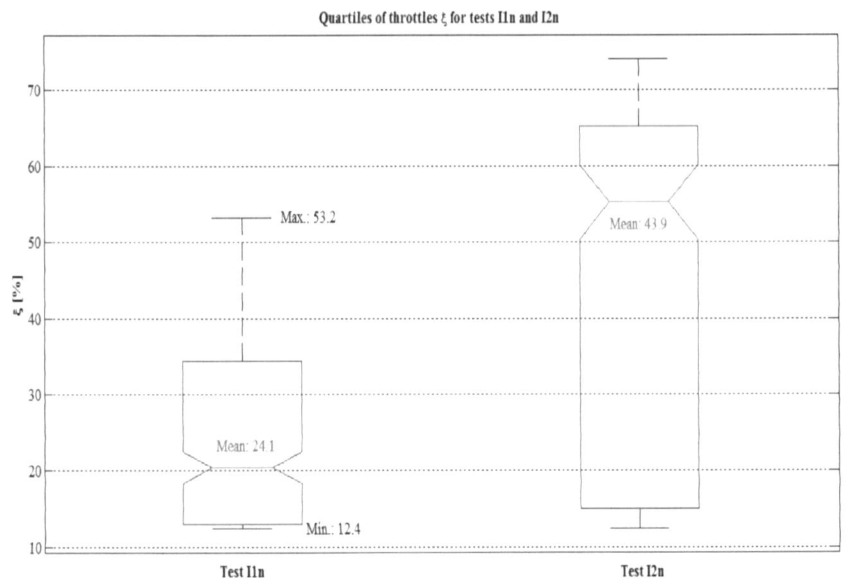

Figure 1 - Quartiles of throttles ξ for tests I1n and I2n

Table 1 shows that $F_c > F_t$ and therefore the two means differ significantly as also seen in the figure 1. It results that the throttle ξ for tests I1n and I2n considerably influences the process so both should be used to develop simple regressions models.

In the case of the same factor, throttle ξ but for tests I1n and I3n is obtained values presented in table 3 and figure 2.

Tabel 3. One-Way ANOVA for throttle ξ and tests I1n and I3n

Tipe of variance	Sum of squares *SPA*	Nuber of degrees of freedom *GL*	Variance *D*	Fisher's coefficient (computed) F_c	Fisher's coefficient (table) F_t
Factorial	23.8	1	23.8	0.085	0.771
Residual	143281.2	510	280.9		
Total	143305.1	511			

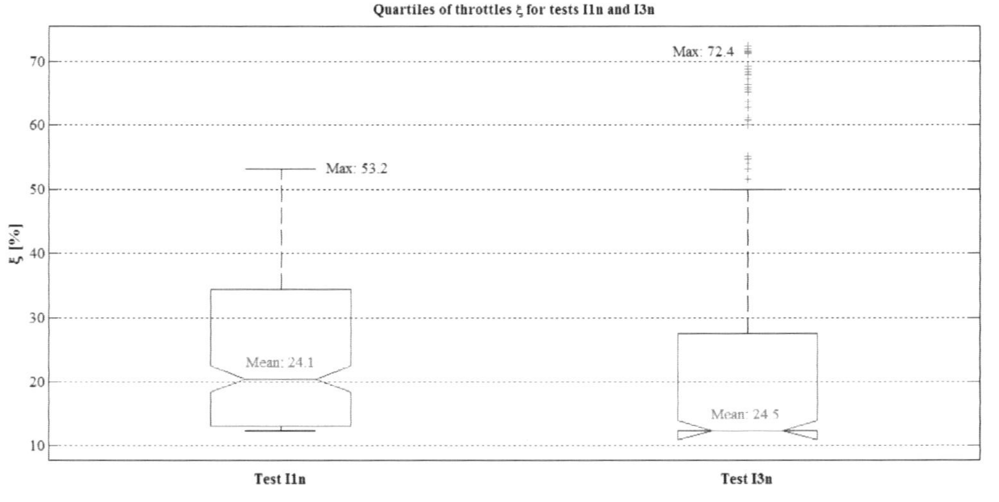

Figure 2 - Quartiles of throttles ξ for tests I1n and I3n

This time, $F_c < F_t$, that means the null hypothesis H_0 with the significance level α is accepted. The two samples have close means as can be seen on the figure 2. Taking into account the above mentioned exemple (where semples of tests I1n and I2n were considered), the finel conclusion is that the sample of throttle ξ for test I3n does not significantly influence the modeling process of the vehicle dynamics, so it can be eliminated when determining the mathematical model (by simple regression).

Similar analysis can be conducted for all tests of a specific factor and for all parameters.

5. Conclusions

The process of modeling vehicle dynamics on a statistical basis requires prior data preparation.[5] A useful tool with good effectiveness is One-Way ANOVA. It is a way to find out if a string of values or experiment results are significant for the process of modeling. In the case of vehicle dynamics it helps us to make the best data base for each type of model .

Also, it has some limitations. It requires a large amount of calculations. One-Way ANOVA tells us that at least two groups are different from each other. But it won't tell us which groups are different.[3] Both, One-Way ANOVA and analysis of correlation used together provides useful information and leads to the construction of data bases necessary for the construction of high precision models.

References

[1] C.O. Ilie, Statistical Modeling of Vehicle Dynamics, Military Technical Academy, Bucharest, 2008

[2] B Laszlo, D Voicu 2016 "Developing a measuring system able to determine the mechanical stress on a rectangular cross – section bar", *MTA Review,* Vol. XXVI, No. 1, pp. 47-52

[3] https://www.analyticsvidhya.com/blog/2018/01/anova-analysis-of-variance/

[4] L Z Rumsiski, Prelucrarea matematica a datelor experimentale, Editura Tehnică, București, 1974

[5] F Popescu, F Enache 2013 Training of RBF Neural Networks: a comparative overview, "Mircea cel Batran" Naval Academy Publishing House, *Scientific Bulletin*, Constanța, Romania

IOP Conf. Series: Earth and Environmental Science **172** (2018) 012029 doi:10.1088/1755-1315/172/1/012029

CFD approach used for modelling hydrodynamic analysis and motion characteristics of a remotely operated vehicle

A Nedelcu[1], O Tărăbuţă[2], C Clinci[3], G Ichimoaiei[4]
[1] PhD Student, " Mircea cel Bătrân" Naval Academy, Constanta, Romania
[2] Assistant professor PhD., " Mircea cel Bătrân" Naval Academy, Constanta, Romania
[3] Lecturer PhD, " Mircea cel Bătrân" Naval Academy, Constanta, Romania
[4] Lecturer PhD, " Mircea cel Bătrân" Naval Academy, Constanta, Romania

andra.nedelcu@anmb.ro
octavian.tarabuta@anmb.ro
catalin.clinci@anmb.ro
gheorghe.ichimoaiei@anmb.ro

Corresponding autor: andra.nedelcu@anmb.ro

Abstract. Towed underwater vehicles represents essential tools for providing safe access to the underwater world with many applications including ocean research, naval operations, inspection and repair different of undersea structures. Recently, there is a necessity to use smaller underwater vehicles, tethered or untethered by ship in rivers, lakes or oceans. This underwater vehicle represents an integral part of scientific equipment to explore the seas and oceans. Remotely operated vehicle (ROV) is one of these vehicles which consists of a surface body, connecting cable and underwater vehicle. The dynamic behavior of Remotely operated vehicle (translational and rotational motion or velocity, also sum of all forces acting on the system) is numerically investigated. This paper provided a model used for simulation and modeling approach to obtain the hydrodynamic characteristics of an underwater vehicle using computational fluid dynamic (performed by the Ansys CFX). This is the preliminary design stage where extensive hydrodynamic test facilities are not available.

1. Introduction

The remotely operated vehicles are used mostly in the exploration of ocean resources. They come to help humans to investigate in a safety mode the deep and hostile underwater environments without their physical presence. They have different applications including naval operations (Banerjee and Do, 1995; Ohkusu et al., 1987; Park et al., 2003), ocean research and exploration (Boe et al., 2013; Feng and Allen, 2004), acoustic surveying (Buckham et al., 2003), assembly, inspection and repair of undersea structures (Kamali and Khojasteh, 2016; Khojasteh, 2015). ROV is a towed underwater vehicle. The system is composed of the underwater vehicle, which is connected to the water surface on a vessel, buoy or platform, a handling system and other associated power supplies. The vehicle is linked to the ship using an umbilical cable. The vehicles can very depend on size: from small vehicles with video cameras for simple observation up to complex systems, which can have several and different equipment's, tools and other manipulators. Take into consideration the importance of vehicles applications nowadays has been made many applications and experiments to examine the dynamics of underwater vehicle, towing cable and surface vehicle.

2. Computational fluid dynamic analysis of an experimental underwater body vehicle

In this paper, the principal objective is to establish and determinate a CFD methodology for the computation of the drag and lift forces on body vehicle. The behavior of these forces gives the hydrodynamic coefficients of the body vehicle that are required for its trajectory simulation. As a result, an accurate estimation of the hydrodynamic lift and drag forces is critical in studying body vehicle trajectories regarding a variety of operating conditions. Since no experimental measurements of these hydrodynamic forces have not been performed in the present work, it is important to validate the CFD methodology with experimental data and associated phenomena by means of computer-based simulation.

The hydrodynamic forces on the body vehicle depend on the body construction. We considered the body vehicle used for CFD analysis for estimating the drag and lift forces for different speed combinations (in Figure 8). The design particulars are presented in Table 1.

Table 1 Design particulars of the ROV

Physical property	Value	Units
Length overall	1200	[mm]
Diameter of body	300	[mm]
Diameter of propellers	120	[mm]
Number of propellers	5	[piece]
Length of wings	165	[mm]
Number of wings	4	[piece]
Diameter of tail	90	[mm]

The preliminary computer-aided design (CAD) model of an ROV will be created using SolidWorks software. Through CAD model we obtained the hydrodynamic parameters like hydrodynamic damping, pressure and forces. For this, we use Ansys CFX software.

In order to achieve the ROV modeling approach, Figure 1 presents the overall approach from numerical modeling to control system design.

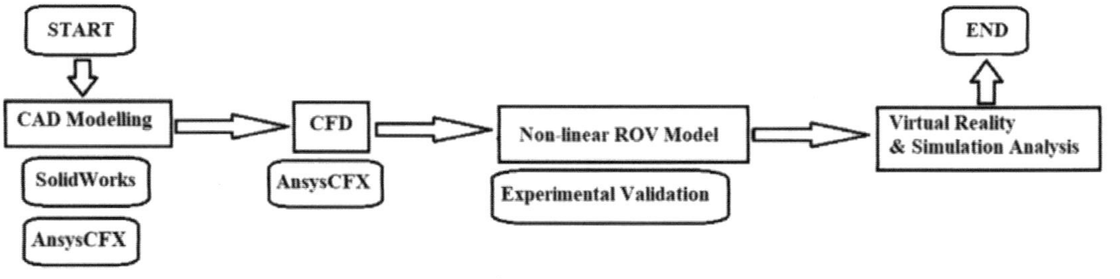

Figure.1. Chart of proposal systematic computation of ROV model for virtual reality

In the Figure 2 it is presented the geometrical model (front, left, top and right side) of the ROV vehicle after modeled by SolidWorks. The design particulars are presented in Table 1. The ROV presents six degrees of freedom for the body position and speed.

The computational domain around the experimental hull body shown in Figure 3 it is represented by a sphere with a radius of 7.5 m and 1 767,15 m^3 volume.

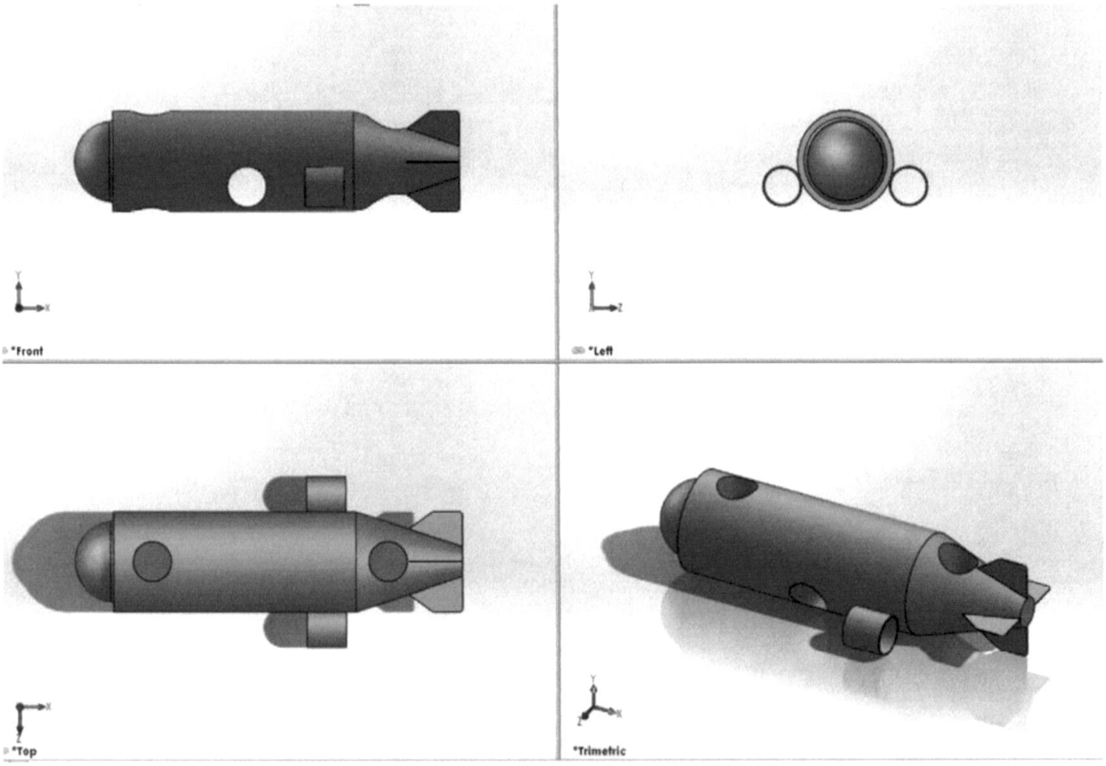

Figure. 2. ROV model in SolidWorks

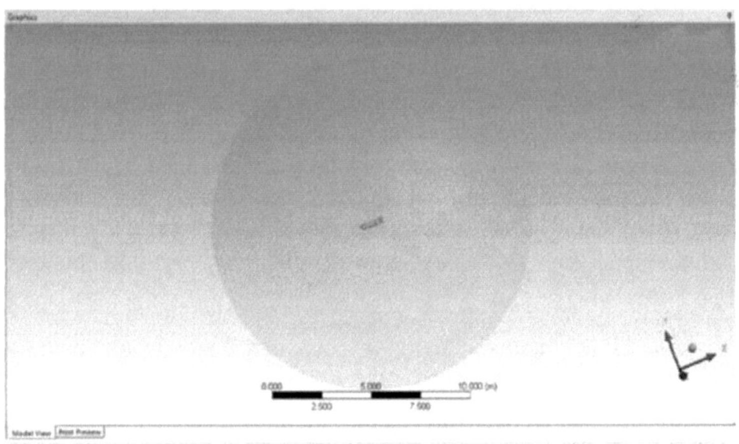

Figure. 3. Computational domain

The vehicle is situated along OX axis. We consider the domain (the sphere) full of water with the follow condition presented in Tabel 2:

Tabel 2. Fluid domain data

Fluid domain data	Value	Units
Density	997.0	[kg/m^3]
Molar Mass	18.02	[kg/kmol]

Specific Heat Capacity	4181.7	[J/kgK]
Reference Pressure	1	[atm]
Reference Temperature	25	[C]
Dynamic Viscosity	8.899E-4	[kg/ms]
Thermal Conductivity	0.6069	[W/mK]
Absorption Coefficient	1.0	[1/m]
Thermal Expansivity	2.57E-04	[1/K]

To simplify the dynamic equation derivation of ROV, we consider the following properties for the model:

- Operational at slow speed (approximate 2m/s);
- The vehicle body must be rigid and fully submerged in water (without wave and current disturbance);
- The tether (or umbilical cable) dynamic attached to ROV body is not considered.

3. Theoretical issue

3.1. Navier-Stokes model for laminar motion

In the following will be related a few physical, mathematical and numerical aspects about the Navier-Stokes equations system.

Physical aspects.

The movement of a fluid is determined by the interaction of many parameters. An essential role is played by the ratio of two types of forces: inertial forces and respectively, friction forces. This report is highlighted by the Reynolds number. Depending on Reynolds number, we can distinguish three flow regimes: laminar regime (for small Reynolds number), turbulent regime (high Reynolds number) and transient regime (between laminar and turbulent flow regimes). In conformity with flow regime, the appearance of movement is very different. For example, for a laminar flow, when the friction forces are dominant, the motion has a perfectly ordered appearance. Contrariwise, for a turbulent flow, when the inertial forces are dominant, the aspect of the movement become completely chaotic, because this chaotic character is mostly highlighted by large variations in time and space of all motion parameters (speed, pressure field, etc.).

The movement can reach different values according to the flow regime. In case of a laminar flow, when the friction forces are dominant, the motion has a perfectly ordered appearance, for a turbulent flow, (dominant inertial forces), the aspect of the movement becomes completely chaotic. The chaotic character is highlighted by large variations in time and space of all motion parameters (speed, pressure field, etc.).

Mathematical aspects.

The Navier-Stokes non-stationary equations represent a system of non-linear partial-derivate equations of the second order. Nonlinearity is a result of inertial forces. If inertial (destabilizing) forces are small according to friction (viscous diffusion plays a stabilizing role), the nonlinearity is relatively weak, and the respective equations can be numerically integrated without further hypothesis. According to the above, this case corresponds to the laminar flow regime.

Numeric aspects.

The basic problem in the numerical calculation of turbulent flows is the coexistence of structures characterized by time scales, velocities and lengths of different orders of magnitude: from scales corresponding to the general convection movement, to scales with several orders of smaller size characteristic of turbulent or viscous diffusion.

Navier-Stokes equation for momentum conservation

The viscous stresses in the momentum equation can be related to the rates of linear deformations of the fluid element. The rates of linear deformations are presented by the following formulae (Achkinadze et al, 2007):

- Linear elongating deformations:

$$\varepsilon_x = \frac{\partial u}{\partial x}\ ;\ \varepsilon_y = \frac{\partial v}{\partial y}\ ;\ \varepsilon_z = \frac{\partial w}{\partial z} \tag{1}$$

- Linear shearing deformations:

$$\vartheta_x = \frac{1}{2}\left(\frac{\partial w}{\partial y} + \frac{\partial v}{\partial z}\right);\ \vartheta_y = \frac{1}{2}\left(\frac{\partial u}{\partial z} + \frac{\partial w}{\partial x}\right);\ \vartheta_z = \frac{1}{2}\left(\frac{\partial v}{\partial x} + \frac{\partial u}{\partial y}\right) \tag{2}$$

According to the Newton`s hypothesis, the viscous stresses are represented as follow:

$$\tau_{xx} = 2\mu\frac{\partial u}{\partial x}\ ;\ \tau_{yy} = 2\mu\frac{\partial v}{\partial y}\ ;\ \tau_{zz} = 2\mu\frac{\partial w}{\partial z} \tag{3}$$

$$\tau_{xy} = \tau_{yx} = \mu\left(\frac{\partial v}{\partial x} + \frac{\partial u}{\partial y}\right)\ ;\ \tau_{zx} = \tau_{xz} = \mu\left(\frac{\partial u}{\partial z} + \frac{\partial w}{\partial x}\right)\ ;\ \tau_{yz} = \tau_{zy} = \mu\left(\frac{\partial w}{\partial y} + \frac{\partial v}{\partial z}\right) \tag{4}$$

In the fallow, we substitute the above expressions for viscous stresses into the momentum equations. For the x-, y- and z- momentum, one can write down:

$$\rho\frac{Du}{Dt} = -\frac{\partial p}{\partial x} + \frac{\partial}{\partial x}\left(2\mu\frac{\partial u}{\partial x}\right) + \frac{\partial}{\partial y}\left[\mu\left(\frac{\partial v}{\partial x} + \frac{\partial u}{\partial y}\right)\right] + \frac{\partial}{\partial z}\left[\mu\left(\frac{\partial u}{\partial z} + \frac{\partial w}{\partial x}\right)\right] + S_{Mx} \tag{5}$$

$$\rho\frac{Dv}{Dt} = -\frac{\partial p}{\partial y} + \frac{\partial}{\partial x}\left[\mu\left(\frac{\partial v}{\partial x} + \frac{\partial u}{\partial y}\right)\right] + \frac{\partial}{\partial y}\left(2\mu\frac{\partial v}{\partial y}\right) + \frac{\partial}{\partial z}\left[\mu\left(\frac{\partial w}{\partial y} + \frac{\partial v}{\partial z}\right)\right] + S_{My} \tag{6}$$

$$\rho\frac{Dw}{Dt} = -\frac{\partial p}{\partial z} + \frac{\partial}{\partial x}\left[\mu\left(\frac{\partial u}{\partial z} + \frac{\partial w}{\partial x}\right)\right] + \frac{\partial}{\partial y}\left[\mu\left(\frac{\partial w}{\partial y} + \frac{\partial v}{\partial z}\right)\right] + \frac{\partial}{\partial z}\left(2\mu\frac{\partial w}{\partial z}\right) + S_{Mz} \tag{7}$$

The equations are named Navier-Stokes equations, and they represent the form of momentum equations that is the most advantageous for the finite volume method formulation.

3.2. Reynolds equation for turbulence motion

According with theory, the Navier-Stokes Equations for Laminar Flow of Real Fluids represent the dynamic equilibrium equations between the inertial unitary forces on the one hand and the external unitary forces (e.g. mass, pressure and friction), on the other hand, applied to a fluid particle. This equation also applies to turbulent movements, only respect the condition that we have done the average of each term, so we can write:

$$\frac{1}{T}\int_0^T\left(\frac{\partial u}{\partial t} + u\frac{\partial u}{\partial x} + v\frac{\partial u}{\partial y} + w\frac{\partial u}{\partial z}\right)dt = \frac{1}{T}\int_0^t\left(f_x - \frac{1}{\rho}\frac{\partial p}{\partial x} + v\Delta u\right)dt \tag{8}$$

The group of terms representing the unitary convective force of inertia can be written as:

$$u\frac{\partial u}{\partial x} + v\frac{\partial u}{\partial y} + w\frac{\partial u}{\partial z} = \frac{\partial(u.u)}{\partial x} + \frac{\partial(v.u)}{\partial y} + \frac{\partial(w.u)}{\partial z} \tag{9}$$

By making the derivatives of the second member of equation, the term $u\left(\frac{\partial u}{\partial x} + \frac{\partial v}{\partial y} + \frac{\partial w}{\partial z}\right)$ is null resulted from the equation of continuity. Consider the fact that the mass forces are constant, and that the density is also constant (incompressible fluid) and the mediation rule applies. The equation becomes:

$$\frac{\partial \bar{u}}{\partial t} + \frac{\partial\overline{(u.u)}}{\partial x} + \frac{\partial\overline{(v,u)}}{\partial y} + \frac{\partial\overline{(w.u)}}{\partial z} = f_x - \frac{1}{\rho}\frac{\partial \bar{p}}{\partial x} + v.\Delta\bar{u} \tag{10}$$

$$\frac{\partial \bar{u}}{\partial t}+\frac{\partial(\bar{u}.\bar{u})}{\partial x}+\frac{\partial(\bar{v}.\bar{u})}{\partial y}+\frac{\partial(\bar{w}.\bar{u})}{\partial z}=f_x-\frac{1}{\rho}\frac{\partial \bar{p}}{\partial x}+\nu\Delta\bar{u}-\frac{\partial(\overline{u'u'})}{\partial x}-\frac{\partial(\overline{v'u'})}{\partial y}-\frac{\partial(\overline{w'u'})}{\partial z} \quad (11)$$

If we execute the derivation in the left part of the equation and considered the continuity equation $\frac{\partial \bar{u}}{\partial x}+\frac{\partial \bar{v}}{\partial y}+\frac{\partial \bar{w}}{\partial z}=0$, it follows:

$$\frac{\partial \bar{u}}{\partial t}+\bar{u}\frac{\partial \bar{u}}{\partial x}+\bar{v}\frac{\partial \bar{u}}{\partial y}+\bar{w}\frac{\partial \bar{u}}{\partial z}=f_x-\frac{1}{\rho}\frac{\partial \bar{p}}{\partial x}+\nu\Delta\bar{u}+\frac{1}{\rho}\left[\frac{\partial}{\partial x}(-\rho\overline{u'u'})+\frac{\partial}{\partial y}(-\rho\overline{v'u'})+\frac{\partial}{\partial z}(-\rho\overline{w'u'})\right] \quad (12)$$

The first term in the left-part of equation represents the local unitary force of inertia, the next three represents the unitary convective force of inertia. The first term in the right-part of equation is the mass unitary force, the second, the unitary pressure force, the third, the unitary viscosity force and the last, the unit forces due to the turbulent pulses.

Analogous equations are obtained in the directions Oy and Oz which together with the continuity equation represent Reynolds equations for turbulent motion.

3.3. Shear Stress Transport

Computational Fluid Dynamics (CFD) is a way of analyzing fluid flows numerically on a computer by a set of algebraic equations. The algebraic equations are obtained by discretizing the partial differential equations (PDEs), which may be momentum, conservation of mass, energy etc. The solution is attained at discrete points and hence the computational domain needs to be discretized into discrete areas or a volume - which is the mesh or grid. For this reason, the grid resolution is of great importance. For instance, in order to fully describe the physics of the flow close to a wall, the grid resolution need to be very precise and fine.

During the simulation, the Shear Stress Transport (SST) model in CFD was using software ANSYS. In the areas close to the surface of the ROV, the URANS equations will be used in conjunction with the k−ω SST (Shear Stress Transport) model of Menter. This model is basically a combination of the k −ω model by Wilcox and the standard k −ε model by Jones and Launder. Close to the cylinder surface, i.e. in the near-wall region, the original k−ω model is used. In the free shear layers and in the outer wake region, the standard k −ε model is used.

The k-x SST model is a common model for predicting the flow separation under adverse pressure gradient. The k-ω based SST model (Shear Stress Transport) accounts for the transport of the turbulent shear stress. Furthermore, the SST gives highly accurate predictions of the onset and the amount of flow separation under adverse pressure gradients (eddy-viscosity).

On the other hand, the BSL model combines the advantages of the Wilcox and the k-ε model. At the same time, still fails to properly predict the onset and amount of flow separation from smooth surfaces. The principal reason is that both models do not account for the transport of the turbulent shear stress. This is the reason that this results in an over-prediction of the eddy-viscosity. The proper transport behavior can be obtained from the eddy-viscosity:

$$v_t=\frac{a_1 k}{\max(a_1\omega, SF_2)} \quad (13)$$

Where $v_t=\mu_t/p$

F_2 is a blending function similar to F_1 function, the second function restricts the limiter to the wall boundary layer. During the underlying assumptions are not correct for free shear flows, S is an invariant measure of the strain rate.

We expected to have a turbulent movement in the fluid domain. In this case, the temperature is fixed at 25°C and the water is modeled and considered as an incompressible fluid. It is important to set

IOP Conf. Series: Earth and Environmental Science **172** (2018) 012029 doi:10.1088/1755-1315/172/1/012029

the fluid domain to be a large one, to analyze the damping force acting on ROV in CFD. The side, top, and bottom of fluid domain are modeled using free slip wall boundary conditions. The conditions used are presented in Tabel 2.

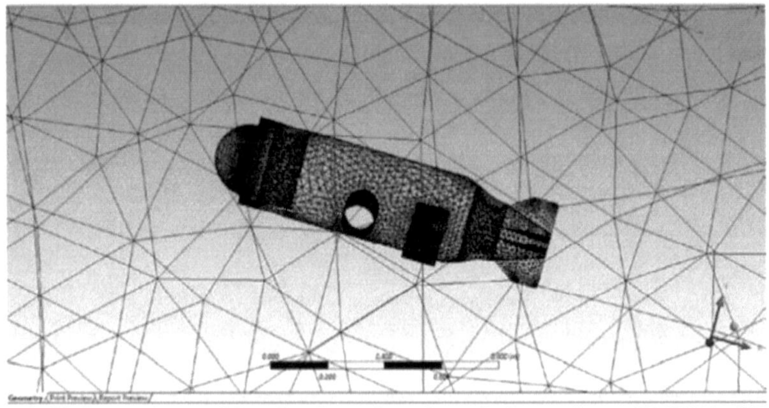
Figure 4. Volumetric meshing for flow domain and close view of ROV in 3D view

In this case, the physical state of the solution domain has been represented by a set of boundary conditions. The Default domain present 2 boundary: the vehicle and the domain. The body vehicle present no slip wall and smooth wall roughness. The domain present the boundary type opening, a subsonic flow regime and for turbulence a Fractional Intensity 0.05 and Eddy Viscosity Ratio of 10.

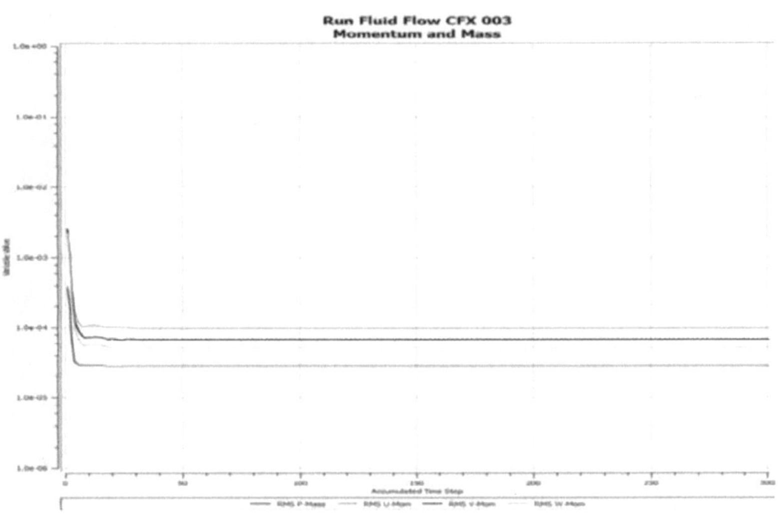

Figure 5. Momentum and Mass

Figure 5 shows the moments and mass of vehicle accumulated in 300 time step when the model was towed in the surge direction. We observe a decrease of value from 1.0e-025 to 1.0e-045 for the first 10 accumulated time step. After the first 10 step, the graphic present a constant value, around 1.0e-0.45. During the simulation, we see few curves variations of momentum and mass.

IOP Conf. Series: Earth and Environmental Science **172** (2018) 012029 doi:10.1088/1755-1315/172/1/012029

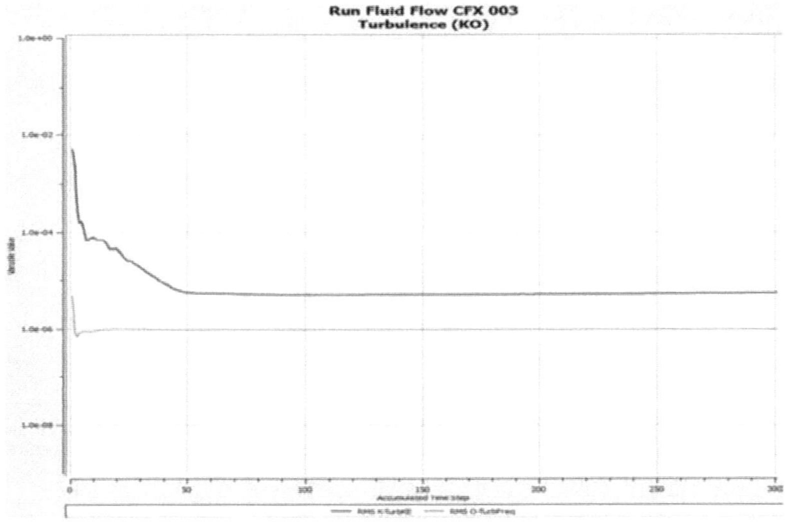

Figure 6. Turbulence

Figure 6 shows the convergence criteria for turbulence of vehicle. The value decrease for the first 12, respectievly 50 steps, and then is constant.

Before performing the CFD, the mesh size needs to be defined properly as the shape of the boundary. It is recommended that the mesh size to be sufficiently small to capture the geometry of the ROV. For example, in this case, the number of elements represented approximately 756330 elements . The fluid domain has around 135403 nodes. The volume is 2232.9 m^3. The 3D view of volumetric mesh of the flow domain around ROV it is represented in Figure 4. In Table 3 are presented the extens of mesh.

Table 3. Extents of mesh

Extends	Value	Units
min x, max x	-7.49959, 8.71459	[m]
min y, max y	-8.10808, 8.10817	[m]
min z, max z	-8.10813, 8.10831	[m]

Figure 7.- Pressure of ROV and streamlines

In Figure 7 is presented the pressure around the vehicle. From figure, result a high-pressure region at the front of vehicle. As observed from simulation, there is certainly flow separation on the ROV. In Figure 7, the vehicle performs an immersion movement. As observed in Figure 7, the wakes

IOP Conf. Series: Earth and Environmental Science **172** (2018) 012029 doi:10.1088/1755-1315/172/1/012029

are formed at the rear and at the top of the ROV body when the turbulent/laminar flow past the vehicle. The minimum pressure has a value estimated at -4556.19 Pa situated at the bottom of the front, and a maxim pressure around 2709.41 Pa. When the vehicle performs a horizontal motion, reaches a minimum value of -3217.52 Pa, and a maximum value of 1952.23 Pa. We also observed the presence of streamlines. Those depend on the movement of vehicle : in horizontal plan or in vertical (immersion) plan. In Figure 7, the vehicle is in immersion plan.

In Figure 8 it is represented the graphic with the pressure distribution measured in equatorial plan of vehicle.

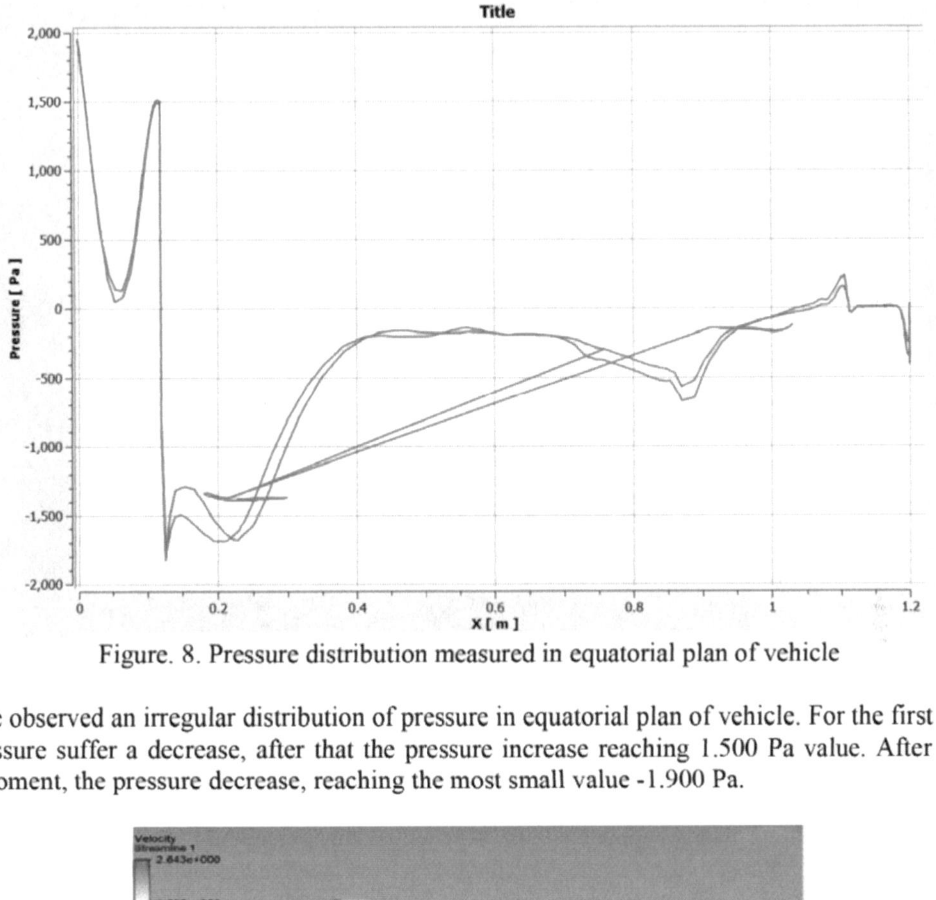

Figure. 8. Pressure distribution measured in equatorial plan of vehicle

We observed an irregular distribution of pressure in equatorial plan of vehicle. For the first 0.1 m, the pressure suffer a decrease, after that the pressure increase reaching 1.500 Pa value. After the increase moment, the pressure decrease, reaching the most small value -1.900 Pa.

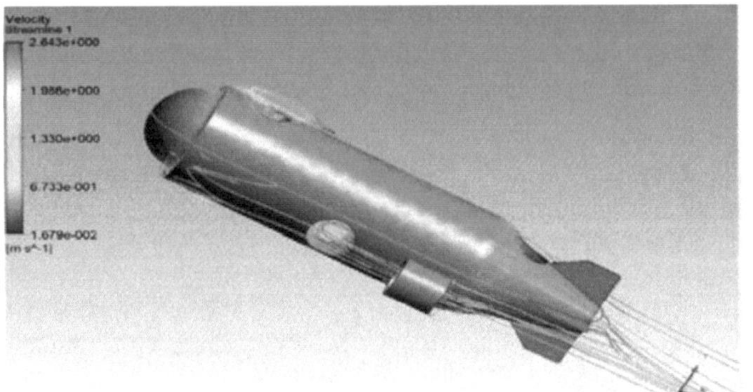

Figure.9. Velocity of ROV and streamlines

A maxim value reaches a value of 2.64 m/s, and a minimum value of 0.016 m/s. In Figure 9, the vehicle movement is in horizontal plan.

At the end of simulation, the software Ansys CFX perform the following information:

Table 4. Final results

Variable Name	min	max
Density	9.97E+02	9.97E+02 |
Specific Heat Capacity at Constant Pressure	4.18E+03	4.18E+03 |
Dynamic Viscosity	8.90E-04	8.90E-04 |
Thermal Conductivity	6.07E-01	6.07E-01 |
Static Entropy	0.00E+00	0.00E+00 |
Velocity u	-1.36E+00	2.49E+00 |
Velocity v	-1.99E+00	1.94E+00 |
Velocity w	-1.82E+00	1.96E+00 |
Pressure	-3.09E+03	2.00E+03 |
Turbulence Kinetic Energy	5.29E-06	2.32E-01 |
Turbulence Eddy Frequency	1.83E+00	1.62E+05 |
Eddy Viscosity	5.99E-08|	1.24E+00 |
Temperature	2.98E+02	2.98E+02 |
Wall Scale	2.26E-04	5.31E+02 |
Wall Distance	1.32E-06	3.25E+01

Figure.10 The vehicle

The overall vehicle image up to now is shown in the Figure 10. Based on the shape design and arrangement of equipment on body, the design is made. In the future, the authors will examine and will test the vehicle for different speed and comparing the numerical simulation with the ROV.

4. Conclusion

In this paper was presented a systematic modeling of the hydrodynamic dragging forces of a remotely operated vehicle using Ansys CFX. The simulated result is going to be verified in the following with the experimental tests in basic and sea water. Future works could improve the accuracy of the CFD results by comparing the numerical simulation with the ROV using real-time adaptive identification approach in sea trial.

References

[1] Committee on Autonomous Vehicles in Support of Naval Operations Naval Studies Board Division on Engineering and Physical Sciences, *Autonomous Vehicles in Support of Naval Operations,* 2005;

[2] Robert D. Christ, Robert L. Wernli, Sr., *The ROV Manual,* A User Guide for Remotely Operated Vehicles, Second Edition;

[3] Thor I. Fossen, *Models for Ships, Offshore Structures and Underwater Vehicles*, 2011;

[4] Gianluca Antonelli, *Underwater Robots: Motion and Force Control of Vehicle-Manipulator Systems*, Springer Tractsin Advanced Robotics, New York, 2006,
[5] Fossen TI (1994) Guidance and control of ocean vehicles, Wiley, New York;
[6] Banerjee, A.K., Do, V.N., 1995. Deployment control of a cable connecting a ship to an underwater vehicle. Appl. Math. Comput. 70 (2), 97–116.
[7] Ohkusu, M., Koterayama, W., Kashiwagi, M., 1987. Hydrodynamics of a depth controlled towed vehicle. Oil and gas fields of Persian Gulf. Available from: ⟨⟨http://meed.com/ ⟩⟩; (accessed 11.08.15).
[8] Park, H., Jung, D., Koterayama, W., 2003. A numerical and experimental study on dynamics of a towed low tension cable. Appl. Ocean Res. 25 (5), 289–299.
[9] Boe, C., Rodriguez, J., Plazaola, C., Banfield, I., Fong, A., Caballero, R., Vega, A., 2013. A hydrodynamic assessment of a remotely operated underwater vehicle based on computational fluid dynamic–Part 1–numerical simulation. CMES: Comput. Model. Eng. Sci. 90 (2), 165–177.
[10] Fang, M.-C., Hou, C.-S., Luo, J.-H., 2007. On the motions of the underwater remotely operated vehicle with the umbilical cable effect. Ocean Eng. 34 (8), 1275–1289.
[11] Buckham, B., Nahon, M., Seto, M., Zhao, X., Lambert, C., 2003. Dynamics and control of a towed underwater vehicle system, part I: model development. Ocean Eng. 30 (4), 453–470.
[12] Kamali, R., Khojasteh, D., 2016. Investigation of dynamic behavior of a towed underwater vehicle at Persian Gulf. In: Proceedings of the 24th Annual International Conference on Mechanical Engineering-ISME, Yazd, Iran.
[13] Khojasteh, D., 2015. Dynamic Modeling of a Towed Undersea Body and Dynamic Studyof a Point Absorber Wave Energy Converter (Master Thesis). Shiraz University, Shiraz, Iran.
[14] http://www.marine-knowledge.com/types_of_ships/an-underwater-robot/
[15] http://www.sciencedirect.com/
[16] SolidWorks software
[17] AnsysCFX software

Water interference effect on ship due to square shaped object shielding

Ionuț Cristian SCURTU, Cătălin CLINCI, Adrian POPA

Naval Academy "Mircea cel Batran", Constanta, Romania

E-mail: ionut.scurtu@anmb.ro

Abstract. This scientific paper presents performed numerical 2D Forces resulted from pressure variation due to interference with a square shaped object are analyzed by computer fluid dynamics module ANSYS. Using the 2d geometry we created a three-dimensional mesh around the ship with water speed v=ct and variable shielding are used for analysis. Interaction is computed on 2D model for distances between objects 0.2L…0.9 L, where L is the length of the ship. Various positions of the square shaped object will determine different case as presented in the fluid flow simulation in a restricted domain simulating open sea operation. The pressure for each case can be easily found in CFD-Post and interference is presented for ship operation in open sea.

1. Introduction

Water speed v=ct and current variable incidence angle are used for analysis in Workbench parametric simulation. Interaction is computed on 2D model for distances between objects 0.2L…0.9 L, where L is the length of the ship. The main goal is to determine the numerical water interference effect on ship performed by computational tools provide appropriate approaches for calculations of the complex 2D presented model characteristics. The main value of the paper is that the obtained results with the realized numerical flow simulation can be compared with the experimental data measures in ship operation near shielding objects.

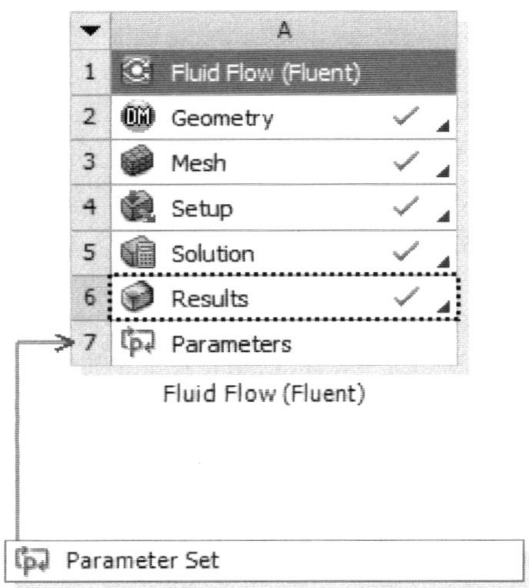

Figure 1. Ansys Parametric simulation

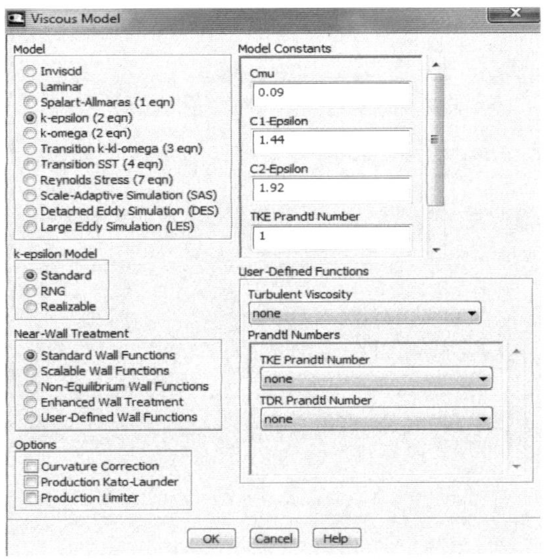

Figure 2. Viscous model used in simulation steps

2. Mathematical Model

The performed numerical simulation is governed by the continuity, Navier-Stokes (momentum conservation) and energy equations. But the flow is turbulent, that's why it was used Reynolds Average version of their equations. Therefore it was Standard k-ε Model turbulence model. In the beginning, it has six variables to solve for: 3 components of velocity, pressure, temperature, and kinematic eddy viscosity. The equations are:

$$\frac{\partial \rho}{\partial t}+\frac{\partial}{\partial x_i}(\rho u_i)=0 \quad (1)$$

$$\rho\left(\frac{\partial U_i}{\partial t}+\frac{\partial}{\partial x_j}(U_i U_j)\right)=\frac{\partial P}{\partial x_i}+\frac{\partial}{\partial x_j}\left(2\mu S_{ij}-\rho\overline{u_i' u_j'}\right) \quad (2)$$

$$\frac{\partial}{\partial t}(\rho e_0)+\frac{\partial}{\partial x_j}\left[\rho u_j e_0+u_j p+q_j-u_i\tau_{ij}=0\right] \quad (3)$$

Where (1) is Continuity Equation,
(2) is Reynolds Averaged Navier-Stokes equation,
(3) is conservation of energy equation,

Using the approach from [1] the given above equation are converted to algebraic equations. It then solves for our six variables ay each of the cell centers of our mesh. This means that if we have 3.000.000 cells, CFX is going to solve 4.8 million equations to solve the problem.

3. Standard k-ε Model theory ued in calculation by Ansys Software.

Two-equation turbulence models allow the determination of both, a turbulent length and time scale by solving two separate transport equations. The standard k - ε model in ANSYS Fluent falls within this class of models and has become the workhorse of practical engineering flow calculations in the time since it was proposed by Launder and Spalding . Fast calculation and reasonable accuracy for a wide range of turbulent flows explain its popularity in industrial flow and heat transfer simulations. It is a semi-empirical model, and the derivation of the model equations relies on phenomenological considerations and empiricism.

The turbulence kinetic energy, k, and its rate of dissipation, ε, are obtained from the following transport equations:

$$\frac{\partial}{\partial t}(\rho k)+\frac{\partial}{\partial x_i}(\rho k u_i)=\frac{\partial}{\partial x_j}\left[\left(\mu+\frac{\mu_t}{\sigma_k}\right)\frac{\partial k}{\partial x_j}\right]+G_k+G_b-\rho\varepsilon-Y_M+S_k$$

and

$$\frac{\partial}{\partial t}(\rho\varepsilon)+\frac{\partial}{\partial x_i}(\rho\varepsilon u_i)=\frac{\partial}{\partial x_j}\left[\left(\mu+\frac{\mu_t}{\sigma_\varepsilon}\right)\frac{\partial\varepsilon}{\partial x_j}\right]+C_{1\varepsilon}\frac{\varepsilon}{k}(G_k+C_{3\varepsilon}G_b)-C_{2\varepsilon}\rho\frac{\varepsilon^2}{k}+S_\varepsilon$$

In these equations, G_k represents the generation of turbulence kinetic energy due to the mean velocity gradients, calculated as described in *Modeling Turbulent Production in the k-ε Models*.

IOP Conf. Series: Earth and Environmental Science **172** (2018) 012030 doi:10.1088/1755-1315/172/1/012030

Various positions of the square shaped object will determine different case as presented in the fluid flow simulation in a restricted domain simulating open sea operation in a 2D domain 12L X 10B where L is ship length and B is ship breatdh. S1 and S2 dimensions are used to determine square object position relative to ship bow.

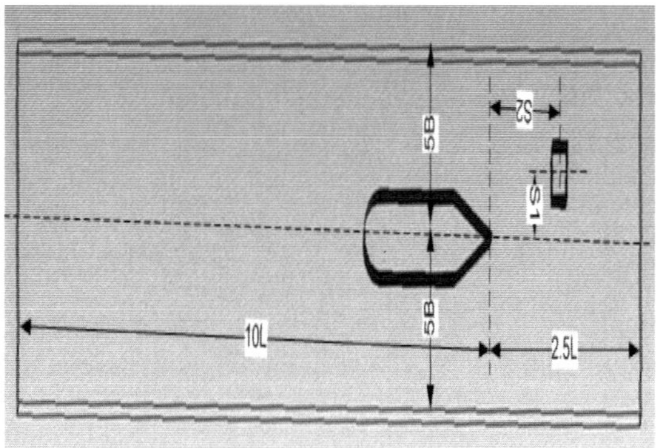

Figure 3. 2 D domain dimensions

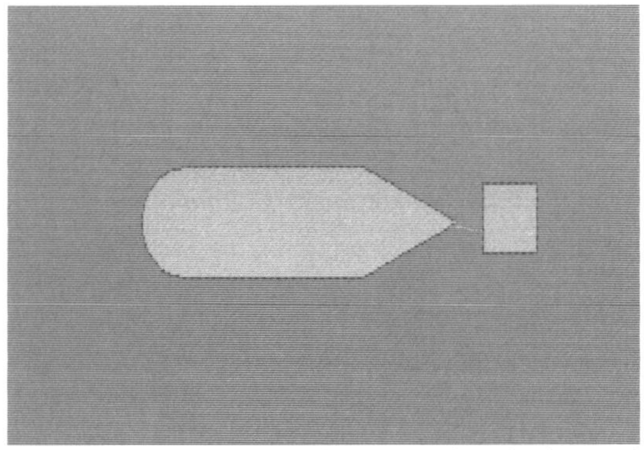

Figure 4. Simetric interference object in front of the ship

The pressure for each case can be easily found in CFD-Post and interference is presented for ship operation in open sea as presented below.

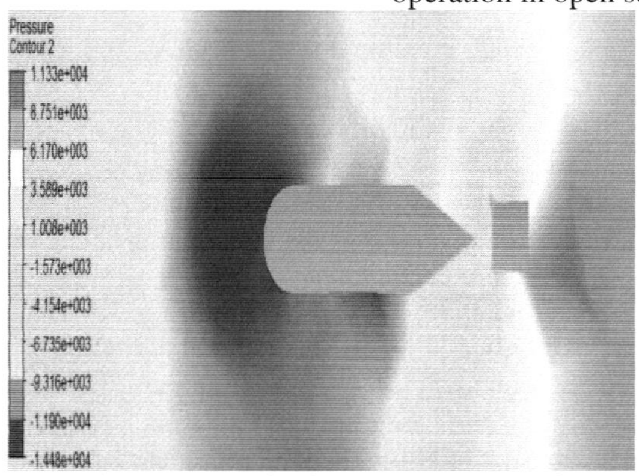

Figure 5. Results for v=8,5 and o=16

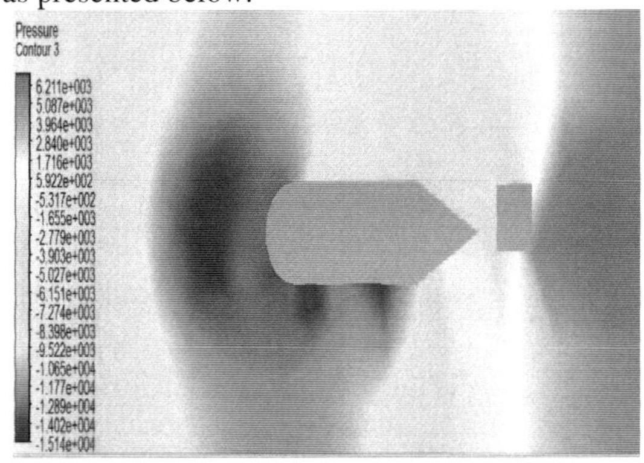

Figure 6. Results for v=8 and o=16

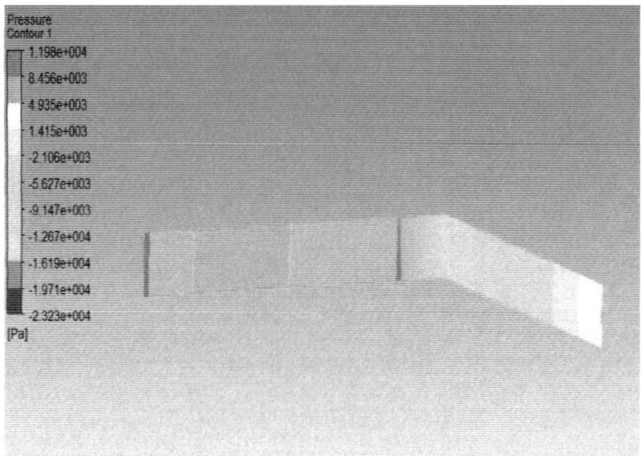

Figure 7. Port side pressure contour

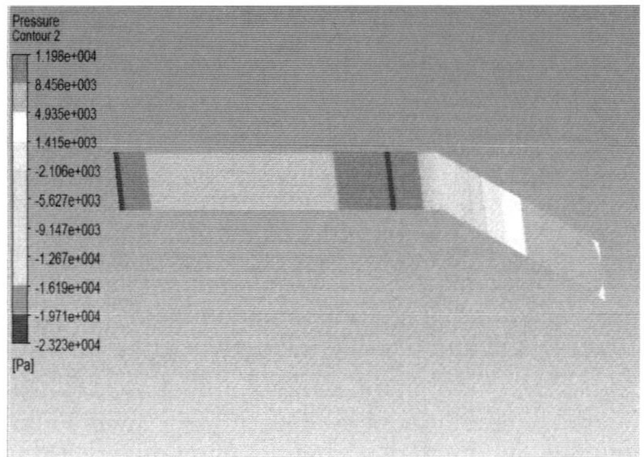

Figure 8. Starboard side pressure contour

IOP Conf. Series: Earth and Environmental Science **172** (2018) 012030 doi:10.1088/1755-1315/172/1/012030

Boundaries not used are set as no friction surfaces for a simulation reflecting interference in open sea. Differences in pressure contour on each side can be easily computed in lateral force.

Figure 9. Bow pressures

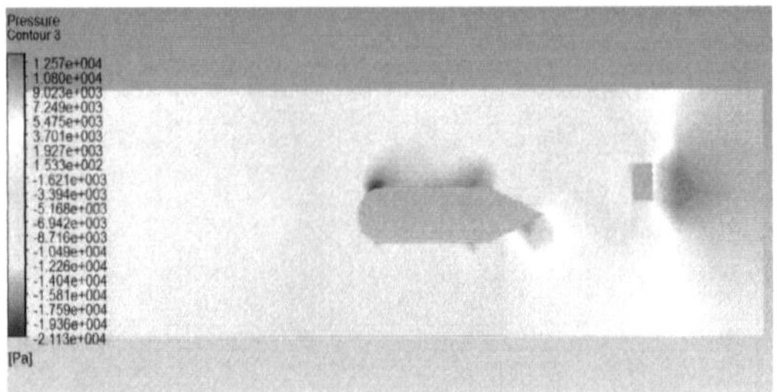

Figure 11. Pressure contours 6 m

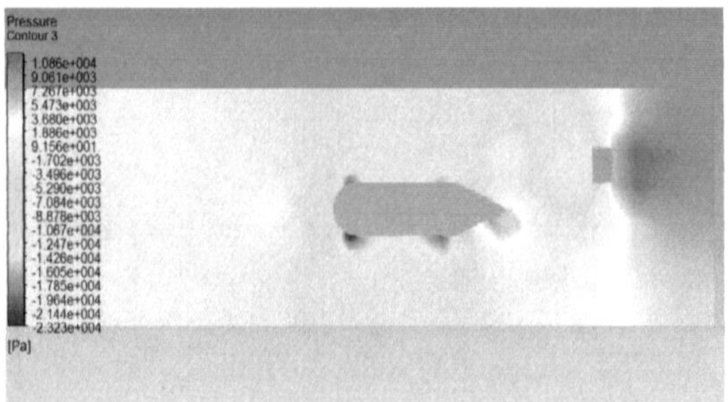

Figure 10. Pressure contours 5 m

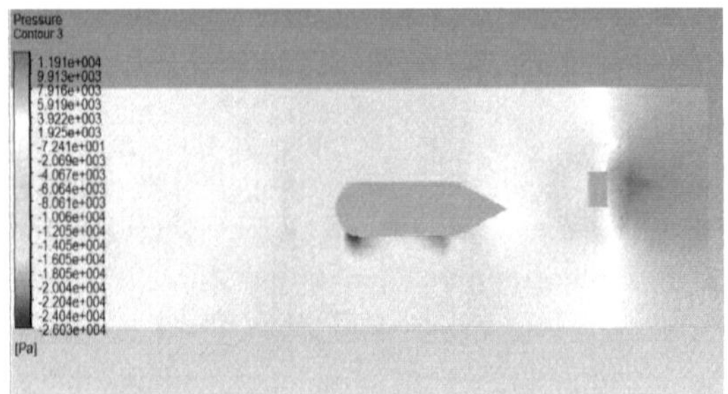

Figure 12. Pressure contours 7 m

Figure 13. Pressure contours 8 m

Figure 14. Pressure contours 10 m

Figure 15. Pressure contours 12 m

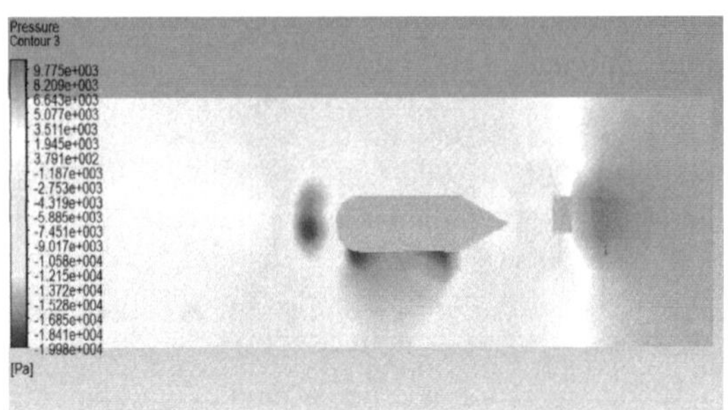

Pressure contours 14 m

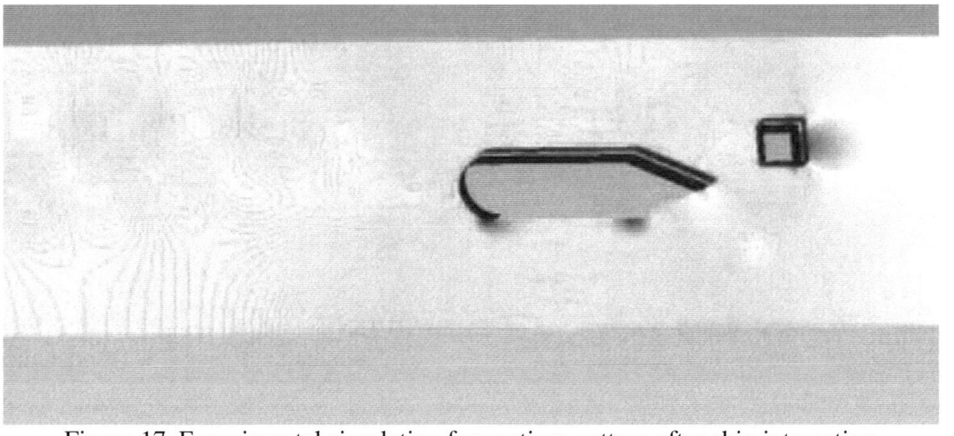

Figure 17. Experimental simulation for vortices pattern after ship interaction

The geometry and the domain around the ship in ANSYS and also the chosen boundary conditions according to ship in open sea operation
Creating an accurate mesh involves a grid generation using appropriate shape cells; here the cell for the mesh is triangular. Essentially, it consists of converting the grid into a format which can be understood by the CFX solver in order to approximate the equations of the fluid mechanics in each cell.

4. Numerical results

Lateral force computed after the simulation is done, CFX give the following values about S1and S2 coefficients: S1=5,6,7,8 m and S2=1,2,4m. Presented results from simulation are showing less interference for higher s2 distance for any s1 value.

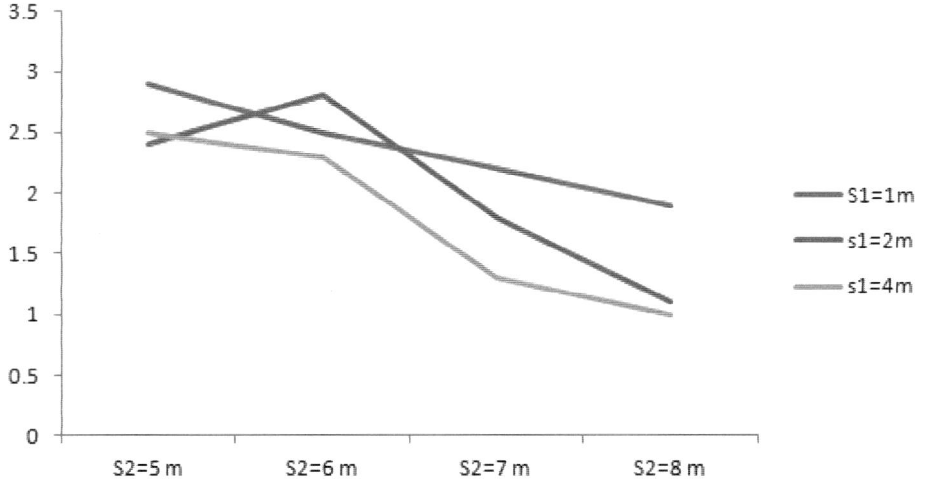

Figure 18. Graphical results for lateral force on generated ship in KN

Results can be improved in simulation for applicative and real operation assessment using real ship design values. The shielding effect on 2D ship shaped model is computed as lateral force on the wall surface representing ship in a no slip boundary.

5. Conclusions

The main conclusion that can be made onto this stage of the study is that observing the result for lateral force shown in figure 18. For more precision it have to be made numerical experiments and it have to be included the technique mesh refinement on a HPC.

The numerical simulation method for 2D viscous flow was formulated using ship interference with a square shaped object. We explicitly reduced number of design point in order to obtain a fast calculation procedure.

Forces resulted from pressure variation due to interference with a square shaped object are analyzed and presented by computer fluid dynamics module ANSYS.

Water speed v=ct and current variable incidence angle are used for analysis. Interaction is computed on 2D model for distances between objects 0.2L…0.9 L, where L is the length of the ship. Various positions of the square shaped object will determine different case as presented in the fluid flow simulation in a restricted domain simulating open sea operation. The pressure for each case is presented in CFD-Post and interference is presented for ship operation in calm water.

References

[1] www.ansys.com
[2] Domnisoru L, Găvan E, Popovici O – Analiza structurilor navale prin metoda elementului finit, Editura Didactica si Pedagogica, Bucuresti 2005, ISBN 973 – 30 – 1075 – 8
[3] Wei Tong, ed. (2010). Wind Power Generation. WIT Press. ISBN 978-1-84564-205-1.
[4] Călimănescu I., Stan L. C., Computer fluid dynamics (CFD), Atom 2016, Conference Paper.
[5] Stan L. Călimănescu I.,. C.A New innovative, 2016, Conference Paper.
[6]http://web.mit.edu/xeviroca/www/rocaMeshForSim.pdf
[7]https://www.padtinc.com/blog/wp-content/ANSYS-Meshing_Blog.pdf
[8]https://www.ozeninc.com/wp-content/uploads/2014/11/MESHING_WORKSHOP_2014.pdf

Establishing the optimal apparent power of electric transformers using the economic criterion

L Pana[1] and D Pasculescu[1]

[1]PhD. Lecturer Eng., Department of Computer and Electrical Engineering, University of Petrosani, Universitatii Str. 20, 332006, Petrosani, Romania
[1]PhD. Associate Professor, Department of Computer and Electrical Engineering, University of Petrosani, Universitatii Str. 20, 332006, Petrosani, Romania

Email: leon_pana@yahoo.com, pdragos_74@yahoo.com

Abstract. The electrical stations (substations) represents the basic subsystems of electric power system and is necessary the modes of operation evaluation and modeling of these with a view to establish, later on, a degree of continuity in electric energy supply of load points or customers connected at these electrical stations or substations. Adoption the optimal solutions of supply with electrical energy of industrial consumers is a very complex problem which involve analyses one of very large number of restrictions and factors, with determinist character quasi-determinate and aleatory. In this paper is presented a method for determinate the optimal apparent power of electric transformers, respective the optimal demand coefficient in the first year of exploitation using the economic criteria and the Matlab –Simulink program.

Key words. Operational states, electric stations, peak load, power loss, apparent nominal power, demand factor

1. Introduction

In this paper for determination the optimal power of electric transformer, respective establishment the initial values from the first year of exploitation (demand factor at minimum that limit $K_{lim.min}$) of annual peaks load are used the economic criteria namely: total up to date annual costs. The economical load from transformer will determinate taking into account a dynamic advance of annual load peak. Correct analyses must take into consideration as costs concerning acquisitioning, assembling and servicing transformer, both energy loss costs produced in operating time of this.

2. Analytical model presentation

In this case the analytical expression of $E_n(S)$, at the charging rate, on the period of *n* years, can be written in the same mode: [1], [2], [7], [10].

$$E_n(S) = \frac{T_n}{T_{20}} \cdot \left[C_T + C_0 \cdot (\Delta P_0 + k_e \cdot \Delta Q_0) + C_k \cdot (\Delta P_{kn} + k_e \cdot \Delta Q_{kn}) \cdot \left(\frac{S_{M1}}{S_{nT}} \right)^2 \right] \quad (1)$$

where:

$$C_0 = C_p + C_w \cdot T_f \cdot T_{20}$$
$$C_k = C_P + C_w \cdot \tau \cdot T_{20} \cdot m_l \quad (2)$$

$E_n(S)$ are the total up to date annual costs summarization on the period of *n* years of operation, in €/year;

T_n is the update quantity of a service life of *n* years and has expression:

$$T_n = \sum_{x=1}^{n} (1+a)^{-x} \quad (3)$$

C_T transformer cost, including workmanship, in €;

C_0 cost of the update values of the stand by loss power on the period of *n* years of operation, in €/kW;

ΔP_0 stand by loss active power of electric transformer, in kW;

C_k cost of the update values of the short circuit loss power unit and in assumption of the rate *l* of increasing the annual load peak, in €/kW;

ΔP_{kn} nominal short circuit active loss power (transformer operating at the nominal load), in kW;

ΔQ_0 stand by loss reactive power of electric transformer, in KVAr;

ΔQ_{kn} nominal short circuit reactive loss power (transformer operating at the nominal load), in kVAr;

C_P specific cost of installed power in equalization electric station, of base, in update values, in €/kW;

C_w specific cost medium of system from kilowatt-hour of loss calculated at the electric station level of HV/MV, in €/kWh;

k_e electric power equivalent, in kW/kVAr;

S_{M1} maximum apparent power, from the first year of operating of transformer, in kVA;

S_{nT} nominal apparent power of transformer, in kVA;

m_l load multiplier and has expression:

$$m_l = \frac{1}{(1+a) \cdot T_n} \cdot \sum_{m=0}^{n-1} \frac{(1+l)^{2m}}{(1+a)^m} \quad (4)$$

in which:

l is the increasing rate of annual load peak in respective period and *a* is the updating rate. In this case, are considered that l=0 and n=20 year, the values of load multiplier is: $m_l = 1$;

τ calculus duration annual technological energy loss, in terms of h/year, and has the expression:

$$\tau = T_{SM} \frac{10000 + T_{SM}}{27520 - T_{SM}} \quad (5)$$

IOP Conf. Series: Earth and Environmental Science **172** (2018) 012031 doi:10.1088/1755-1315/172/1/012031

in which:

T_{SM} represents the duration of annual apparent power utilization, in terms of hr/y;

T_f operating time of electric transformer, in terms of hr.

When designing new substations for generation, transmission or distribution, the new approach should therefore be to look at this as a functional solution, not a puzzle of various equipment.

Stand by loss, short-circuit loss and the transformer cost can be replace, in accordance with hypotheses made, in function of some specific values related to nominal assignment of transformer, at power β, in this way: [4], [5], [9].

$$C_T = \alpha_T \cdot S_{nT}{}^{\beta}$$
$$\Delta P_0 = \alpha_0 \cdot S_{nT}{}^{\beta}$$
$$\Delta P_k = \alpha_k \cdot S_{nT}{}^{\beta} \qquad (6)$$

In like manner can be write the relations from the stand by loss reactive power and in short circuit of electric power transformer: [3], [8]

$$\Delta Q_0 = \lambda_0 \cdot S_{nT}{}^{\beta}$$
$$\Delta Q_k = \lambda_k \cdot S_{nT}{}^{\beta} \qquad (7)$$

where:

β has value 0.66 or 0.75.

In final the analytical expression of load rating are given by:

$$K_{lim.min} = \frac{S_{M1}}{S_{nT}} = \begin{cases} 0.707 \cdot \sqrt{\dfrac{\alpha_T + (\alpha_0 + k_e \cdot \lambda_0)\cdot(C_P + T_f \cdot T_{20} \cdot C_W)}{(\alpha_k + k_e \cdot \lambda_k)\cdot(C_P + \tau \cdot T_{20} \cdot C_W \cdot m_1)}} & \text{if } \beta = 0.66 \\ 0.775 \cdot \sqrt{\dfrac{\alpha_T + (\alpha_0 + k_e \cdot \lambda_0)\cdot(C_P + T_f \cdot T_{20} \cdot C_W)}{(\alpha_k + k_e \cdot \lambda_k)\cdot(C_P + \tau \cdot T_{20} \cdot C_W \cdot m_1)}} & \text{if } \beta = 0.75 \end{cases} \qquad (8)$$

From analytical expression gives by relation (7) results relation of calculus in apparent nominal power of transformer in first time of operating:

$$S_{M1} = \begin{cases} 0.707 \cdot S_{nT} \cdot \sqrt{\dfrac{\alpha_T + (\alpha_0 + k_e \cdot \lambda_0)\cdot(C_P + T_f \cdot T_{20} \cdot C_W)}{(\alpha_k + k_e \cdot \lambda_k)\cdot(C_P + \tau \cdot T_{20} \cdot C_W \cdot m_1)}} & \text{if } \beta = 0.66 \\ 0.775 \cdot S_{nT} \cdot \sqrt{\dfrac{\alpha_T + (\alpha_0 + k_e \cdot \lambda_0)\cdot(C_P + T_f \cdot T_{20} \cdot C_W)}{(\alpha_k + k_e \cdot \lambda_k)\cdot(C_P + \tau \cdot T_{20} \cdot C_W \cdot m_1)}} & \text{if } \beta = 0.75 \end{cases} \qquad (9)$$

For the present electric transformer series produced in Romania, the final load value, in appreciation that get indicated transformer supplanting with a step immediate superior, is given by: [6]

$$K_{lim.max} = 1.6 \cdot K_{lim.min} \qquad (10)$$

In this case results:

$$K_{lim.max} = \begin{cases} 1.13 \cdot \sqrt{\dfrac{\alpha_T + (\alpha_0 + k_e \cdot \lambda_0) \cdot (C_P + T_f \cdot T_{20} \cdot C_W)}{(\alpha_k + k_e \cdot \lambda_k) \cdot (C_P + \tau \cdot T_{20} \cdot C_W \cdot m_l)}} & \text{if } \beta = 0.66 \\ 1.25 \cdot \sqrt{\dfrac{\alpha_T + (\alpha_0 + k_e \cdot \lambda_0) \cdot (C_P + T_f \cdot T_{20} \cdot C_W)}{(\alpha_k + k_e \cdot \lambda_k) \cdot (C_P + \tau \cdot T_{20} \cdot C_W \cdot m_l)}} & \text{if } \beta = 0.75 \end{cases} \quad (11)$$

The infrastructure of a new power company i.e. assets such as generating stations, transformer stations and distribution stations, are valued on the deregulated market not only what the equipment is "worth" but also for the efficiency and return of capital employed. Consequently low operational and maintenance costs as well as high availability are most important criteria.

3. Simulation results

In continuation aid relations (7) and (9), the present series of transformers TTUS-ONAF from the electric station of HV/MV, respective TTU-ONAN of MV/LV has calculated the initial economic loading from the annual load peaks in advance with a constant rate and different values of kilowatt-hour costs C_W as result of figure 1. In figure 2 are presented the optimal apparent power for these series of transformer mentioned above.

The diagrams from figure 1 and figure 2 have traced aid the Matlab program. According can see of diagram, all characteristics from different specific values at the kilowatt-hour loss cost C_W, elapse through the same point D, in which derivate is:

$$\frac{dK_{lim.min}}{dC_W} = 0 \quad (12)$$

In this case can be determining abscissa and ordinate the point D.

The analytical expression from the abscissa of the point D is:

$$\tau_D = \frac{(\alpha_0 + k_e \cdot \lambda_0) \cdot T_f \cdot C_P}{[\alpha_T + (\alpha_0 + k_e \cdot \lambda_0) \cdot C_P] \cdot m_l} \quad (13)$$

In final the analytical expression from the ordinate of point D is obtaining of relation (8) from the specific loss cost $C_W=0$:

$$K_{lim.D} = \begin{cases} 0.707 \cdot \sqrt{\dfrac{\alpha_T + (\alpha_0 + k_e \cdot \lambda_0) \cdot C_P}{(\alpha_k + k_e \cdot \lambda_k) \cdot C_P}} & \text{if } \beta = 0.66 \\ 0.775 \cdot \sqrt{\dfrac{\alpha_T + (\alpha_0 + k_e \cdot \lambda_0) \cdot C_P}{(\alpha_k + k_e \cdot \lambda_k) \cdot C_P}} & \text{if } \beta = 0.75 \end{cases} \quad (14)$$

The economical or optimal power of transformer is in fact, minimum value the constant power on all operational duration which, at transit by transformer produce minimum loss power.

In assumption the minimum expression the theoretic domain of variation at optimal peak loading from figure 2 and figure 3 is limited from two characteristics.

The first characteristic corresponding from $C_W=0$:

$$K_{lim.min.1} = \begin{cases} \lim_{C_W \to 0}\left[0.707 \cdot \sqrt{\dfrac{\alpha_T + (\alpha_0 + k_e \cdot \lambda_0)\cdot(C_P + T_f \cdot T_{20} \cdot C_W)}{(\alpha_k + k_e \cdot \lambda_k)\cdot(C_P + \tau \cdot T_{20} \cdot C_W \cdot m_1)}}\right] = & \text{if } \beta = 0.66 \\ = 0.707 \cdot \sqrt{\dfrac{\alpha_T + (\alpha_0 + k_e \cdot \lambda_0)\cdot C_P}{(\alpha_k + k_e \cdot \lambda_k)\cdot C_P}} & \\ \lim_{C_W \to 0}\left[0.775 \cdot \sqrt{\dfrac{\alpha_T + (\alpha_0 + k_e \cdot \lambda_0)\cdot(C_P + T_f \cdot T_{20} \cdot C_W)}{(\alpha_k + k_e \cdot \lambda_k)\cdot(C_P + \tau \cdot T_{20} \cdot C_W \cdot m_1)}}\right] = & \text{if } \beta = 0.75 \\ = 0.775 \cdot \sqrt{\dfrac{\alpha_T + (\alpha_0 + k_e \cdot \lambda_0)\cdot C_P}{(\alpha_k + k_e \cdot \lambda_k)\cdot C_P}} & \end{cases} \quad (15)$$

respective, in the case of the second characteristic can be writing:

$$K_{lim.min.2} = \begin{cases} \lim_{C_W \to \infty}\left[0.707 \cdot \sqrt{\dfrac{\alpha_T + (\alpha_0 + k_e \cdot \lambda_0)\cdot(C_P + T_f \cdot T_{20} \cdot C_W)}{(\alpha_k + k_e \cdot \lambda_k)\cdot(C_P + \tau \cdot T_{20} \cdot C_W \cdot m_1)}}\right] = & \text{if } \beta = 0.66 \\ = 0.707 \cdot \sqrt{\dfrac{(\alpha_0 + k_e \cdot \lambda_0)\cdot T_f}{(\alpha_k + k_e \cdot \lambda_k)\cdot \cdot \tau \cdot m_1}} & \\ \lim_{C_W \to \infty}\left[0.775 \cdot \sqrt{\dfrac{\alpha_T + (\alpha_0 + k_e \cdot \lambda_0)\cdot(C_P + T_f \cdot T_{20} \cdot C_W)}{(\alpha_k + k_e \cdot \lambda_k)\cdot(C_P + \tau \cdot T_{20} \cdot C_W \cdot m_1)}}\right] = & \text{if } \beta = 0.75 \\ = 0.775 \cdot \sqrt{\dfrac{(\alpha_0 + k_e \cdot \lambda_0)\cdot T_f}{(\alpha_k + k_e \cdot \lambda_k)\cdot \cdot \tau \cdot m_1}} & \end{cases} \quad (16)$$

At same the time with increasing the cost of kilowatt-hour C_W, in term of €/kWh, optimal peak load degree decrease or increase according the annual number of hour utilization at the maximum load (T_{SM}) is elder or less than $T_{M.D}$ defined by following relation:

$$T_{M.D} = 10^4 \cdot \sqrt{\frac{\tau_D}{8760}} - 1240 \quad (17)$$

If in equations (1) are considered value of β=0, results in final the following equation:

$$K_{lim.min} = \frac{S_{MI}}{S_{nT}} = \sqrt{\frac{\alpha_T + \alpha_0 \cdot (C_p + C_w \cdot T_f \cdot T_{20})}{\alpha_k \cdot (C_p + C_w \cdot \tau \cdot T_{20} \cdot m_1)}} = \sqrt{\frac{C_T + \Delta P_0 \cdot (C_p + C_w \cdot T_f \cdot T_{20})}{\Delta P_k \cdot (C_p + C_w \cdot \tau \cdot T_{20} \cdot m_1)}} \quad (18)$$

In table 1 are presented the input data and the simulation results with Matlab program for transformers analyzed. The values for $K_{lim.\,min}$ are indicate in table 1 and respective in figure 2 and values for S_{MI} is presented in figure 3. Know that the powers required of consumers supplied at a certain transformer are in permanence variation, after a certain load-duration graphic.

Table 1. Input data and the results simulation for transformers analyzed.

S_{nT} [MVA]	U_1/U_2 [kV/kV]	ΔP_0 [kW]	ΔP_k [kW]	u_k [%]	i_0 [%]	[€]	β = 0.66	β = 0.75	β = 0
								$K_{lim.min}$	
10	20/6	12.5	65	10	0.9	85000	0.51	0.59	0.75
16	110/20	18	94	11	0.8	130000	0.50	0.84	0.70

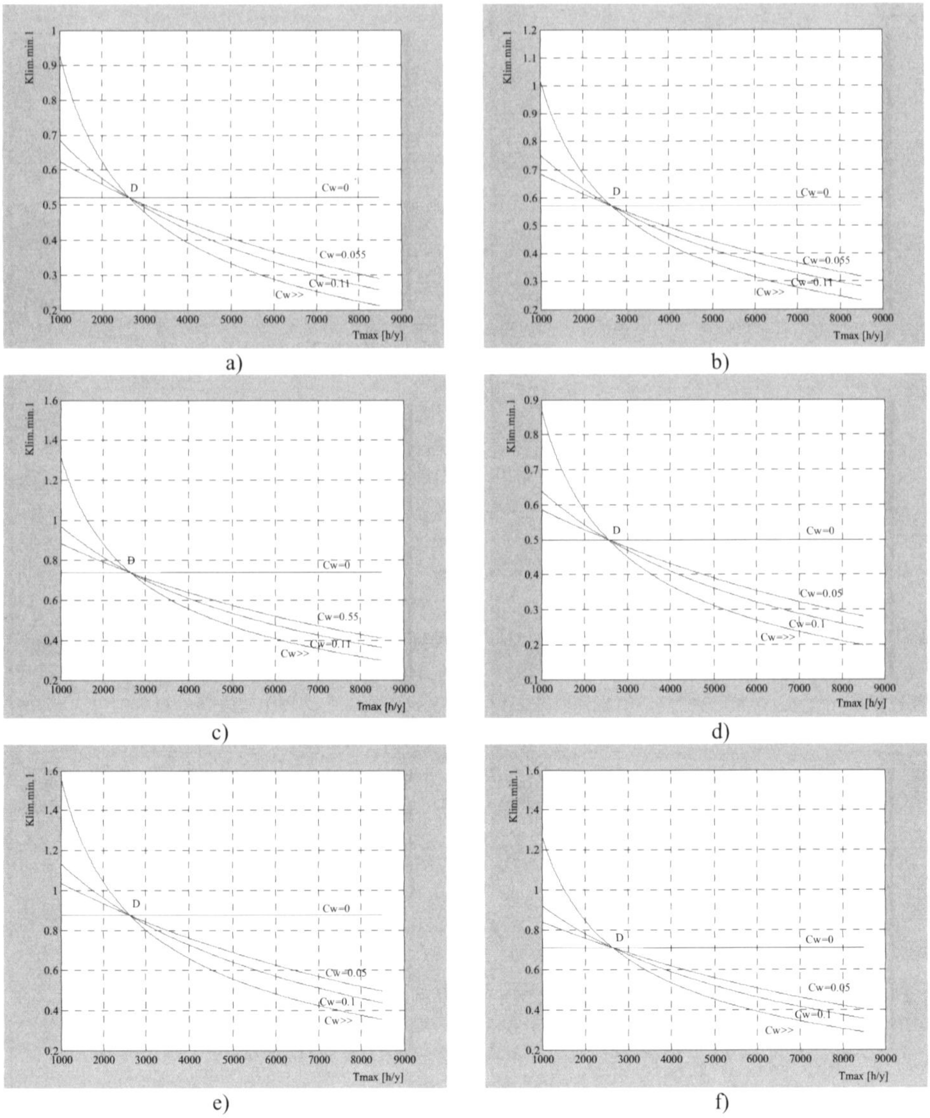

Figure 1. Economic loads from annual peak assignment determinate on the strength criteria from the present series of transformers TTUS-ONAF of HV/MV and TTU-ONAN of MV/LV, with increasing rate l=0.03

a) S_{nT} = 10 MVA, β=0.66 b) S_{nT} = 10 MVA, β=0.75
c) S_{nT} = 10 MVA, β=0 d) S_{nT} = 16 MVA, β=0.66

e) S_{nT} = 16 MVA, β=0.75 f) S_{nT} = 16 MVA, β=0.

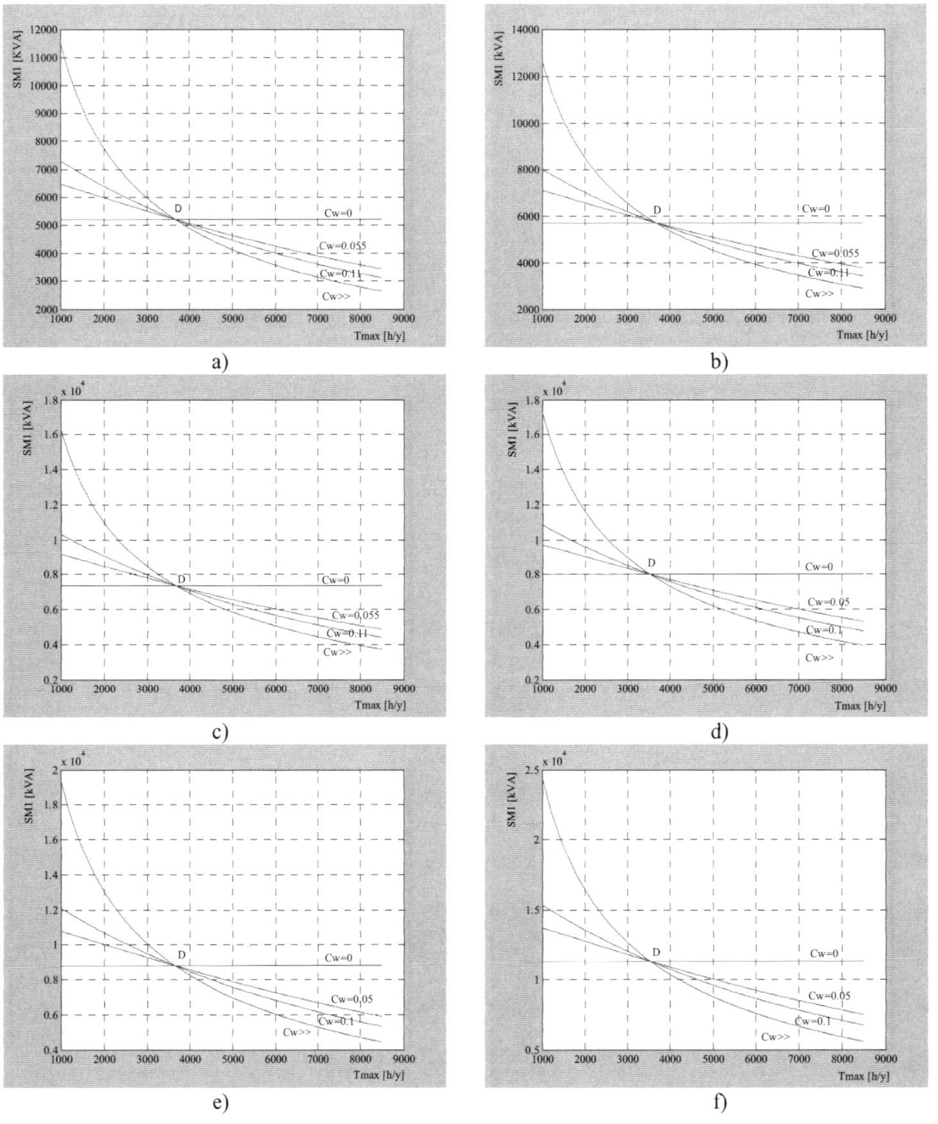

Figure 2. Optimum apparent power in the first year of exploitation from annual peak assignment determinate on the strength criteria from the present series of transformers TTUS-ONAF of HV/MV and TTU-ONAN of MV/LV, with an increasing rate l=0

a) S_{nT} = 10 MVA, β=0.66 b) S_{nT} = 10 MVA, β=0.75
c) S_{nT} = 10 MVA, β=0 d) S_{nT} = 16 MVA, β=0.66
e) S_{nT} = 16 MVA, β=0.75 f) S_{nT} = 16 MVA, β=0.

Investment costs in terms of economic of electric station, differ more enough from case at case. Value costs used in results simulation are following as:

- The medium cost of kilowatt, with which been assimilated specific cost of installed power in standard power station: C_P = 1250 €/kW
- Medium prices of energy with which are assimilated the specific costs of lost energy at the three steps of distributions are:
- At HV (110 kV) $C_w = 5.0\ 10^{-2}$ €/kWh
- At MV (6-20 kV) $C_w = 5.5\ 10^{-2}$ €/kWh
- At LV (0.4 kV) $C_w = 7.5\ 10^{-2}$ €/kWh

4. Conclusions

Load grade analysis in case of transformers made, as a rule, in many goals:

- For checking at transformer capacity, existing in exploitation, of to make face needs (out of overloading over the received at value and duration)
- For comparing the mode of operation, accomplished in relation to optimal regime of existing units
- For analysis of replacing opportunity, on economical efficiency criteria of transformer by another had different technical characteristics (nominal power, stand by loss power and at short circuit, short circuit tensions, etc.)

Strength on the of graphics presented in figure 1 are determined optimal loading coefficients at the annual peak load for an increasing annual rate equal with zero, in terms of time of utilization at peak of the loading and differently value the specific cost of kilowatt - hour of energy loss established at the electric station level.

In activity interval an electric transformer can be kept in the same station long as maximum annual loads are owned in delimited domain of demand factors.

If in time of exploitation transformer the annual maximum load decrease at minimum values ($K_{lim.min}$) is recommended in this case, as in most a year, the respective transformer is replaced with a unit of less nominal apparent power.

In analog mode are determinate the economic demand factor for actually series of transformers in oil of MV/LV.

5. References

[1] Pana L 2006 *Tehnici de optimizare in sistemele electroenergetice miniere* Editura Focus Petrosani
[2] Ionescu T G and Pop O 1998 *Ingineria sistemelor de distrbutie a energiei electrice* Editura Tehnica Bucuresti
[3] Heathcote M J 2007 *J & P Transformer Book* Newnes: Oxford UK
[4] Degeneff B C 2005 *3-Power Transformers* In the Electrical Engineering Handbook Chen W-K Ed. Academic Press: Burlington VT USA
[5] Rashid N 2011 *Short-Time Overloading of Power Transformers* Royal Institute of Technology KTH (Kungliga Tekniska Högskolan): Stockholm Sweden
[6] Willis H L 2004 *Power Distribution Planning Reference Book* CRC Press: Boca Raton FL USA
[7] Siemens AG 2014 *Transformers - Power Engineering Guide* 7th Ed Siemens Energy Sector: Erlangen Germany
[8] Amoiralis E I, Tsili M A and Georgilakis P S 2008 *The state of the art in engineering methods for transformer design and optimization: a survey* Journal of Optoelectronics and Advanced Materials 10: 1149-115
[9] Georgilakis P S 2009 *Spotlight on Modern Transformer Design* Springer London UK
[10] Guzun D B, Gal S A 2013 *Centrale, Statii si Retele Electrice-CSRE Sistemul de servivii aferent* Editura AGIR Bucuresti

IOP Conf. Series: Earth and Environmental Science **172** (2018) 012032 doi:10.1088/1755-1315/172/1/012032

USING THE GNU RADIO SOFTWARE IN THE OFDM MULTICARRIER COMMUNICATIONS

Florenţiu DELIU[1], Nicolae BADARA[1], Ovidiu CRISTEA[1]
Naval Academy "Mircea cel B trân", Constanţa

Email: florentiu.deliu@anmb.ro

Abstract: Data transmission in HF and UHF radio channels require modern types of modulations. OFDM and Multicarrier are modulation types that have been imposed on 4G and Wi-Fi data transmission channels in 900Mhz, 1800Mhz and 2400Mhz beams. In this paper we are studying the use of these modulation types to increase the data transmission speed in HF and UHF maritime bands.

1. Introduction. The SDR soft defined radio allows the use of universal hard drives for the reception and broadcasting modules for radio frequency signals. The Universal Peripheral Radio Software uses the SDR to construct the interface between the PC and the radio domain, that is, the hard interface of the interface. Future emitters and receivers will be USPR.

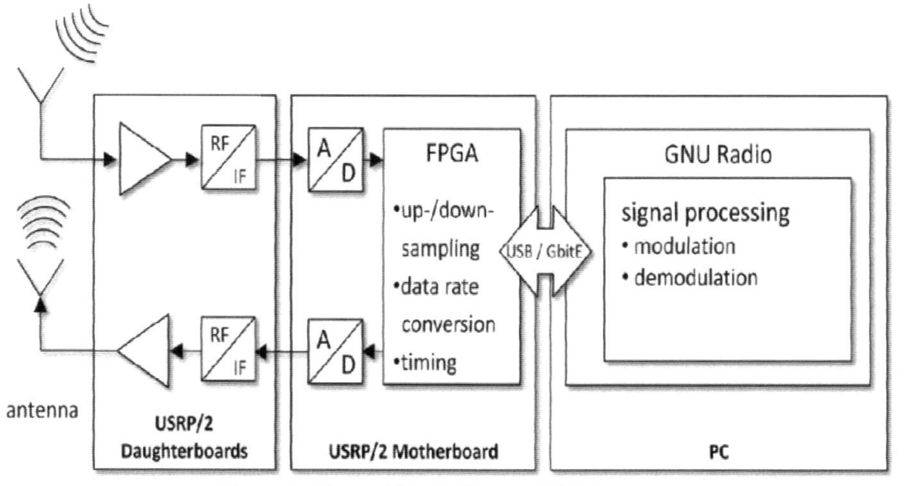

Figure 1. General block diagram USRP

GNU Radio is an open source programming environment that schedules these interfaces. Another programming environment is LabView, but it is not open source, and the license price is for the time

IOP Conf. Series: Earth and Environmental Science **172** (2018) 012032 doi:10.1088/1755-1315/172/1/012032

being prohibitive. But they are very productive USRP manufacturers like Ettus, National Instruments whose products supports both environments. To achieve the OFDM modulation transceiver we use NI USPR 2901 with parametric techniques:

Transmitter:

Frequency range 70 MHz to 6 GHz
Frequency step <1 kHz
Maximum output power (Pout) 20 dBm
Gain range1 89.75 dB
Gain step 0.25 dB
Frequency accuracy2 2.5 ppm
Maximum instantaneous real-time
bandwidth
56 MHz
Maximum I/Q rate3
Streaming4 15 MS/s
Burst
One channel 61.44 MS/s
Two channels 30.72 MS/s
Digital-to-analog converter (DAC)5 12 bits

Receiver:

Frequency range 70 MHz to 6 GHz
Frequency step <1 kHz
Gain range6 76 dB
Gain step 1.0 dB

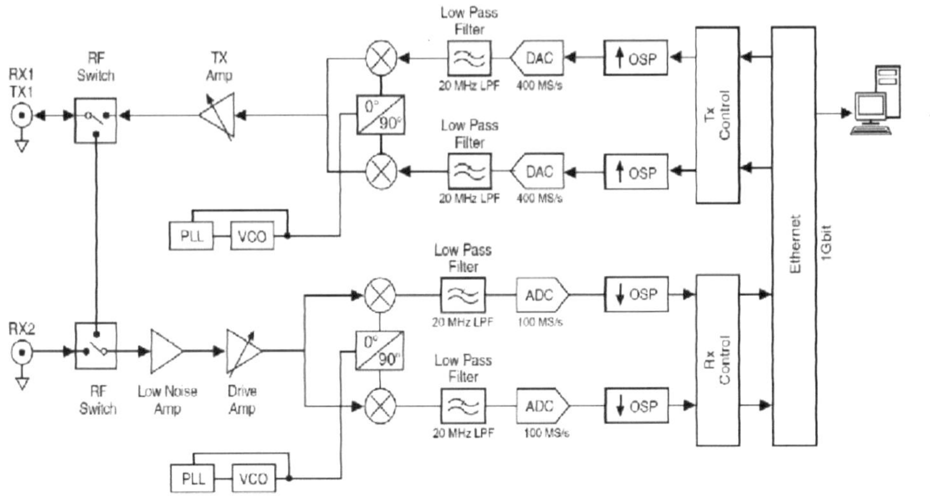

Figure 2. Block diagram NI USPR 2901

The basic principle of OFDM is to divide data at a high rate from a data source into N lower rate streams. These flows are PSK or QAM modulated. These streams are simultaneously transmitted on N orthogonal subparts by summation to obtain an OFDM symbol. Mathematically, if X [n] n = 0, 1 ,. . . , N - 1, represents the complex input of the subpopulation n, N is the number of sub - carriers, and T the duration of the symbol can be written as:

$$X[n] = \sum_{k=0}^{N-1} x(K)e^{J\frac{2\pi}{N}} = \sum_{k=0}^{N-1} x(k)\left[\left(c \quad \frac{2\pi}{N}\right) + j\!: \left(c \quad \frac{2\pi}{N}\right)\right] \tag{1}$$

for n=0,1.........N-1

The receiver performs the inverse operation from the emission, X [k] for k = 0, ...N. The signal is filtered, converted to digital signal using an analogue-to-digital converter (A / D). Sample time stream serial stream is converted to parallel streams by the S / P converter, DFT is used to convert time domain data into the frequency range:

$$X(k) = \frac{1}{n}\sum_{n=0}^{N-1} x(n)e^{-J\frac{2\pi}{N}} = \frac{1}{N}\sum_{n=0}^{N-1} x(n)\left[\left(c \quad \frac{2\pi}{N}\right) - j\!: \left(c \quad \frac{2\pi}{N}\right)\right] \tag{2}$$

for k=0,1......N-1

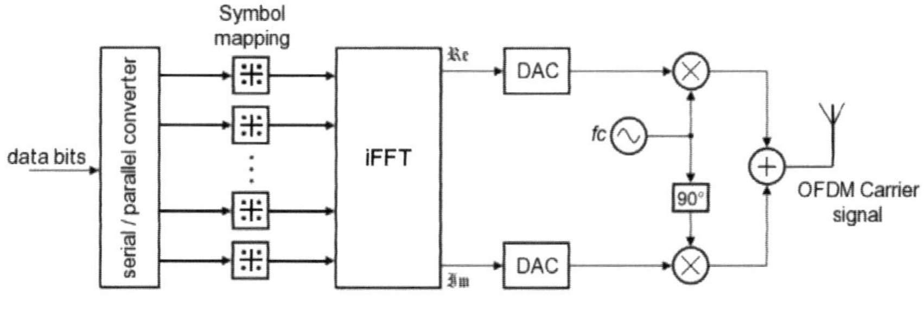

Figure 3. OFDM demodulator modulator diagram

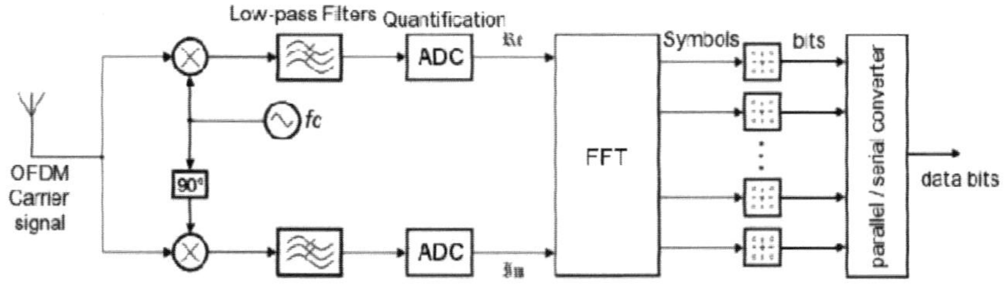

Figure 4. Implementing OFDM Transmitter with Functional Blocks GNU Radio.

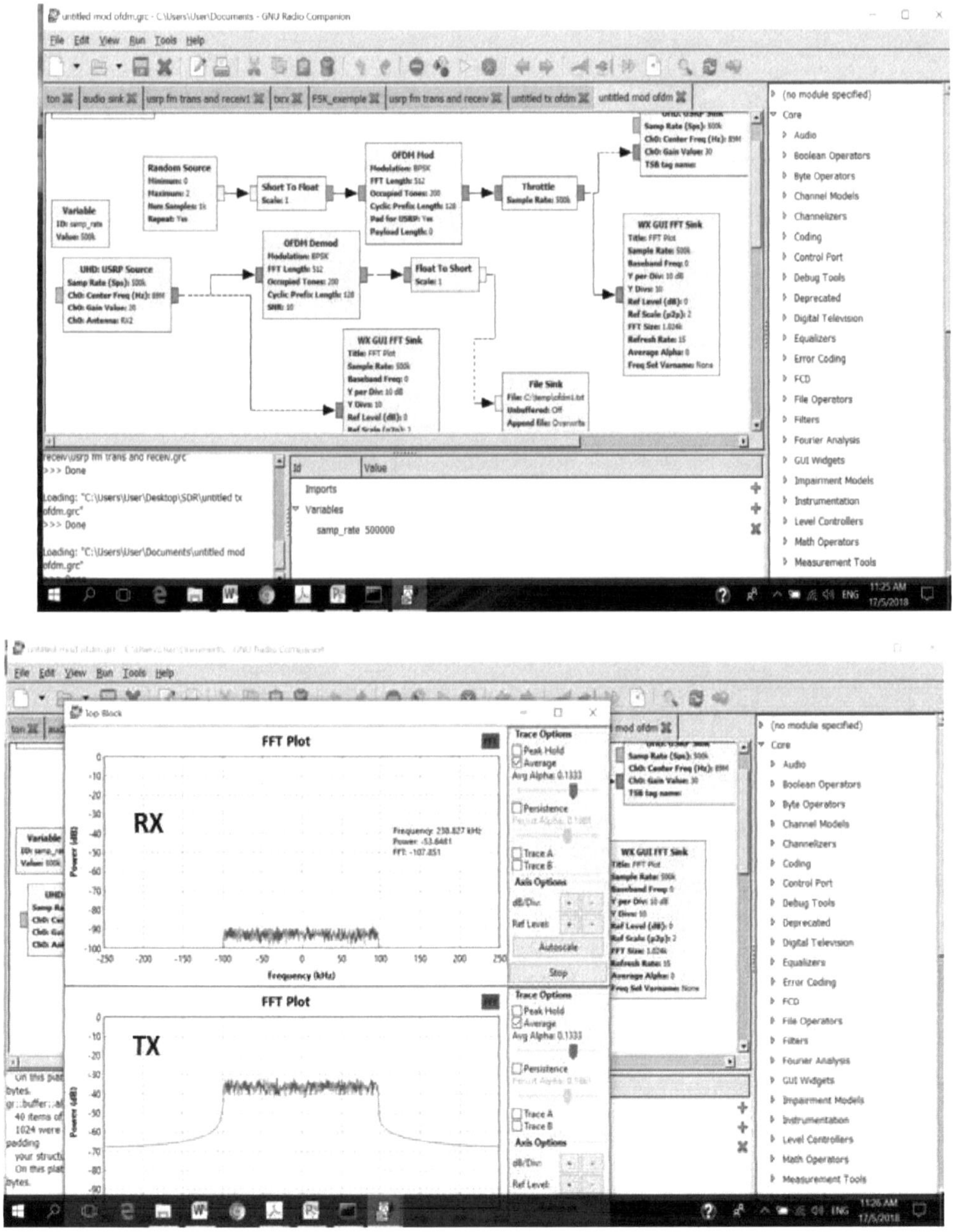

Figure 4. Implementing the OFDM receiver with GNU Radio functional blocks.

Conclusions

Experiments conducted with the NI USRP-2901 and the GRU RADIO platform for the development of an OFDM modulated communications channel have shown that using these low-cost devices with an OPEN SOURCE platform can be a technical solution for many applications in the field of maritime and terrestrial communications.

Random Source from the transmitter scheme randomly selected 1K binary values that were transmitted in the 89 MHZ frequency. The OFDM modulator uses 200 tones and 128 Cyclic Prefix. In the sink file in the receiver were written 1K binary values. During the study there were no communication errors. The study can be deployed in communications networks in the VHF maritime band, shipboard, presenting main advantage increase stability in terms of propagation and decoding errors.

References

1] S. Karande and P. N. Kota, 2017, "Analytical performance of OFDM transceiver on SDR and GNU radio-companion," *International Conference on Communication and Signal Processing (ICCSP)* , pp. 0001–0005.

[2] C. D. R. Rodriguez, G. P. Leguizamon, and C. S. Fajardo, 2017 , "Processing threshold in an IEEE 802.11a/g/p receiver over GNU radio: A fuzzy logic application," in *IEEE Symposium Series on Computational Intelligence (SSCI)*, pp. 1–8.

[3] J. Talukdar, B. Mehta, K. Aggrawal, and M. Kamani, 2017 , "Implementation of SNR estimation based energy detection on USRP and GNU radio for cognitive radio networks," in *2017 International Conference on Wireless Communications, Signal Processing and Networking (WiSPNET)* , pp. 304–308.

[4] K. Aggrawal, M. Kamani, and K. Vachhani, 2017 ,"Analysis of GSM air interface using DVB-T receiver and GNU radio," in *International Conference on Trends in Electronics and Informatics (ICEI)*, pp. 635–640.

[5] G. Swetha and B. N. Bhandari, 2017, "Energy detection spectrum sensing on DPSK modulation transceiver using GNU radio," in *2nd International Conference for Convergence in Technology (I2CT)*, 2017, pp. 974–978.

[6] www.ni.com/documentation/specs-uspr-2901

IOP Conf. Series: Earth and Environmental Science **172** (2018) 012033 doi:10.1088/1755-1315/172/1/012033

Comparison of two thermodynamic combustion models

A Sabau

Maritime University of Constanta, 104, Mircea cel Batran, St.9000663, Constanta, RO
adrian.sabau@cmu-edu.eu

Abstract. The paper aims at making a comparison between two models of the process of thermodynamic combustion, the fuel rate model and the Wiebe's function model, used for the prediction of the engine performance. The paper shows a practical method that enables a basic calibration of the constants encountered in the mathematical models when there are no concrete experimental data for the specified engines, but there are data for similar engines. The models have opposite starting hypotheses in the frame of the combustion process modelling and there is the advantage of a large number of well documented case studies in the reference literature. Both models use the Annand and Pflaum approximations for the heat release assessment, that have particular relations for the piston, the linear cylinder and the cylinder cover. The chemical mechanism use stoichiometric reactions taking into account the dissociation.

1. Introduction

A major problem of the design stage of the compression ignition engine is represented by the prediction of the engine performance. This task is difficult to accomplish since the number of parameters is rather large and the links between them are non-linear. The same happens to all the other processes that take place in the engine, and especially with the combustion process which is the most complex of them all due to the large number of phenomena that take place simultaneously.

Many combustion models have been developed, most of them being very complex since they require a large number of parameters. Consequently, it becomes impossible to apply this type of model in the early stage of the design because most of the initial data are not known.

The paper presents a theoretical model of the entire cycle for a compression ignition engine by using two combustion models. For this model the author has developed an original computing code, the main feature being the minimum number of the initial parameters taken into consideration.

The theoretical approaches are based on the models of thermodynamic combustion and all other processes are treated in a global manner so that the entire cycle can be actually simulated.

2. Theoretical backgrounds

The inlet process is described using the First Law of Thermodynamics for an open system, the continuity equation and the equations of the pressure of steady flow from orifices.

These equations were solved using different approaches in the flow equation and thermodynamics constants. Finally we can obtain pressure p_{in} and temperature T_{in} at the theoretical end of the inlet process:

$$p_{in} = \frac{p_s}{\sigma}\left(1 - \frac{n^2}{1800\frac{n_a}{n_a - 1}\frac{RT_s}{M_{aer}}} \frac{\left(\frac{\varepsilon - \mu\rho}{\varepsilon - 1}\right)^2}{\phi^2\left(fm\_Vs \cdot n\right)^2}\right)^{\frac{n_a}{n_a - 1}}, \tag{1}$$

$$T_{in} = T_s \frac{p_{in}}{p_s} \frac{\varepsilon}{\varepsilon - 1} \left[1 + \frac{\rho p_{out} T_s \nu \, p_{in}}{p_s^2 T_{out} (\varepsilon - 1)^2 \theta \, n_a} \cdot \left(\varepsilon + (n_a - 1)(\varepsilon - 1)\sigma - \frac{p_{out}}{p_{in}\rho} \right) \right], \tag{2}$$

T_s ; p_s inlet air parameters, T_{out} ; p_{out} exhaust gas parameters, ε compression ratio, θ ratio of temperature increase, ρ scavenge coefficient, ϕ debit and velocity coefficient flow in inlet valve, fm_V_s ratio of section time and piston displacement, $\mu = \frac{p_{out}}{p_{in}} \frac{T_{in}}{T_{out}}$, ν coefficient of inertial inlet, n_a isentropic index, n speed engine [1] .

The compression process may be calculated using the first law of thermodynamics for a closed system.

In order to calculate the changes for p and T, we discretize the volume $dV = V_{j+1} - V_j$ and for every small step we ar use the First Law of Thermodynamics. For this approach we can apply the First Law of Thermodynamics in the form:

$$dQ = \Delta E + dW, \tag{3}$$

where dQ is the heat transfer to the cylinder walls.

If the cylinder wall temperature is T_w and the exposed surface area is A, then the heat transfer is:

$$\begin{aligned} \frac{dQ_t}{dt} &= \alpha_c A(T_{mj} - T_w) + \alpha_r A(T_{mj}^4 - T_w^4) \quad \left[\frac{j}{s}\right] \\ \frac{dQ_t}{d\alpha} &= \left[\alpha_c A(T_{mj} - T_w) + \alpha_r A(T_{mj}^4 - T_w^4)\right]\frac{1}{6n} \quad \left[\frac{j}{\deg}\right] \end{aligned}, \tag{4}$$

α_c convective heat transfer coefficient, α_r radiative heat transfer coefficient.

Anand, Pflaum, Woschni, Eichelberg et. al proposed formula for heat transfer. We used Annand and Pflaum relations [2].

$$\text{Annand:} \begin{cases} \alpha_c = a \dfrac{k}{D} \mathrm{Re}^b \\ \alpha_r = c \end{cases}, \tag{5}$$

$$\text{Pflaum:} \begin{cases} \alpha_c = (k_1 + k_2 \dfrac{p_s}{p_{in}})(k_3 + k_4 e^{k_5 \overline{w_p}})\sqrt{pT} \\ \alpha_r = \sigma_B c_r \end{cases}, \tag{6}$$

Re Reylonds number, K thermal conductivity, $\overline{w_p}$ mean piston velocity, σ_B Stefan Boltzman constant, k_1,.. k_5, c_r coefficients of Pflaum formulae regarding different area type, a, b, c coefficients of Annand's formula.

For a step of time/angle the pressure change from p_j to p_{j+1}. If the volume change is small, the "work" term can be given approximately by

$$dW = pdV = \frac{p_j + p_{j+1}}{2} \left(V_{j+1} - V_j \right). \tag{7}$$

For the volume we used the approaches:

$$V_j = \frac{V_s}{\varepsilon - 1} + \frac{V_s}{2}\left[\left(1-\cos(\alpha_j)\right) + \frac{\lambda}{4}\left(1-\cos(2\alpha_j)\right)\right], \tag{8}$$

which are the first two terms of a Taylor series, where, $\lambda = \frac{S}{2L}$, S and L is the piston stroke and the connecting rod length.

Equation (3) cannot be solved analytically and a numerical solution must be sought. In principle this requires the estimation of T_{j+1} to obtain p_{j+1} at which step of time/angle:

$$p_{j+1} = \frac{V_j}{V_{j+1}} \frac{T_{j+1}}{T_j} p_j . \tag{9}$$

This pressure is substituted in the rearranged equations of the First Law:

$$e = \Delta E + dW - dQ_t . \tag{10}$$

A solution is obtained when $e \to 0\,(|e| < .001)$. The numerical scheme used is the Newton-Raphson method. In this method if $(T_{j+1})_{q-1}$ is the estimated value of T_{j+1} this has the expression [3]:

$$\left(T_{j+1}\right)_q = \left(T_{j+1}\right)_{q-1} - \frac{e_{q-1}}{MCv[(T_{j+1})_{q-1}]} . \tag{11}$$

For a volume change dV the conditions at moment j the beginning of the step will be equal to the conditions at the state $j+1$ at end of the previous volume step. The first estimate of T_{j+1} may be obtained by assuming an isentropic change from the state conditions at T_j thus:

$$T_{j+1} = T_j \left(\frac{V_j}{V_{j+1}}\right)^{\frac{\Re}{Cv(T_j)}} . \tag{12}$$

The internal energy $E(T_j)$ and the specific heats $Cv(T_j)$ are calculated used the gas composition and temperature (we used a polynomial IV degree expression).

The injection, ignition and expansion, the most important processes in the engine cycle will be treated simultaneously.

The ignition reaction was modelled using the stoichiometric equation:

$$\begin{cases} C + O_2 = CO_2 \\ H_2 + \frac{1}{2} O_2 = H_2O , \\ S + O_2 = SO_2 \end{cases} \tag{13}$$

and reaction of CO_2 dissociation,

$$\delta CO_2 \rightarrow \delta CO + \frac{\delta}{2} O_2, \tag{14}$$

where $K_{Pco}\sqrt{\frac{p}{v}} = \frac{1-\delta}{\delta\left(\frac{1}{2}\delta\right)^{\frac{1}{2}}}$, and $\delta = \sqrt[3]{\frac{2}{K_{Pco}^2}\frac{v}{p}}$, or $\delta = \delta^{'}\sqrt[3]{\frac{v}{p}}$, using : $\begin{cases} v = v + \frac{1}{2}v \cong 1 \\ \delta^{'} \cong 10^{-7}(T-1400)^2 \end{cases}$ [3].

The mass of CO obtained from dissociation is too small and we considered its mass as being negligible but it is very important for heat balance when the temperature raises over 1400 K.

$$dQ_d = \delta \cdot dm_{CO_2} \cdot Q_{CO} \text{ [j]}, \tag{15}$$

where Q_{CO}=279231 [kj/kg].

The fuel was approximated as a chemical pure substance having the formula: $C_nH_mO_r$. This approach allowed us to calculate the products of combustion according to the reaction:

$$C_nH_mO_r + (n+\frac{m}{4}-\frac{r}{2})O_2 + \frac{79}{21}(n+\frac{m}{4}-\frac{r}{2})N_2 \Rightarrow nCO_2 + \frac{m}{2}H_2O + \frac{79}{21}(n+\frac{m}{4}-\frac{r}{2})N_2. \tag{16}$$

The ignition delay time was calculated with the formula proposed by Baert [4]

$$\tau_{aa} = \tau_{ao} + k_1 p^{-n1} \exp\left(\frac{E}{\Re T}\right) + k_2 p^{-n2} \text{ [ms]}, \tag{17}$$

where $\tau_{ao} = 0.36$, k_1=1.154x10^{-4}, $k2$=2550, $n1$=0.75, $n2$=2.45, $E/\Re = 9495$.

The injection law considered was a linear one, but supplementary parabolic and polynomial (second degree) have been taken into account.

The fuel instantaneous vaporisation has also been treated only from the energetic point of view:

$$dQ_v = \left[Lv + c_l(T_s - T_l)\right]dm_{inj} \text{ [j]}, \tag{18}$$

whit Lv latent vaporisation heat; c_l specific heat; T_s vaporisation temperature; T_l temperature of fuel.

During the combustion period, the heat release was obtained by using two models:

- the Wiebe functions model,
- the fuel rate Whitehouse-Way model.

The first model proposes an empirical formula for the rate of heat release based on the analysis of the experimental rate of heat release diagram according to the expression:

$$\frac{dQ_b}{d\alpha} = a_p \frac{\Delta Q_{pr}}{\Delta \alpha_p}\left(M_{pr}+1\right)\left(\frac{\alpha}{\Delta\alpha_{pr}}\right)^{M_{pr}} e^{-a_p\left(\frac{\alpha}{\Delta\alpha_{pr}}\right)^{M_{pr}+1}} + a_d \frac{\Delta Q_d}{\Delta\alpha_d}\left(M_d+1\right)\left(\frac{\alpha}{\Delta\alpha_d}\right)^{M_d} e^{-a_d\left(\frac{\alpha}{\Delta\alpha_d}\right)^{M_d+1}}. \tag{19}$$

For the automatic calibration of the constants we used experimental correlations [5]. This expression takes into account dependencies of constants to the timing injection, considering that, this is the most important parameter. Usually $a_p=a_d=6.9$.

The numeric calculation expressed heat release in a small step according to the formula:

$$dQc=(dQ_b d\alpha)m_{inj}\ [\mathrm{j}], \tag{20}$$

This formula permits the heat release calculation for any time step of the calculus.

The second method calculates heat release using two expressions for the fuel preparation and reaction rates [6] .

The fuel preparation rate is given by:

$$\frac{dm_{pr}}{d\alpha}=km_{inj}^{1-x}m_u^x p_{O_2}^m\ \ [\mathrm{kg/deg}], \tag{21}$$

the reaction rate by:

$$\frac{dm_b}{d\alpha}=\frac{60k' p_{O_2}}{n\sqrt{T}}e^{\frac{-E}{T}}\int\left(dm_{pr}-dm_b\right)\ [\mathrm{kg/deg}]. \tag{22}$$

m_{inj} [kg] total mass of injected fuel, m_u [kg] mass of unburned fuel; $m_u=m_{inj}-\int(dm_{pr})d\alpha$, k and m constants, p_{O_2} [bar] partial pressure of oxygen, E [j] activation energy, k' constant, n engine speed.

The numeric calculations using the fuel rate model consider that:

$$(m_{inj})_{i-1}=\sum_{1}^{n-1}\frac{dm_{inj}}{d\alpha}\Delta\alpha\ . \tag{23}$$

Similarly, the total fuel prepared and burnt in the cylinder is given by:

$$(m_{pr})_{i-1}=\sum_{1}^{i-1}dm_{pr}\Delta\alpha\ , \tag{24}$$

$$(m_b)_{i-1}=\sum_{1}^{i-1}dm_b\Delta\alpha\ . \tag{25}$$

In principle average values of m_{inj} and m_u for step would be the logical values, but this demands a knowledge of the conditions at the end of the step, which can only be obtained by an iterative procedure. If the change in the values during the step is small enough, relative to the absolute values, which is usually true for the step size used in these calculations, the initial values could be used.

For the first step of the process, when m_{inj} and m_u are very small, this condition is not valid and use of initial values alone would underestimate preparation rate. For this reason, as m_{inj} is usually known and specified as data, the final value of m_{inj} for the step may be used [6].

Then in equation (21):

$$m_{inj}=(m_{inj})_i \text{ and } m_u=(m_{inj})_i-(m_{pr})_{i-1}\ , \tag{26}$$

gives the preparation rate during the step as:

$$\left(\frac{dm_{pr}}{d\alpha}\right)_i = k(m_{inj})_i^{1-x}\left[(m_{inj})_i - (m_{pr})_{i-1}\right]^x p_{O_2}^m \text{ [kg/deg]}, \tag{27}$$

and the total fuel prepared

$$(m_{pr})_i = (m_{pr})_{i-1} + \left(\frac{dm_{pr}}{d\alpha}\right)_i \Delta\alpha \text{ kg]}. \tag{28}$$

For the fuel burnt during this step we use the reaction rate equation (22)

$$\frac{dm_b}{d\alpha} = \frac{60k' p_{O_2}}{n\sqrt{T}} e^{\frac{-E}{T}}\left[(m_{inj})_i - (m_b)_{i-1}\right] \text{ [kg/deg]}. \tag{29}$$

The fuel burnt including that burnt in the step under consideration is now:

$$(m_b)_i = (m_b)_{i-1} + \left(\frac{dm_b}{d\alpha}\right)_i \Delta\alpha \text{ [kg]}. \tag{30}$$

If $(m_b)_i < (m_{pr})_i$the burning rate is lower due to the conditions in cylinder, then the fuel consumed in the step is assumed to be:

$$dm_c = (m_b)_i \Delta\alpha \text{ [kg]}. \tag{31}$$

If $(m_b)_i > (m_{pr})_i$, the burning rate is higher than the assumed rate, the process is controlled by the rate of preparation and fuel consumed in the step is:

$$dm_c = (m_{pr})_i \Delta\alpha \text{ [kg]}. \tag{32}$$

Using the mass of fuel consumed in each step of the cycle calculation we may determinate the value of the heat release dQ_c:

$$dQ_c = dm_c Q_i \text{ [j]}, \tag{33}$$

where Q_i is the lower specific heat value of the fuel.

The first law of thermodynamics for combustion and expansion process in accord whit the rearranged formula is:

$$e = \left[E_{j+1}(T_{j+1}) - E_{j+1}(T_s)\right] - \left[E_j(T_j) - E_j(T_s)\right] + dW - dQ_c - dQ_t - dQ_v - dQ_d, \tag{34}$$

where T_s is the reference temperature for the chemical reaction and the others terms in the equation has the form previously expressed.

A solution is obtained when $e \to 0(|e| < .001)$. The numerical method used for solving the first law expression is Newton-Raphson. According to this method if $(T_{j+1})_{q-1}$ is the estimated value of T_{j+1} it is possible to conduct an iterative scheme to solve the equations and this is done by the expression:

$$\left(T_{j+1}\right)_q=\left(T_{j+1}\right)_{q-1}-\frac{e_{q-1}}{MCv[(T_{j+1})_{q-1}]}. \tag{35}$$

In this case the pressure and the temperature at the end of the step have the following formula:

$$p_{j+1}=\frac{m_{j+1}}{m_j}\frac{V_j}{V_{j+1}}\frac{T_{j+1}}{T_j}p_j, \tag{36}$$

$$T_{j+1}=T_j\left(\frac{V_j}{V_{j+1}}\right)^{\frac{\Re}{Cv(T_j)}}+\frac{dQ_c}{m_jC_v\left(T_j\right)}. \tag{37}$$

The expansion stroke was solved simultaneously with the combustion process as a continuation of the last one with $dQ_c=0$.

The exhaust process was globally treated and the corresponding pressure was constant (the initial parameter). The temperature of the residual gases which was initially approximated was calculated simulating the expansion as a polytropic process with a constant exponential coefficient, ending at the inlet pressure. The temperature was corrected until the imposed error was less then 1%.

Mechanical efficiency is a very important parameter for the estimation of the engine performance. The lost friction power was modelled based on Millington & Hartles and Whinterbone & Tenant's expresions [7].

Approaches for cycle calculation. The inlet and exhaust strokes have 180 revolution degrees and an imposed pressure value sets as initial data. The injection timing is initially set and has a constant value. The computation has as goal to obtain the engine dimensions according to the imposed engine power and the correction of the necessary residual gas temperature or to estimate the output power for the imposed engine dimensions. The optimization consists in variating the initial parameters so that in the end we obtain the performance indicators.

Numerical simulation was done by using an original code developed in Matlab MatWorks environment. Matlab has an impressive library of functions for every part of the simulation program: mathematical functions, functions for solving transcendent equations or differential equations, functions for manipulating data and for graphical representations. The code is developed in an open architecture, only the initial data imput has a minimal interface for data protection and flexibility. The rest of the program is in fact a collection of functions called by a main routine.

The data result is not filtered and for this reason it is mandatory to interpret results with additional attention. It is possible to obtain a non-physical value duet o the numerical algorithm and cut-off errors. To facilitate the data comparison between the result values and the experimental data the code has a library of interpolation function.[8].

3. Results

Results of the cycle calculations are presented using the diagrams below. Figure 1 is shown the cycle diagram (p(α)) for a four stroke direct injection AT 25 Sulzer engine.

In Figure 2 to 6 we present comparative results calculation based on Wiebe's function combustion model and fuel rate Whitehouse-Way's model by using Annand's formula for the heat transfer .

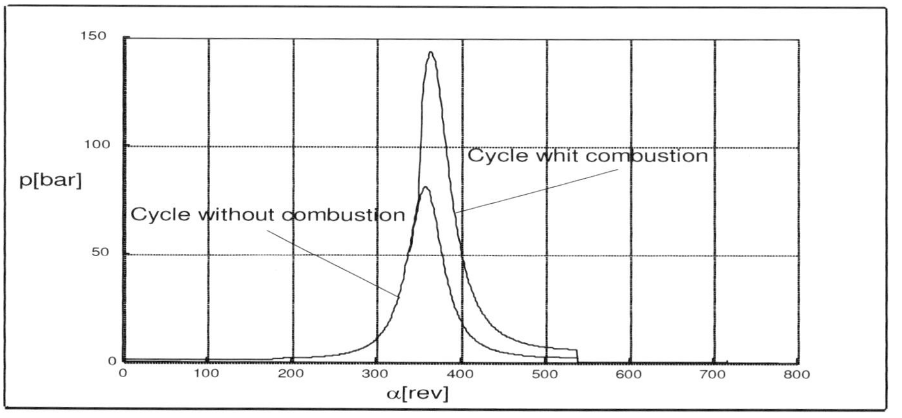

Figure 1 p(α) diagram

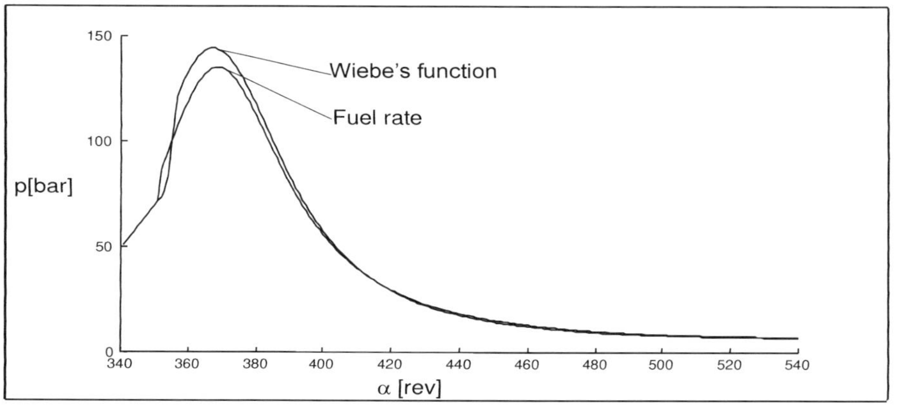

Figure 2. p(α) combustion and expansion

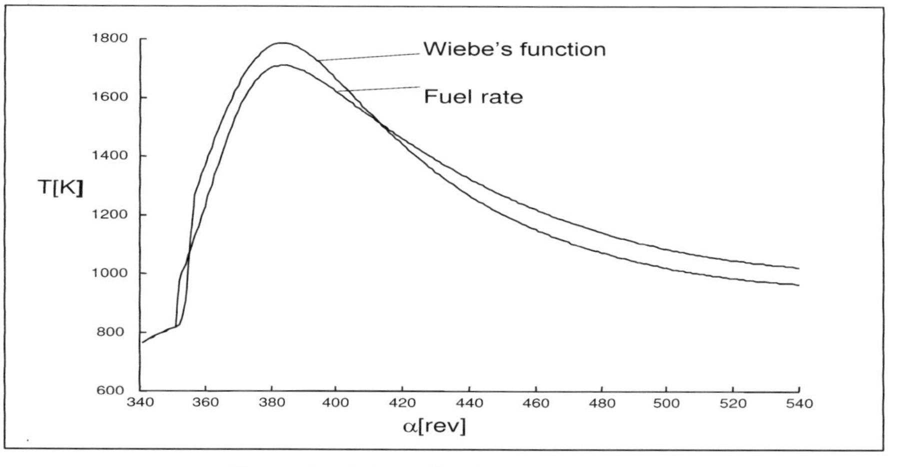

Figure 3. T(α) combustion and expansion

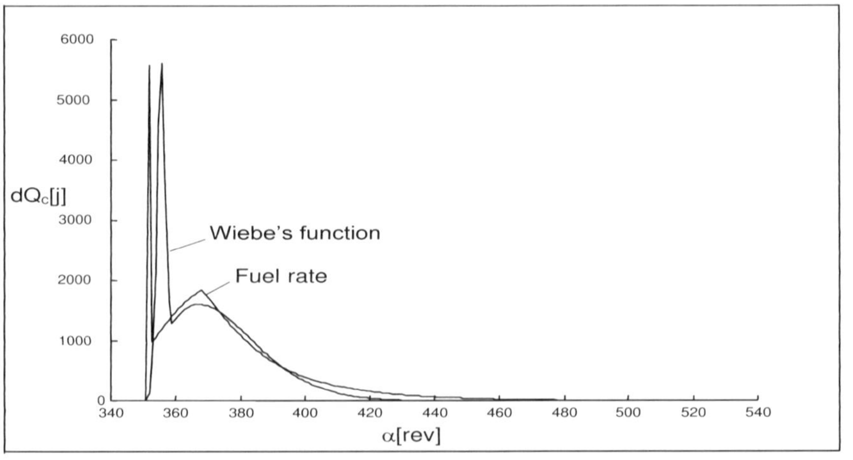

Figure 4. Heat release

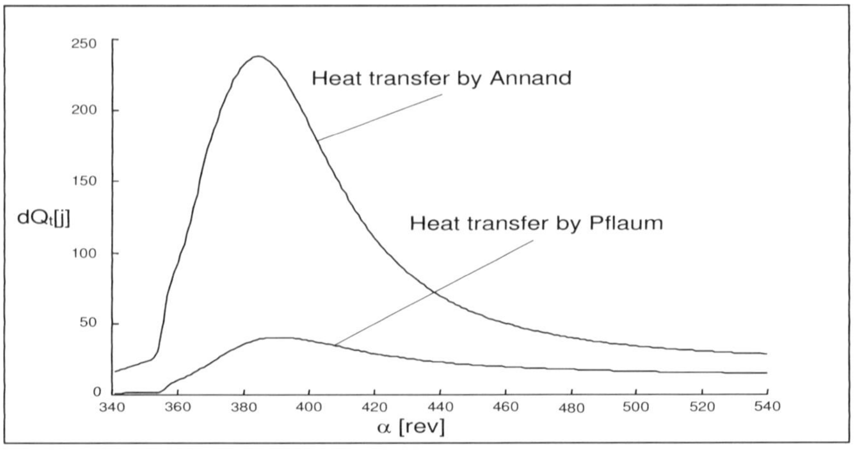

Figure 5. Heat transfer

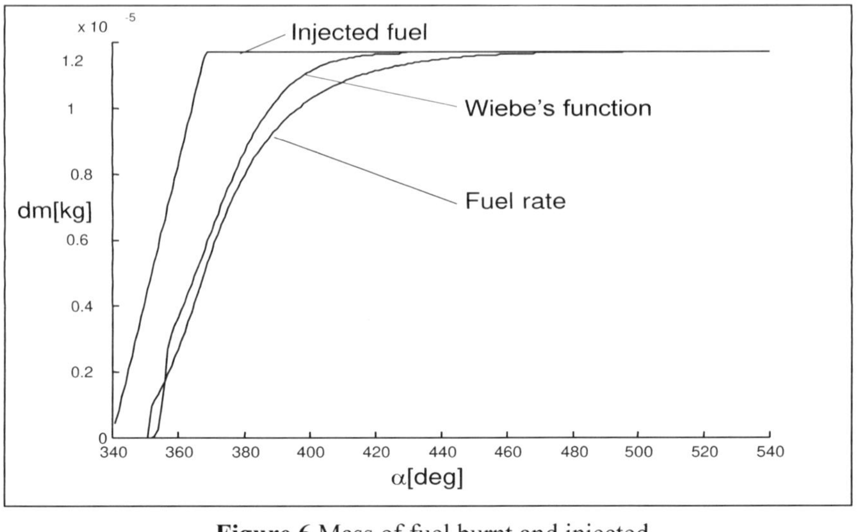

Figure 6 Mass of fuel burnt and injected

The heat transfer pattern for both methods Annand and Pflaum formulations is presented in Figure 5. Pflaum formula gives a small heat transfer rate and, consequently, not satisfactory results. This is the reason why we used only Annand's formula for the calculus.

The code results were verified whit experimental data p(α) diagram at 1000 rev/min and 100% (nominal regime). The real indicated diagram (Figures 7 and 8) was available for this checking which was compared with the computed indicated diagram.

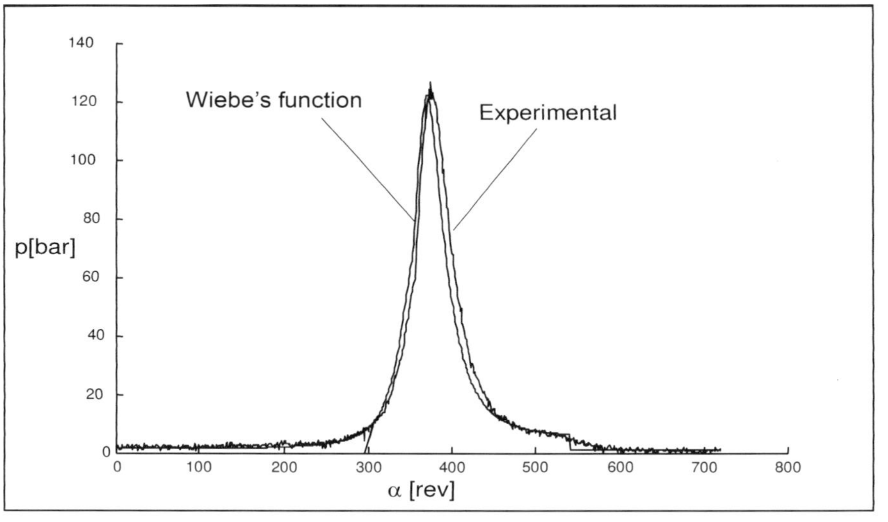

Figure 7. Indicated Wiebe's function and experimental diagram.

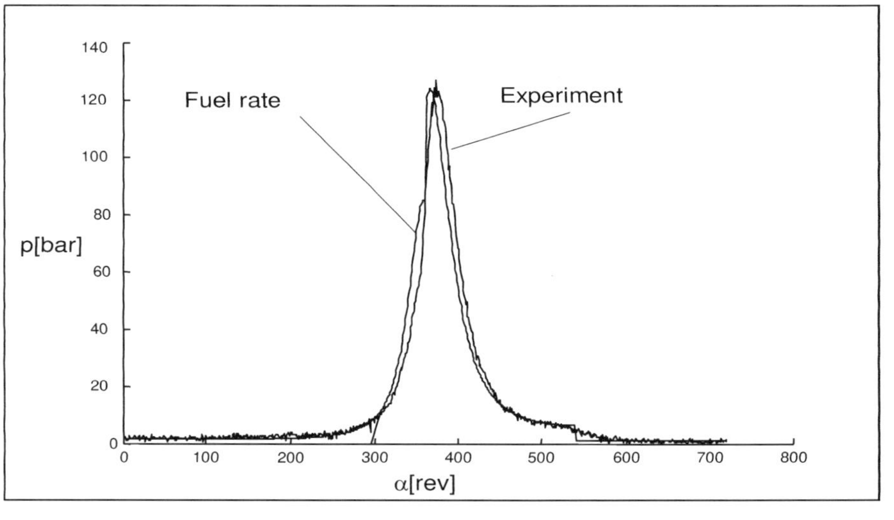

Figure 8. Indicated fuel rate and experimental diagram.

The numerical calculus with automatic calibrated Wiebe's function offered a satisfactory match between the diagram patterns. We must underline on the moderate differences of the amplitudes. The percentage errors are bot maximum 2% for pressure, 1% for power and 1.5% the fuel consumption.

For the fuel rate model the result is not so good, since the percentage errors are of maximum 4.3% for pressure, 2% for power and 1.5% for the fuel consumption. In addition, the shape of the diagram is slightly different and the pressure peak is not in the same position.

IOP Conf. Series: Earth and Environmental Science **172** (2018) 012033 doi:10.1088/1755-1315/172/1/012033

4. Conclusions

The models used for the combustion process are enough accurate for the assessment of the performance parameter (power, efficiency and fuel consumption) for a compression ignition engine at nominal regime.

The Wiebe's function combustion model is more flexible than the fuel rate combustion model and offers a higher accuracy. The flexibility and accuracy of this model is based on a larger number of constants that must be set up as initial data. Instead the calibration of constants is a very difficult task. With a good calibration the model may offer good predictability for regimes close to the nominal one. By using Wiebe's function and proper experimental data it is possible to model any regime, but with limited accuracy and predictability.

The fuel rate combustion model has a smaller number of constants and it offers a quick solution, but with lower precision. Fuel rate models have very limited flexibility and predictability.

The code obtained from the present study is a useful tool in the early stages of design. It has the capacity to predict the running parameters and performance indicators of the compression ignition engine. The precision of the calculus depends on the initial data and can be improved if there are calibration data for that particular engine.

Even if the numerical results are accurate enough it is not possible to immediately select the final construction solution. We can only make a global evaluation and indicate the optimization directions which should be followed because the code involves many simplified hypotheses, empirical constants in the mathematical expressions and a minimum of initial parameters (these are good for the design stage but prove insufficient for a final optimal solution).

The present theoretical approaches offer appropriate solutions for the nominal regime but only satisfactory solutions for the other engines. This observation underlines on the fact that some initial parameters are not constant and are not optimal for the partial engine loads.

References

[1] Arama, C. Grunwald B. 1966 *Motoare cu ardere interna. Procese si caracteristici* (Editura Tehnica Bucuresti)
[2] Benson R.S. Whitehouse N.D. 1983 *Internal combustion engine* (Pergamon Press London)
[3] Buzbuchi N Manea L Dinescu C 1997 *Motoare navale. Procese si caracteristici* (Editura Didactica si Pedagogica Bucuresti)
[4] Dugal, V.K. Kuoo T.W. Mukerjee A., Singhal A.K. 1985 Three-dimensional modeling of in cylinder processes in DL Diesel Engines (*SAE)* Paper 840227
[5] Heywood J.B. 1988 *Internal Combustion Engine Fundamantals* (McGraw-Hill Book Company)
[6] Ramos J.I. 1989 *Internal Combustion Engines Modeling* (Hemisphere Publishing Corporation, New York)
[7] Winterbone, D.E., Tennant, D.W.H. 1981, The Variation of Friction and Combustion Rates During Diesel Engine Transients (*SAE)* Paper 810339
[8] Oanta E. Danisor A. Tamas R. 2016 *Study regarding the spline interpolation accuracy of the experimentally acquired data* (*ATOM VIII Constanta*) 1001007 doi: 10.1117/12.2242996

IOP Conf. Series: Earth and Environmental Science **172** (2018) 012034 doi:10.1088/1755-1315/172/1/012034

Unmanned Surface Vessel for Marine Data Acquisition

Vasile Dobref[1], I Popa[2], P Popov[3] and I C Scurtu[4]

[1]Professor PhD eng , Naval Academy "Mircea cel Bătrân", Constanța, RO
[2]Lecturer PhD eng , Naval Academy "Mircea cel Bătrân", Constanța, RO
[3]Lecturer PhD, Naval Academy "Mircea cel Bătrân", Constanța, RO
[4]Lecturer PhD eng , Naval Academy "Mircea cel Bătrân", Constanța, RO

E-mail: petrica.popov@anmb.ro

Abstract. This paper brings to the fore, the construction of a remote controlled naval platform, without crew on board, and its possibilities to transmit or record data in inaccessible areas or in marine areas in which human life would be endangered. Although these unmanned platforms are not a novelty, they being used mostly for military, the solution presented is a Romanian premiere, similar to international standards, but at much lower cost, with the advantage that certain environmental action is totally insignificant. It was presented only a range of applications (video data, forecasts and submarines), the platform built in the "Mircea cel Batran" Naval Academy, allowing many improvements based on equipment that can be mounted on board.

1. Introduction

Depending on the objectives, many of the techniques and technologies that are currently used for marine survey, are causing irreversible damage to the natural environment of the current ecosystems. Regarding at the use of USV (Unmanned Surface Vehicle) vehicles type, unmanned, remote controlled or guided on imposed routes, they have the advantage of the human absence, taking information that may be relevant as research data [1].

The remote-controlled vehicles have been built since 1940 and were used in military operations for demining marine areas. Unmanned vehicles have also been used for the environmental research, and for taking water samples after the release of the famous radioactive atomic bombs.

The main features of USV are:

- Autonomy: the ability to operate over a long period of time in order to extend the investigated area, this can be extended by using non-conventional energies.
- Reduced risk: avoiding endangering or loss of human lives and damage the picturesque of sea.
- Displacement: due to the small USV size this can slip easily through natural or artificial obstacles.
- Economy: low power consumption and reduced number of maintenance and operation team, assigns to USV an advantage in terms of human resource and energy costs.
- Profile: USV is operating at sea with low acoustic and electromagnetic signature, which represents a minor risk for the marine environment.

The most important feature of a USV is that the operator remains in a safe environment while the vehicle fulfils its mission in various environments with high risk, even hostile or inaccessible for the large ships.

The domains serviced by USV's can be classified as [2]:

1. Education and Science, with an emphasis on education (applications and practice), research and de-velopment, oceanography (marine research and data transmission, collecting water samples and biological), ecology, etc.

2. Emergency services, related to police operations (search for evidence, observation of interest, monitoring of polluted areas), border police operations (observation, search and rescue, inspection of unidentified objects etc.).

3. Military, the main missions are: research and observation of navigation districts; action against marine mines (the rapid establishment of a safe transit zone for the fleet by surveying the area and destruction of possible mines with missiles and mines launched from USV), antiterrorism (the capacity to observe, query and even destroy the unidentified boats), antisubmarine (USV have performing equipment on board radiolocation and hydrolocation able to discover and track submarines etc.).

Existing autonomous vehicle models are developed in the presence, as follows:

- Autonomous surface vehicle Cat C-4 (Figure 1a)-developed to take samples of water and assess the aquatic environment on coastal zones [3].

- Autonomous surface vehicle WAM-V (Figure 1b). It was built in 2013 by the Office of Naval Research USA and has been used by Google to map the coast waterfront of San Francisco. It can be mounted / dismounted in 7 parts of two people in less than 3 minutes [4].

- Autonomous surface vehicle Rafael Protector. The vehicle is also part of a group of autonomous vehicles in the US Navy (Figure 1c), with the difference that is larger and is intended to be included in some combat missions. Rafael Protector USV is a high-speed craft, built with a rigid and resistant body. It is a modular vehicle that has on board weapons system and can be adapted to different types of missions such as: surveillance, reconnaissance, antiterrorism and electronic warfare [5].

b.

a. c.

Figure 1 a. C-Cat 4 b. WAM-V c. Rafael Protector

2. Elements of design for Marine Engineering Faculty - USV 01

Table The design elements that give mission and performance requirements of USV were generated by potential product capabilities. This elements, also, provides the basis for the design and construction of on board systems. The requirements of the mission provide first information about tasks that will be

IOP Conf. Series: Earth and Environmental Science **172** (2018) 012034 doi:10.1088/1755-1315/172/1/012034

fulfilled launching the USV. Performance requirements are derived from USV's specific missions and establish the capabilities for meeting the requirements envisaged (speed, maneuverability, endurance, autonomy, etc.). These requirements are the basis of vehicle characteristics development and associated systems, while being essential elements for performing the analysis of cost-effectiveness and that must allow further development with knowledge evolution in the field.

To obtain these items, we conducted an experimental model that we tested by various methods of transmitting data and various modern integrated systems related to propulsion and maneuverability.

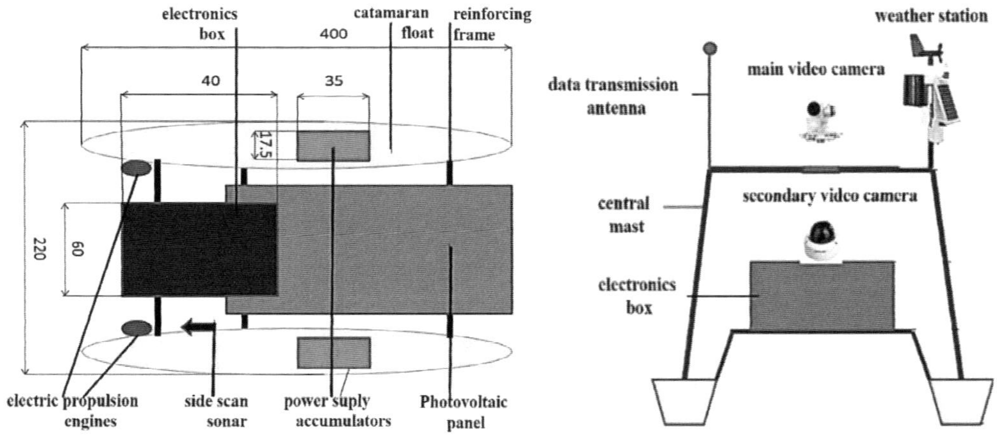

Figure 2 Marine Engineering Faculty-USV 01

The experimental model is a representative for an autonomous vehicle surface, in order to investigate experiment and demonstrate the possibility of building a USV Class X (Figure 2). Design and construction of the prototype was done in the Research Laboratory of the Faculty of Marine Engineering in the Naval Academy in Constanta, name Marine Engineering Faculty-USV 01.

USV's body was designed as a hydrodynamic body, composed of a lightweight material, namely glass fiber reinforced with aluminum structure. In order to achieve this body it was necessary to use a very light material for reduction of the total mass and obtaining a positive value of buoyancy, also, taking into account the weight of all equipment necessary to accomplish missions. The design of the body that revealed USV's capabilities was established as catamaran boat type with two identical immersed bodies, called "twin-hulls' houses muonted on basic structure represented by a rigid alumium frame. The choice of the following body were taken into account the benefits, specific to "twin-hulls" bodies, as follows:

• the catamaran can sail in shallow waters due to smaller draft, the displacement of USV being distributed between the two hulls;

• space for equipment and renewable energy systems is more generous compared to other types of boats;

• has a high stability against the waves from all directions, very difficult sunk;

• aerodynamic and hydrodynamic shape lead to low drag resistance, thus the USV can reach great speeds relative to another type of boat at the same size.

IOP Conf. Series: Earth and Environmental Science **172** (2018) 012034 doi:10.1088/1755-1315/172/1/012034

The main element to be considered for body design was the study of "optimum distance" between immersed hulls, because this type of structure can create an own waves system, affecting structural strength during operation and can also considerably increase the advancing resistance of the vessel.

Technical features designed for generic platform, are:

- Length: 4[m];
- Body weight: 60 [kg];
- Propelling force: two electric motors- 2 x 18[daN];
- Displacement- 130 [kg], max 200 [kg];
- Autonomy – 3 h;
- Draught - 0.11 [m];

The starting point in building the catamaran was made up of two kayak bodies fiberglass profiled, representing hull type "twin-hull" wich are secured by rigid structure of aluminum bars.

The stability calculations of USV's it was done by calculation, which showed that this type of platform is an optimal solution for autonomous naval USV 01 FIM platform. The diagram on the right hull catamaran floating in calm water is shown in Figure 3. The straight hull diagram of catamaran floating, for calm waters, was based on the values calculated in the Autoship program.

The sizes in the diagram have the following meanings:

Awj - surface of floating area;

XFj- the abscissa of the geometric center of the floating surface;

Ilj– momentum of longitudinal inertia;

Itj - momentum of transverse inertia;

CWj – coefficient of float surface.

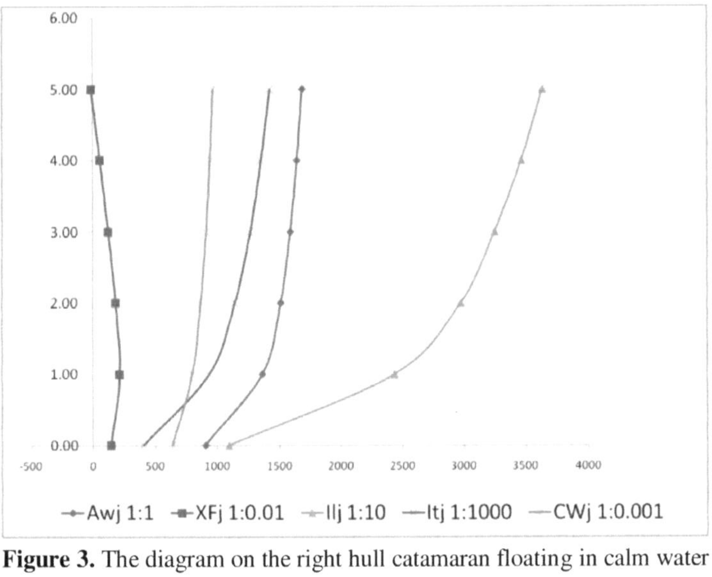

Figure 3. The diagram on the right hull catamaran floating in calm water

The straight hull diagram and stability calculation of the ship have demonstrate the special nautical qualities, proving that it can safely sail even under unfavorable condition.

The total displacement of the vessel:

$$\Delta = \Delta_{ng} + \sum m_i = 130[\text{kg}] \tag{1}$$

Where: Δ_{ng} is displacement of the vessel without load and $\sum m_i$ is embarked masses of the vessel.

Using ρ – sea water density, it was calculated V – hulls, and from straight hull diagram was established the vessel draft at this volume of hulls, following to extract Il and $\overline{KB}$ (the center line of straight hulls) terms for appropriate floating.

$$V = \frac{\Delta}{\rho} = 126000\ [cm^3] \quad (2)$$

$$\overline{KB} = 6[cm]$$

$$Il_{(cat)} = 402000\ [cm^4]$$

Centre of gravity is calculated using the formula:

$$\overline{KG} = \frac{KG^* \cdot \Delta_{ng} + \sum_{i=1}^{n} m_i \cdot z_i}{\Delta_{ng} + \sum_{i=1}^{n} m_i} = 2{,}6\ [cm] \quad (3)$$

For calculating the center of gravity quota, it was specified the following size:

- metacentric radius cross:

$$\overline{BM_T} = \frac{Il}{V} = 3{,}19\ [cm] \quad (4)$$

- - transverse metacenter height:

$$\overline{KM_T} = \overline{BM_T} + \overline{KB} = 9{,}19\ [cm] \quad (5)$$

- - transverse metacentric height:

$$\overline{GM_T} = \overline{KM_T} - \overline{KG} = \overline{BM_T} + \overline{KB} - \overline{KG} = 6{,}59\ [cm] \quad (6)$$

- Reports between dimensions:
  - The ratio between length L and theoretical width B is: L / B = 1.8. This value corresponds to high stability and low speed for single hull vessel, for "twin hulls" catamaran type we have a good speed compared to boats of the same size.
  - The theoretical ratio of the height H and draft d is: H / d = 22.8. This report demonstrates that the USV has a high stability.
  - The ratio of the theoretical length L and the height H is: L / H = 2.5. This value is indicative for a high longitudinal strength.

In conclusion, this size ratios demonstrates the main advantages of catamaran type vessel: stability, ability to keep the sea and good maneuverability. The practical achievement of such a body is shown in **Figure 4.**

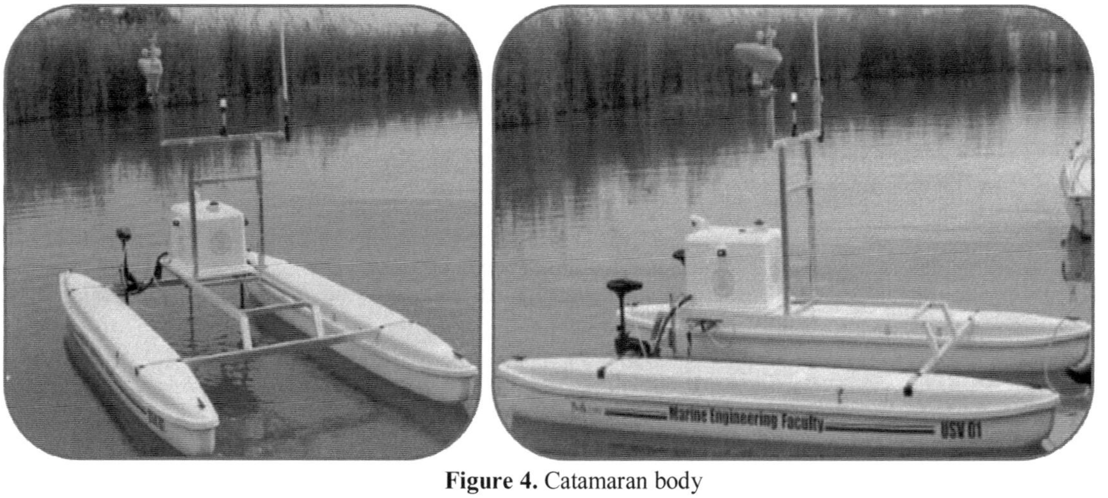

Figure 4. Catamaran body

The propulsion system is intended to move the vehicle forward and backward. Governance body is also included in this scheme and propulsion government belong to the same system, type AZIPOD (azimuthal thrusters).

Under this system are designed to drive contained the following equipment:

- two electric motors Minn Kota Endura C2 New;
- two brushless DC Drive Controller;
- two Futaba receiver 6-J;

The chosen electric motor type was Minn Kota Endura C2, an improved version of the model C1, which develops a thrust of up to 18.1 [kg], the electric power being achieved from 12 [V] alkaline accumulators. The energy autonomy based on chosen accumulators is about 3 hours at a speed of 4 [Nd].

Brushless DC (Drive Controller) is an electronic circuit used in order to vary the speed of the electric motor and to act as a dynamic brake.

Regardless of the used type, a DC, generally, interprets control information not as mechanical movement, such in the case of an actuator, but rather in a way that varies the switching speed of field effect transistors (FET). This allows a variation much smoother and more precise of the motor speed in a more efficient manner than the mechanical type with resistive coil.

The propulsion command is done by the command post of the craft and serves to transmit commands to the thrusters and actuators (Figure 5). The link from the command post to craft, which is designed to transmit commands to the thrusters and actuators vehicle is running with a portable transmitter type Futaba 6EX. The propulsion system has installed a Futaba receiver (Figure 5) for receiving radio signals transmitted by the console when USV is running in manual mode.

Futaba 6EX is a very simple but very powerful remote control of motors and actuators and has both transmitter and receiver own. The operating frequency is programmable from 2.4 [GHz] to 6 channels. The range of the system can be reached in good weather conditions with no obstacles between antennas to 3-4 [km].

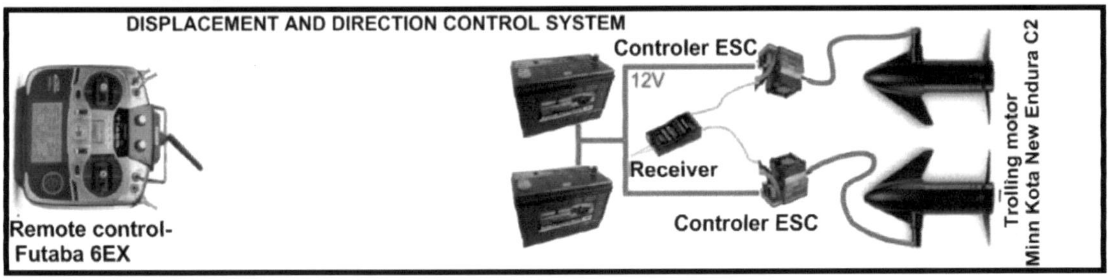

Figure 5. The propulsion command

3. Sensors and data acquisition equipment

Following the analysis, which it was based on data volume and speed transmission, it was concluded that the best connection for data transmission is the radio link in 2.4 [GHz] frequency spectrum (figure 6).

A critical factor limiting range of this connection is closely related to EIRP parameter (Equivalent Isotropically Radiated Power) - radiated power density of an ideal lossless antenna in the direction of maximum gain. Our legislation limits the value of EIRP, on amateurs' regime, at 100 [mW] equivalent to 20 [dBl] maximum gain.

The chosen equipment for making the radio link between the vehicle and the shore station is METAL 2SHPn router that serves to transmit all the data collected by sensors at the post command.

This wireless device, waterproof, operates in 2.4[GHz] frequency. In junction with a very large gain antenna, the working distance can reach distances of 30-40 [km]. We use a 6 [dBI] antenna which will give a range of 6-7 [km] to meet our operational requirements.

In accordance with the mission that USV have to perform the autonomous platform was equipped with various equipment and sensors, as follows:

- ***Camcorders***. The video received from USV's board is captured using two cameras. Main camera that aims mainly carrying out the reconnaissance and surveillance of the vehicle and has a 3-axis freedom of motion, being able to focus on a specific object image. The secondary camera has two axes freedom of motion and the main role is to provide information about navigation status and any obstacles that vehicle encounters on his way.
- ***Side Scan Sonar***. Sonar equipment which gives the underwater images, near the vehicle. It aims to discover navigation hazards and to discover the potential targets for military or civilian parts, like: marine mine, enemy divers, underwater vehicles or small groups of fish. The vehicle is equipped with a Raymarine Wi-Fish sonar type with adobe Down Vision. This small sonar is used more fisheries, but has a very good application on our missions and requirements
- ***Weather Station*** WS-1440-IP OBSERVER. USV-fitted weather station is a small station sizes, powered by three rechargeable AAA batteries, using solar energy. This station has the capability to measure 15 different parameters, of which the most important are: temperature, moisture, atmospheric pressure, wind direction, wind speed, solar radiation, etc.
- ***GPS*** – UBX-G7020 module. The receiver onboard autonomous vehicle is the MOD-GPS type. The device which receives both the GPS satellite system and the GLONASS provides a precision of 10-15m. The receiver is integrated into the microprocessor on board that transmits by the radio link the received position to be displayed on the display from the shore command station.

The current position will be displayed on the ENC (Electronic Nautical Chart) so that the operator will have a clear picture about the position, direction and speed of a moving vehicle.

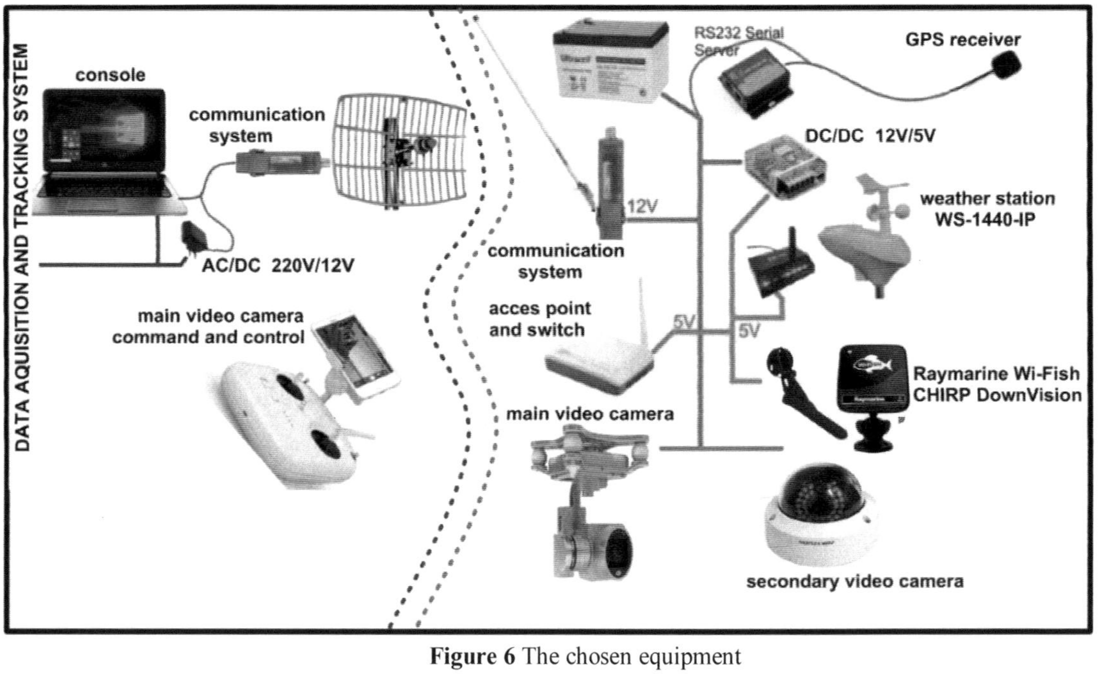

Figure 6 The chosen equipment

4. Conclusion

The remote-controlled surface vehicles represents a very good solution with a very low pollution impact on the environment and marine picturesque. Beginning with the coming years, we have faith that we have to diversify research capabilities for USV. Autonomous surface vehicles will provide additional information, providing better surveillance especially on the open sea, where the loss of human life is unacceptable, but also inland. Through its very small draft, it can operate in areas where it is impossible to send a manned craft. It can also be used in emergency situations such as monitoring the areas contaminated by radiological, chemical or nuclear agents. Another advantage is the short time for operating compared to a manned ship, including their increased speed and maneuverability.

This autonomous platform may have an important contribution for maintaining and carrying out the daily tasks, such as monitoring compliance with maritime laws, collect information and data necessary to identify vessels with illegal activities (piracy, drug traffic, illegal immigration), fisheries and pollution monitoring potential for unattended areas. The many type of mission it can perform, indicate multiple ways of product development. In this respect, existing equipment may also include additionally: video cameras (standard infrared thermal imagers), navigation lights, sensors, radar, sonar, and different types of weapons, both lethal and non-lethal.

The main difficulty faced by USV developers is a challenging physical environment. The vehicle is subjected to the influence of sea currents, wind, waves, temperature and corrosion. Communications are in some cases intermittent and data from sensors are provides with considerable delay. Deployment, operation and recovery of the vehicle are time consuming and expensive, and its dynamic control is a challenge for the operator. Also propulsion is expensive and sometimes slow and operability period is limited by environmental conditions.

Regarding the potential risk in terms of marine environment pollution this is virtually nonexistent, USV's operation is done without been resented by marine surface or underwater ecosystems.

References

[1] Dobref V., Popa I., Popov P., Sintea S., *Study for functional model of autonomous surface naval platforms,* Navy Sectorial Program for Research and Innovation Proiect no.151/2016, "Mircea cel Batran" Naval Academy, Constanta, Romania 2016.

[2] J. Lambert and J. Manley, "*Development of Unmanned Maritime Vehicle (UMV) Standards, An Evolving Trend,*" Sea Technology, Vol. 48, No. 12, December 2007

[3] *** http://asvglobal.com/product/c-cat-4/

[4] Bogart F.; Djapic V.; Galdorisi G.; Prijic C.; Roche J.; Gututala K.; Swigert B.; Dong W.- *Heterogeneous Autonomous Mobile Maritime Expeditionary Robots: Maritime Information Dominance* Naval Engineers Journal, Volume 126, Number 4, 1 December 2014, pp. 87-91(5)

[5] Thomas J. Pastore, Ronald T. Kessel *Unmanned surface vessels for surface and subsurface threats in harbours: background and practical lessons,* UDT Europe, Undersea Defence Technology Europe, Cannes, 9-11 June 2009

[6] Bidoae I., Sârbu N., Chirică I., Ionaş O - *Guidelines ship design theory*, University of Galaţi, Romania, 1986

IOP Conf. Series: Earth and Environmental Science **172** (2018) 012035 doi:10.1088/1755-1315/172/1/012035

Load Profile of Typical Residential Buildings in Bulgaria

Nicola Mihaylov, Boris Evstatiev, Seher Kadirova, Tzvetelin Gueorguiev, Tsvetelina Georgieva, Aleksandar Evtimov

8 Studentska Str., University of Ruse Angel Kanchev, 7017 Ruse, Bulgaria
mihailov@uni-ruse.bg

Abstract. In the current market design, the increasing use of renewable energy sources for electricity generation leads to new challenges in balancing supply and demand. While households are responsible for 29% of total electricity demand in Europe, a good understanding of their consumption and load profiles is missing. Households have become one of the most crucial factors shaping the management of developments towards sustainability. The development and implementation of effective policies for promoting energy efficiency in the household sector has been an emerging target of the EU. Recent analyses of Bulgarian households show this type of housing as the most statistically significant variable to impact electricity savings. This study deals with the statistical analysis of electric consumption. Statistical evaluations of electricity consumption in a typical apartment building have been obtained in the research. Mathematical models of the investigated loads have been developed. The research is based on the periodic results and the object of investigation is a typical residential block of flats in Ruse, Bulgaria. The average daily energy consumption during the weekdays and weekends is presented and analysed in the paper.

1. Introduction

After the socio-economic changes in Bulgaria, there have been significant variations in the structure of electricity consumption. The energy consumption by households has gradually increased greenhouse gas concentrations in the atmosphere and is considered the main source of global warming. Because of global warming and the accompanying climate change, conserving electricity is imperative. Clearly, even a small reduction in the energy consumption of buildings can have appreciable economic and ecological impacts for the society [1]. Under the Energy Efficiency Directive, all EU countries are required to use energy more efficiently at all stages of the energy chain, from production to final consumption. On 30 November 2016 the Commission has proposed an update to the Energy Efficiency Directive, including a new 30% energy efficiency target for 2030, and measures to update the Directive to make sure the new target is met [2]. These documents have also been transposed into national documents (different laws, programs, etc.) Preliminary studies show that the number of studies related to the above mentioned problems is limited, making it difficult to achieve the desired goals.

The development and implementation of effective policies for promoting energy efficiency in the household sector has been one of the main targets of the EU [3]. Energy sustainability and environmental preservation have become worldwide concerns with the many performances of climate change and the continually increasing demand for energy [4].

Energy consumption is increasing in all sectors due to the rapid industrial development and population growth. The potential for household energy efficiency improvements is still high. Households, as well as the people living there, are very different and one of the important factors is to assess a consumer's behavioural aspects [5]. Usually, efficiency is defined as the actions that generate the highest outputs for a definite set of resources [6]. Various studies have been done in the field of impact of user behaviour on the total amount of energy consumption in the households [7].

Nowadays, the operation of various systems and infrastructure is based on the use of electricity, which suggests that we could not imagine life without electricity. Unfortunately, in practice, even very high efficiency of electrical equipment does not guarantee expected savings. Often people do not know or do not want to change anything in their daily routine. Inefficient use of electricity probably will lead to even more inevitable negative consequences in the future. Therefore, increased attention should be paid on finding new innovative solutions for increasing the consumers' motivation to save energy. Smart meters are contributing not only to optimizing the operation of electricity systems, but also lead to a socio-economic platform for the promotion of interactions between suppliers and consumers [5].

In Bulgaria the proportion of households and services in final energy consumption reached 37.4 % in the recent years. Since most of the energy in these two sectors is spent on heating, the influence of the characteristics of the buildings in Bulgaria on the level of energy efficiency also increases. In the future, this impact will be commensurable with the impact of the energy intensive industries. Bulgarian households fail to reduce significantly their energy consumption, despite rising energy prices and the slow growth in their income. They continue to improve their energy comfort at the expense of satisfying other necessities. If the trend of faster growth in energy prices compared to the income of households remains the same and if they do not receive any external support to improve the characteristics of buildings and appliances, it can be expected that the energy comfort of Bulgarian households will decline in the context of already reduced expenses to meet other needs. Reduced consumption of liquid fuels is caused by reduced use of private cars, due to the rapid increase in oil prices. Practically, this means fewer opportunities to travel and necessity to consider practical alternatives to preserve the mobility of Bulgarian citizens [8].

Bibliographic studies have shown that a number of educational and research organizations have been actively involved in the management of electricity consumption in public utilities. But due to differences in climatic features, the heat supply systems and the culture of behaviour in individual countries, it is not possible to use directly the results obtained for the purposes of the efficient energy consumption in Bulgaria. Therefore, specialized studies for investigation of the dynamics of electricity consumption are required. Based on this data statistical assessment of the results can be implemented.

The aim of the paper is to investigate the typical load profile of residential buildings and customers in Bulgaria. This would allow to reliably analyse the investments in renewable energy sources, energy efficiency measures, etc.

2. Materials and methods

The method used to assess the load profile of a block of flats, is explained bellow. Firstly, for each day of the year the cumulative energy consumption at the nth hour of the day E_n, is generated according to:

$$E_n = \sum_{k=1}^{N} E_k^{hour}$$

where N is the number of apartments in the block of flats, and E_k^{hour} is the energy consumption of the kth apartment for a certain hour of the day.

Then the average energy consumption for every month of the year E_n^{mnt} is obtained for each hour of the day:

$$E_n^{mnt} = \sum_{l=1}^{28,29,30\ or\ 31} E_l \Big/ 28,29,39\ or\ 31$$

The experimental data is further analyzed through dividing it into weekdays and weekends. The average daily energy consumption E_n^{avg} throughout the day is estimated by using the formula:

$$E_n^{avg} = \sum_{m=1}^{K} E_m \Big/ K$$

where K is the number of weekdays or weekend days and n is again the hour of the day. The confidence interval is assessed for normal distribution with a 95% probability, based on the variation throughout the different months.

Finally, the normalized load profile for each hour of the day is estimated according to:

$$E_n^{norm} = 24.\sum_{n=1}^{K} E_n^{avg} \Big/ \sum_{p=0}^{23} E_p^{avg}$$

where p varies between 0 and 23 and represents the different hours of the day. The normalized load profile allows to visualize it on a scale whose average is 1. Such representation allows to easily apply the obtained dependencies for different consumers by multiplying them by the investigated average consumption.

3. Results and discussion

The object of the investigation is a typical block of flats in Ruse, Bulgaria. It consists of two entrances, each with 6 floors and 24 apartments, which is owned by the Municipality of Ruse. Since 2013 all the apartments have electric meters whose readings are being automatically stored in a database on hourly basis. The block is not connected to the central heating system of the town, so during the winter months the heating is done mainly by using electrical energy.

The collected data for the past years is used in the present study to analyse the load profile. The average monthly load profile for each month of the year is presented in Figures 1 and 2. It has a minimum in May, June and September and the maximum values are reached during the winter months from December to March.

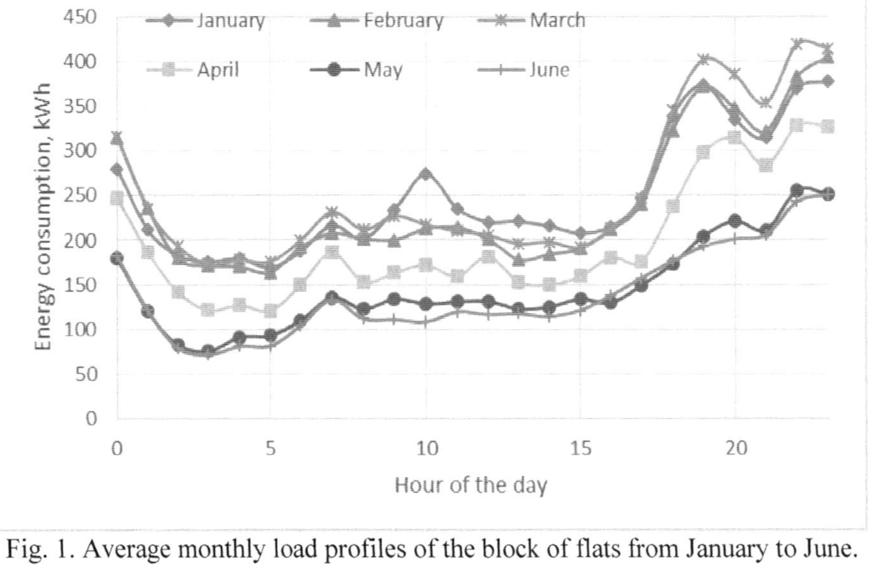

Fig. 1. Average monthly load profiles of the block of flats from January to June.

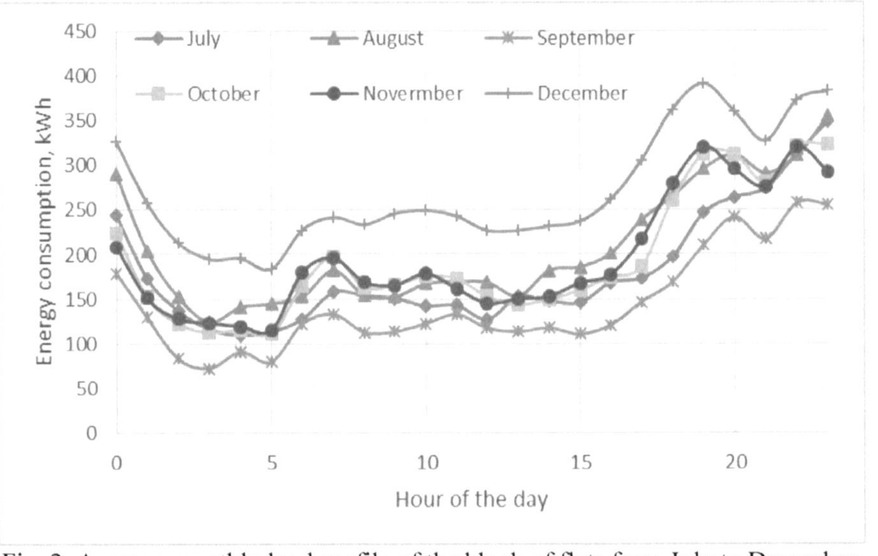

Fig. 2. Average monthly load profile of the block of flats from July to December.

The average daily energy consumption during the weekdays and during the weekends is presented in Figures 3 and 4 respectively. The figures also show the confidence interval for the different months with a 95% probability. As expected, during the weekdays (Fig. 3) there is a peak of the energy consumption at 7:00h when people wake up and prepare to go to work. The consumption starts to increase again after 18:00h when people return home and has peaks at 20:00h and 23:00h.

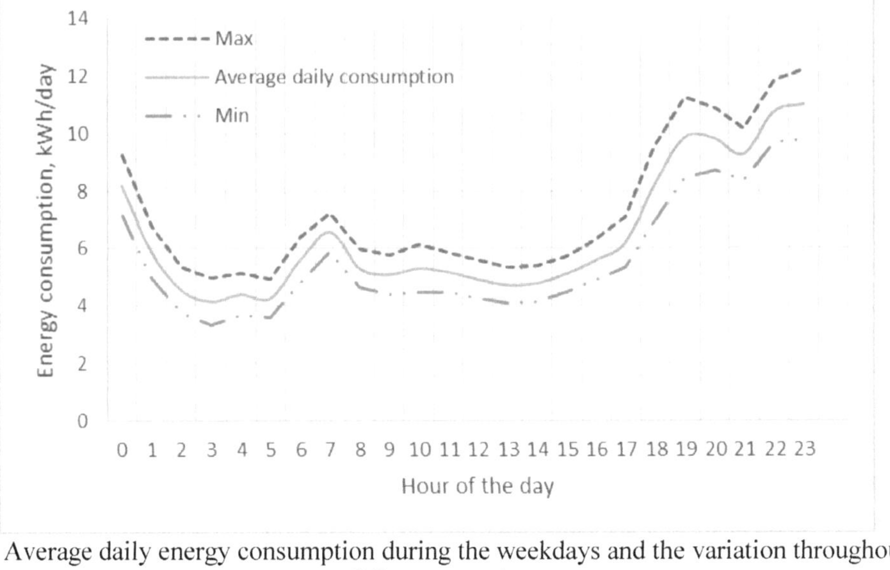

Fig. 3. Average daily energy consumption during the weekdays and the variation throughout the different months.

The situation is slightly different during the weekends (Fig. 4), when the morning peak is at approximately 11:00h, which is explained with the fact that people wake up later than usual. The other two peaks are again at 20:00h and 21:00h.

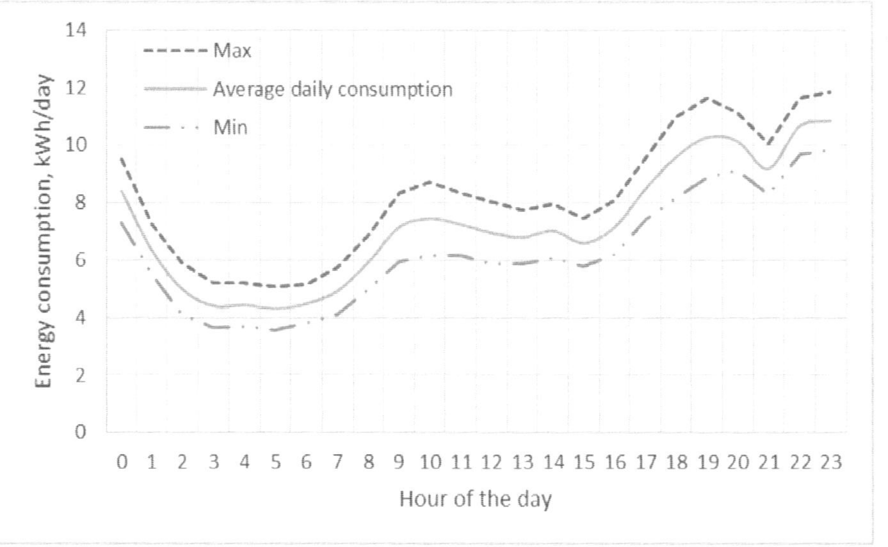

Fig. 4. Average daily energy consumption during the weekends and the variation throughout the different months.

The load profile has been normalized for three situations: during the weekdays, during the weekends, and on average (Fig. 5). The average one has one morning peak at 7:00h, caused by the weekdays, another smaller peak at 11:00h, caused by the weekends, and two significantly higher peaks, at 20:00h and 23:00h.

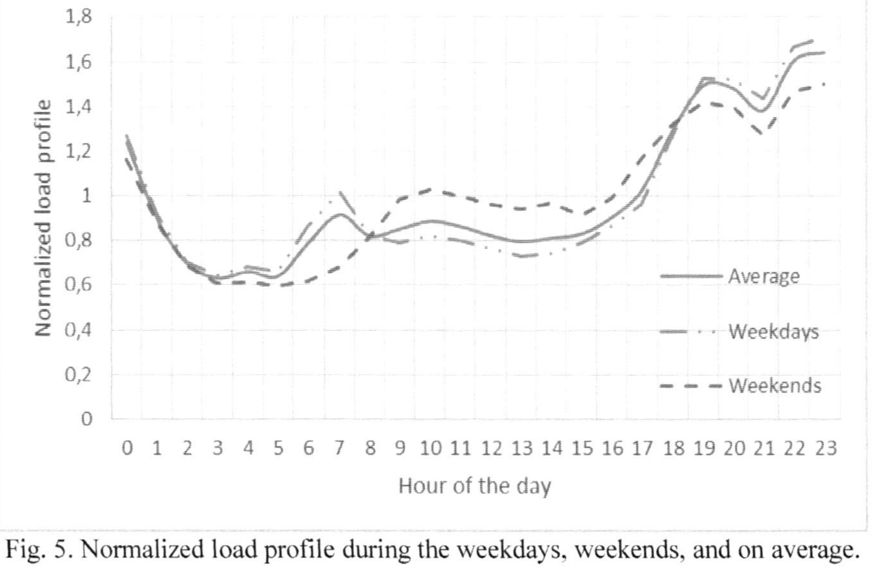

Fig. 5. Normalized load profile during the weekdays, weekends, and on average.

The obtained results show some similarities with studies in other countries [7], yet the peaks are happening in a different timeframe. On the one hand, in Bulgaria, the noon peak from the weekends is not so pronounced, and is approximately 1 hour earlier. On the other hand, the evening peak in Bulgaria is significantly higher and longer. Such a difference can be explained with the different culture and lifestyle.

4. Conclusions

This study presents an analysis of experimental data from 48 apartments in a Bulgarian block of flats, located in the city of Ruse. Hourly data of the energy consumption of each apartment since 2013 have been used for the analysis.

The obtained results demonstrate that during the weekdays there is one peak, early in the morning, around 7:00h, and then other two peaks at 20:00h and 23:00h when people spend their time at home.

During the weekends, the morning peak shifts to 11:00h, and the evening peaks remain the same. The average normalized load profile has been obtained in this study. The load profile has four peaks – at 7:00h, 11:00h, 20:00h, and at 23:00h, and the last two peaks are significantly higher. The results show that a substantial amount of the energy (more than 50%) is consumed during the dark hours of the day when there is no solar light.

The results provide some interesting data about the lifestyle in Bulgaria and show some differences from similar studies conducted in other countries. These results can be used to design renewable energy systems at residential buildings, to perform cost-benefit analysis of different energy saving investments, to design smart grids, and others.

5. Acknowledgment

The study was supported by contract of University of Ruse Angel Kanchev, № BG05M2OP001-2.009-0011-C01, "Support for the development of human resources for research and innovation at the University of Ruse Angel Kanchev". The project is funded with support from the Operational Program "Science and Education for Smart Growth 2014 - 2020" financed by the European Social Fund of the European Union.

References

[1] Chou JS and Ngo NT, 2016, Smart grid data analytics framework for increasing energy savings in residential buildings, *Automation in Construction*, Vol. 72, Part 3, pp. 247 – 257.

[2] https://ec.europa.eu/energy/en/topics/energy-efficiency/energy-efficiency-directive

[3] Laicane I, Blumberga D, Blumberga A and Rosa M, 2015, Comparative multiple regression analysis of household electricity use in Latvia: using smart meter data to examine the effect of different household characteristics, *International Scientific Conference "Environmental and Climate Technologies – CONECT 2014"*, Energy Procedia ,Vol. 72, pp. 49 – 56.

[4] Tuballa ML and Abundo ML, 2016, A review of the development of Smart Grid technologies, *Renewable and Sustainable Energy Reviews*, Vol. 59, pp. 710 -725.

[5] Poznaka L, Laicane I, Blumberga D, Blumberga A and Rosa M, 2015, Analysis of electricity user behavior: case study based on results from extended household survey, *International Scientific Conference "Environmental and Climate Technologies – CONECT 2014"*, Energy Procedia, Vol. 72, pp. 79 – 86.

[6] Camaraa N, Xub D and Binyetc E, 2018, Enhancing household energy consumption: How should it be done?, *Renewable and Sustainable Energy Reviews*, Vol. 81, Part 1, pp. 669-681.

[7] Ghaemi1 S and Brauner G, 2009, User behavior and patterns of electricity use for energy saving, *6. Internationale Energiewirtschaftstagung an der TU Wien IEWT*.

[8] Sustainable Energy Development Agency, http://www.odyssee-mure.eu/publications/national-reports/energy-efficiency-bulgaria.pdf .

IOP Conf. Series: Earth and Environmental Science **172** (2018) 012036 doi:10.1088/1755-1315/172/1/012036

Building and modeling a sustainable expert system, using UML language

Violeta N OPRIȘ
Ph.D. Student, The Military Technical Academy, Faculty of Military Electronic and Information Systems, Bucharest
E-mail: violeta.opris@yahoo.ro

Abstract. The Unified Modeling Language (UML) is the successor of multitudes of oriented analysis and design methods. This model saves energy, resources and also secures the required infrastructure. It is also the result of a process of introducing standardization into object-oriented analysis and design. The paper concentrates to build UML diagrams and to highlight the structure of modules for the expert system. The UML diagrams are useful at a certain point in the system development and have a certain utility in the research paper. This new approach ensures that the expert system is successful and sustainable in the future. Integral elements of sustainable expert systems are research and innovation activities. In this context, guided by system user needs, UML is used in an iterative and incremental approach based on abstraction levels. The personal contribution in this paper is modeling the proposed expert system, using UML diagrams. These are used to represent and visualize modeling elements for an intrusion detection expert system.

1. Introduction

This research paper aims to demonstrate the value of the proposed solution to improve the current infrastructure of expert systems, automate complex processes, monitor networks, intrusion detection, generate reports, increase the performance of the solution by using new information technologies and improve access to available services through system functionalities.

The integrated expert system includes online service components, downloadable and recorded intrusion services, event schedules, the ability to integrate with other systems, the ability to build an integration pilot with external systems, a knowledge base, user interface, automatic and remediation. The expert system is based on an algorithm and a knowledge base.

An important goal is that access to the expert system by users is safely done. The users are assigned to a module and have configured rights and access roles, the group they belong to and the access password. As the organizations have tried hard to monitor and optimize their impact within society, the researchers aim to reduce the impact of businesses on social and natural environment.

The expert system also includes a reporting and statistics module, making it a particularly useful tool for assessing and detecting intrusions. Another goal of intrusion detection activity is to prevent the consequences of intrusion by implementing an effective security control expert system.

The expert system benefits using UML language are:

- availability 24/7 hours;
- knowledge base for storing information and resolving problems;
- competitive access to information and event planning;
- detecting suspect elements that can be considered intrusions;
- automatic remediation and keeping the history of the performed operations;
- managing the detected intrusions;
- tracking possible attacks;
- engine deduction for exploiting the knowledge base;
- monitoring network events;
- fundamental management decisions through reports;
- determining the deviation from normal behaviour;
- assigning tickets and solving problems;
- evidence of events and their storage in a secure location;

- blocking network attacks;
- interfacing with other expert systems.

In this context, the modeling tool responds to security requirements and supports various standards recognized by the organizations [10]. The expert system model has traceability tools that offer total visibility to all interventions. In terms of data storage, securing the data contained in the database is accomplished on several levels: connectivity, authentication, storage and encryption of the information. Living and non-living natural expert systems ensure organization's survival if it shows responsibility for the environment and lead to renewable energy sources [7].

The expert system is based on a framework that will be divided into three essential layers, with a concise objective on the developed application [9].

The presentation layout is the component that has a direct user connection, being provided through the web interface of the expert system. The level of presentation of the expert system is represented by the graphical interface. The application layer is the component that encompasses the logic of the application. The first level is the security filter and the second level of filtering is logging. The expert system model is built on the MVC (Model-View-Controller) framework, thus separating business data and logic from the user interface [10]. The data layer is the database. It will have two components: the operational database and the journaling database.

This research paper has the objective of modeling the expert system using the UML language, which is capable of identifying, managing, monitoring and preventing real-time attacks and network failures on cloud computing infrastructure [5]. As a deep economic reason, migrating to Cloud is uncertain. Taking advantage of technology may lead to higher consumption of it, but in time it shows potential in order to become an organizational strategy itself [8].

Adoption of Cloud Computing has favourable economic impact, while human development speeds the process. Technology improves life quality, reduces long-term costs and aligns economic positive impact with social and environmental issues [8].

The paper is divided in five sections. In the first section, we present the introduction of the research paper, the second section contains a review of the most important previous work and design methods for expert system. In the third section we present an overview of the expert systems, in the fourth part we describe the expert system model using UML language and in the last section we present the conclusions of our work.

2. Previous work. Design methods for expert system

This section has as main objective understanding the concept of modeling. Modeling the expert system ensures effective communication between those involved in the system development [4]. In documenting the expert system to detect intrusions, modeling is considered a learning process.

In this paper, graphical patterns are created using the various UML diagrams to get a clear picture of the main components of the expert system.

Approaching the expert system through UML diagrams leads to various advantages. An important advantage is the quick understanding of the connections between the expert system and the external environment and the non-inclusion of technical aspects of implementation in the analysis phase. Likewise, there are two categories of system models in practice: the logical model that focuses on how the system works and the physical model in which the way the system will be implemented is presented.

The modeling of the system is useful because the diagrams will be modified to be mapped to the requirements of the projected system. A feature of modeling is the flexibility to ensure the consistency of system representation.

For the development of the system, the interactive modeling process is used, which leads to the probability of detecting errors from the analysis phase, eliminating nonconformities and detecting possible omissions [4].

The intrusion detection expert system on cloud computing infrastructure is a specialized IT system that allows computerized monitoring of computer security in order to manage network traffic [5].

The key features of the expert system are:

- it is a technology-based IT solution that ensures the security of the monitored systems regardless of location and time: browser based solution;
- innovative: an innovative tool for identifying, managing intruders in a network, especially with regard to vulnerabilities;
- ensures the entire flow of intrusion monitoring activity in a network;
- accessibility: friendly, accessible, graphical user interface, labels and messages that provide the user with ease in navigation and operation, general knowledge of operating software applications;
- high availability: in terms of trading / multiuser / competitor as well as in terms of each component;
- data security and integrity: access is automatically based on credentials;
- parameterization: the uniqueness of information by eliminating redundant information;
- manageability: in the sense that the system includes mechanisms that enable end-user hierarchy, definition of user / role / group access rights, user logs;
- scalability and flexibility: response to a high degree of parameterization that allows rapid and easy changes within the system, configurable and able to meet the needs of a progressive number of users.

The architecture of the model system has three levels:

- user interface;
- computer system server;
- database.

Business flow activities involve identifying intrusions, panning the desired events, automatically troubleshooting identified issues, communicating between users using the messaging module, scanning events that occur on a network, generating reports, managing organizations and system users, assigning tickets for the IT security team, special administration of the knowledge base, inventory of goods [10].

The modeled expert system in this research paper is a dedicated intrusion detection activation solution, making it an innovative and extremely useful tool.

3. Expert system overview

The concept of developing an expert system is in the attention of the research community. Developing an intrusion detection expert system is a multidisciplinary and complex task. An important factor is the lack of research and general methodology for the development of expert systems managed on cloud computing infrastructures [5].

The system modeled in this paper is an expert system based on artificial intelligence and modeled using the UML language. Artificial Intelligence (AI) is a field of research in computer science. An important definition of AI was given by John McCarthy in 1955: "a car that behaves in a way that could be considered intelligent if it were a man."

Progress of strong AI-based expert systems is in continuous development, almost all simulations of intelligence are based on common rules and algorithms.

In the case of the classical expert systems, different areas of interest and research for AI that have been researched are distinguished:

- expert systems;
- neural networks;
- automatic learning;
- data mining;
- artificial view;

- voice processing;
- processing in natural language.

An essential feature of AI is the ability to learn, with or without external aid, for continuous improvement. Initially, AI was conducted in the field of psychology, which was based on linguistic intelligence. An important example is the Turing test, which consists of a conversation in natural human language with a machine that has been programmed specifically for this test.

Traditional expert systems can be categorized as follows:

- systems that think rationally;
- systems that act as people;
- systems that act rationally.

The concept of modernity of expert systems involves their migration to an innovative research area. Migrating classical (ES) expert infrastructure into cloud computing is a complex process that requires careful and deliberate planning [2]. This process can have unexpected costs, interoperability, security gaps and can create significant obstacles. But cloud computing has benefits and today most organizations use it [5].

The expert system has key features such as performance, easy to understand, trustworthy and highly responsive. It also has training, assisting in decision-making, diagnosis and substitution of human decision makers. Mapping the cloud computing with expert an system requires organizations not to have large data centre. They can store, process and analyse information using cloud computing [5]. This could reduce widespread energy consumption that balances economic and environmental performance. Cloud computing comes with the ability to exchange information on multiple platforms, from different locations, inside and between users [2].

4. Modeling the expert system using UML

Unified Modeling Language is the successor to the multitude of oriented analysis and design methods. The use of the UML language depends on the complexity of the real system to be modeled, the project management and realization method, the type of computer system to be achieved, and the level of detail required in the design phase [6].

In this context, guided by system user needs, UML is used in an iterative and incremental approach, based on abstraction levels.

To model the expert system proposed in this research paper, UML diagrams are used to represent and visualize modelling elements for intrusion detection [1].

4.1. Use Case Diagram

The Use Case Diagram is built for expert system to describe the functionality of the system from the users` point of view (Figure 1). It contains actors and uses case. Actors are system elements that generate or receive events, and a case of use is a sequence of actions triggered when an actor calls the system to complete a process [6].

The use case diagram is also used to specify system boundaries and specifies the interaction that takes place between the actors and the expert system.

IOP Conf. Series: Earth and Environmental Science **172** (2018) 012036 doi:10.1088/1755-1315/172/1/012036

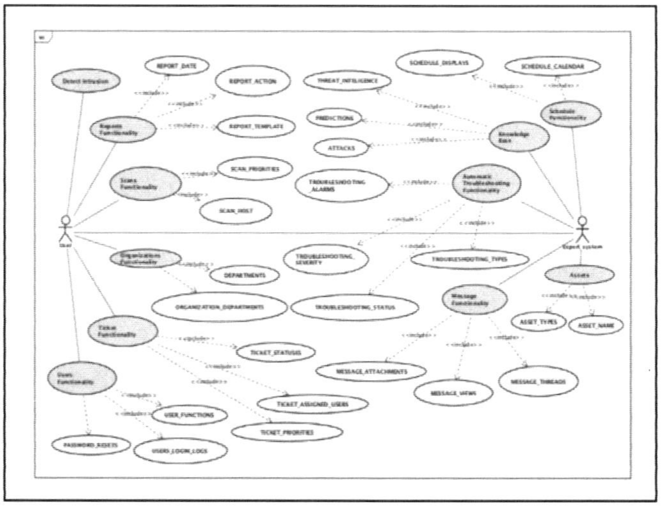

Figure 1. Use Case Diagram

4.2. Class Diagram

The static structure of the expert system, viewed as an ensemble of object classes and relations between classes, is represented by the class diagram [6].

The class diagram presents the structure of the system from a general point of view. In close connection with it, the object diagram highlights objects and links between them [1].

Object diagrams show particular cases and facilitate understanding of complex data structures. Classes correspond to semantics of entities in the real system. A class designates a group of objects that have similar properties, common behaviour, and common relationships with other objects.

4.3. Activity Diagram

By performing the activity diagram, the behaviour of the intrusion detection expert system is described by introducing implementation elements. The diagram of activities of the expert system has in structure the following elements: action, transition, decision [6].

The action from the point of view of the expert system belongs to a stage in the execution of an algorithm. The operations performed within the system can be seen as a chain of actions and detailed in elementary operations [3]. Also, the actions can be grouped into activities to simplify diagrams interacting in the system.

The transition from one action to another is graphically represented by an arrow labelled eventually with the name of the event that determines the transition and the condition of occurrence of the event [1].

4.4. State Diagram

In this paper, the state diagram is used to describe the behaviour of a class's objects through states and events [3].

Using the state diagram for the expert system leads to the following advantages [6]:

- it provides a description of the usage case;
- it is built only for classes with significant dynamic behaviour;
- it models the lifecycle of a single class, highlighting eventual events sent to another class in the system;
- it contains only one initial state and one or more final states determined by the conditions of occurrence of the event.

IOP Conf. Series: Earth and Environmental Science **172** (2018) 012036 doi:10.1088/1755-1315/172/1/012036

4.5. Sequence Diagram

The sequence diagram illustrates interactions between objects across a case scenario (Figure 2). Each object is represented by a rectangle in which the name is entered. The object's line of life is specified by a vertical bar.

Messages are represented by horizontal arrows from the transmitter to the receiver. Assuming that the time runs from top to bottom, the order of reference is given by the position on the vertical axis. The elements presented in the sequence diagram are translated into the class diagram [6].

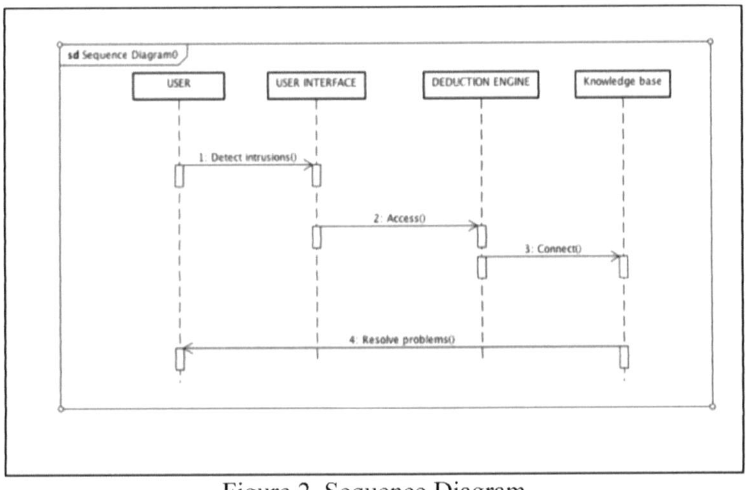

Figure 2. Sequence Diagram

Starting from the initial specifications of the expert system, the proposed solution has a technical architecture consisting of a back-office component that will be exploited by dedicated staff from the entire system administration and a Front Office component, accessible via the Internet to users represented in particular by IT security specialists; a trading component that will handle system authorization requests and an archiving component.

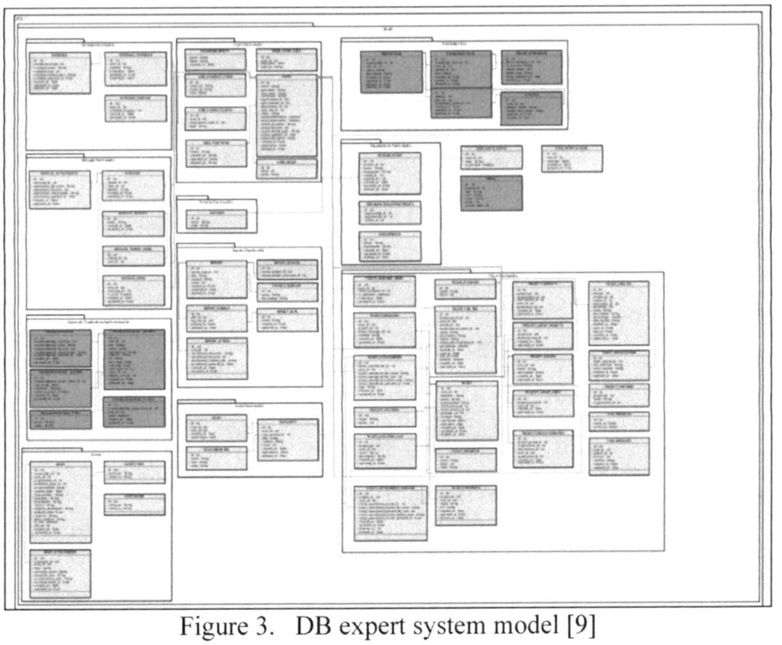

Figure 3. DB expert system model [9]

IOP Conf. Series: Earth and Environmental Science **172** (2018) 012036 doi:10.1088/1755-1315/172/1/012036

The proposed system includes the following modules: event planning, messaging, automatic remediation, inventory, users, reports, scanning, knowledge base, organizations and tickets (Figure 3).

Schedule functionality module provides the ability to customize events in a computer system, network scanning, event monitoring, task assignment.

The Message functionality module is a user-friendly model for easy communication between the organization's departments and the monitoring of the entire infrastructure that is supposed to be innovative.

The Troubleshooting functionality module provides users with automatic task resolution, by the initial software configuration and previous information (Figure 4).

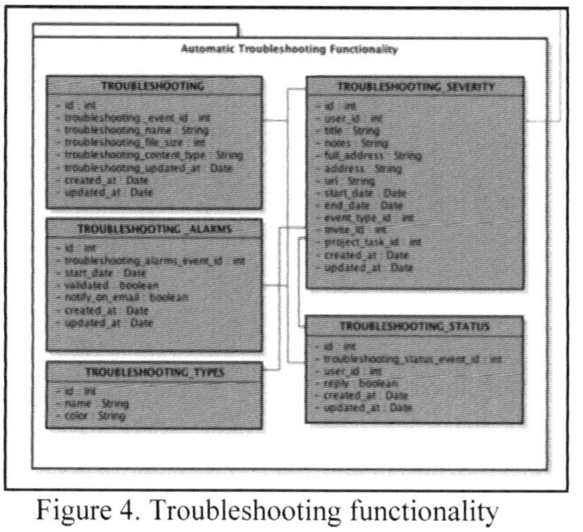

Figure 4. Troubleshooting functionality

The Assets Functionality module is a mechanism that automatically checks inventory of software products for the proper operation of the entire IT system.

The Users Functionality module assumes management of the IT system, which includes user groups configuration, user creation, workstation management.

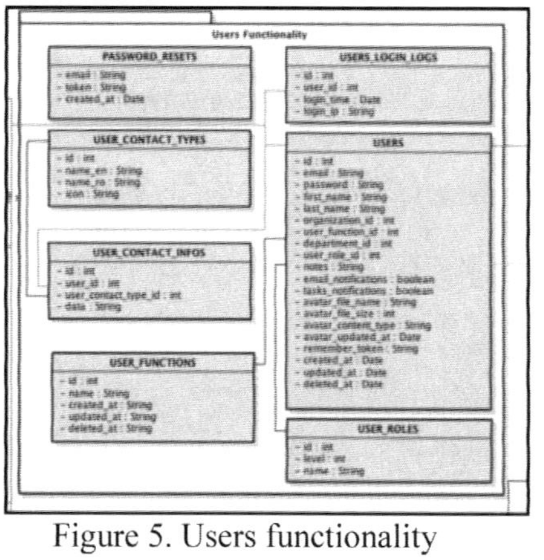

Figure 5. Users functionality

The Reports Functionality module is a particularly useful tool for evaluating and predicting network intrusions and problems.

The Scans Functionality module provides the ability to scan unplanned events that can affect the entire computer system.

The Knowledge Base Functionality module is designed so that each operation is stored in a database, while being the answer to ambiguous solutions to intrusion detection problems (Figure 6).

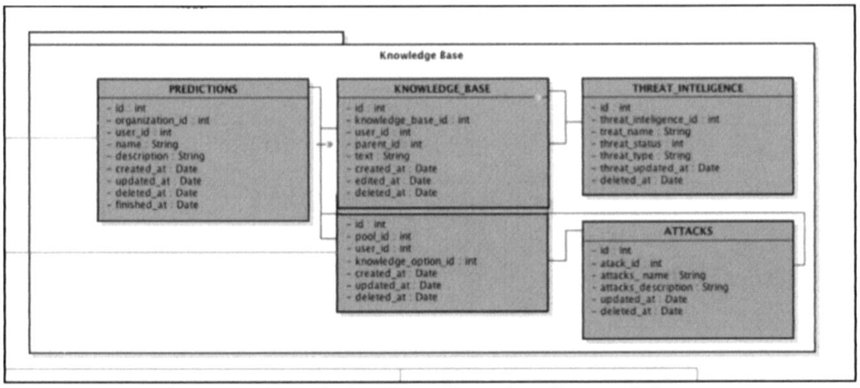

Figure 6. Knowledge Base Functionality module

The Organizations Functionality module facilitates the management of the organizations that have the proposed expert system for development.

The Ticket Functionality module involves the implementation of a full-feature management functionality for the vendor-specific ticketing flow to users specific to the IT department.

5. Conclusions

The final conclusions of the scientific research paper highlight the main contributions, both theoretical and practical.

In the paper, it is described and modeled an expert system for detecting intrusions from a network mapped with cloud computing infrastructure. By analysing UML diagrams, we have noticed that it is essential to understand the expert system interface and modules. Despite the aforementioned findings, the UML diagrams highlight a new model that helps to develop an expert system.

The expert system has a modular structure for optimal use and increased security, a cloud-based infrastructure being a new concept [11]. The proposed IT solution will be provided as a public or private cloud, helping organizations to create security in initiatives to detect network intrusions, also substantially reducing costs.

Considering the benefits of the model, this study provides new approaches of investigating the UML language used for development.

It is difficult for organizations to develop and integrate an expert system into innovative environmental and economic performance tools. Although, it could increase stakeholders trust, but companies have to prove their ability to minimize costs and to improve efficiency.

The solution has many modules, which can be expanded as research and development directions: event planning, messaging, automatic remedies, inventory, users administration, organizations, scanning, tickets, reports, knowledge bases. Of course, this system appears against ever-changing threats, also supporting compliance with existing technologies. Old security software is complex, expensive and ineffective.

This study reveals that the expert system UML model has a significant influence on information security. Starting on-demand scalability from expert system, the paper roughly introduces the new structure and components of the expert system. As a future research work, we plan to design a strong expert system.

IOP Conf. Series: Earth and Environmental Science **172** (2018) 012036 doi:10.1088/1755-1315/172/1/012036

References

[1] Udrică M, Mihăilescu M, Opriş V, Ionită R, Improving The Decisional Process By Using UML Diagrams, Journal of Knowledge Management, Economics and Information Technology, Bucharest 2012

[2] Dumitru L, Eftimie S, Mihailescu M, Nita L, Opris V, Racuciu C, A Novel Architecture for Authenticating Scalable Resources in Hybrid Cloud, 11th International Conference on COMMUNICATIONS, Bucharest, 2016

[3] Mihailescu M, , Opriş V, Beteringhe A, Enhancing UML with Security, 2nd INTERNATIONAL CONFERENCE SEA-CONF, Academia Navală Mircea Cel Bătrân, Constanta, 2016

[4] Paunica M, Opriş V, Opriş M, Tanase G, Financial Performance Improvement Through Sustainable Development, The 12th International Conference Accounting and Management Information Systems, Bucharest, 2017

[5] Opris V, Opriş M, The expert system development technologies in cloud, 3nd INTERNATIONAL CONFERENCE SEA-CONF, Academia Navală Mircea Cel Bâtran, Constanţa, 2017

[6] Sisteme informatice economice – probleme şi solutii, Mioara Udrica, Vatuiu T, Lipcanu A, Varjoghe S, Editura Hamangiu, Universitatea Titu Maiorescu, 2014

[7] Saufi, N.A.A., Daud, S., Hassan, H., 2016, Green Growth and Corporate Sustainability Performance, 7th International Economics & Business Management Conference, 5th & 6th October 2015.

[8] Williams, B., 2012, The Economics of Cloud Computing, Cisco

[9] http://astah.net/editions/professional

[10] https://www.tutorialspoint.com/mvc_framework/mvc_framework_introduction.htm

[11] http://www.ibm.com/cloud-computing/us/en/cloud-solutions/cloud-security

Institute of Physics
Dirac House, Temple Back
Bristol BS1 6BE UK

ISSN: 1755-1307
ISBN 978-1-5108-6746-8
9 781510 867468

CURRAN ASSOCIATES INC.
proceedings.com